Lecture Notes in Computer Science 16150

Founding Editors

Gerhard Goos
Juris Hartmanis

Editorial Board Members

Elisa Bertino, *Purdue University, West Lafayette, IN, USA*
Wen Gao, *Peking University, Beijing, China*
Bernhard Steffen, *TU Dortmund University, Dortmund, Germany*
Moti Yung, *Columbia University, New York, NY, USA*

The series Lecture Notes in Computer Science (LNCS), including its subseries Lecture Notes in Artificial Intelligence (LNAI) and Lecture Notes in Bioinformatics (LNBI), has established itself as a medium for the publication of new developments in computer science and information technology research, teaching, and education.

LNCS enjoys close cooperation with the computer science R & D community, the series counts many renowned academics among its volume editors and paper authors, and collaborates with prestigious societies. Its mission is to serve this international community by providing an invaluable service, mainly focused on the publication of conference and workshop proceedings and postproceedings. LNCS commenced publication in 1973.

Lina Felsner · Thomas Küstner · Andreas Maier ·
Chen Qin · Seyed-Ahmad Ahmadi · Anees Kazi ·
Xiaoling Hu
Editors

Reconstruction and Imaging Motion Estimation, and Graphs in Biomedical Image Analysis

First International Workshop, RIME 2025
and 7th International Workshop, GRAIL 2025
Daejeon, South Korea, September 27, 2025
Proceedings

Editors
Lina Felsner
Technical University of Munich
Garching, Germany

Andreas Maier
University of Erlangen-Nuremberg
Erlangen, Germany

Seyed-Ahmad Ahmadi
NVIDIA
Munich, Germany

Xiaoling Hu
Harvard Medical School
Boston, MA, USA

Thomas Küstner
University Hospital of Tübingen
Tübingen, Germany

Chen Qin
Imperial College London
London, UK

Anees Kazi
Harvard Medical School
Boston, MA, USA

ISSN 0302-9743　　　　　　ISSN 1611-3349　(electronic)
Lecture Notes in Computer Science
ISBN 978-3-032-06102-7　　　ISBN 978-3-032-06103-4　(eBook)
https://doi.org/10.1007/978-3-032-06103-4

© The Editor(s) (if applicable) and The Author(s), under exclusive license
to Springer Nature Switzerland AG 2026

This work is subject to copyright. All rights are solely and exclusively licensed by the Publisher, whether the whole or part of the material is concerned, specifically the rights of translation, reprinting, reuse of illustrations, recitation, broadcasting, reproduction on microfilms or in any other physical way, and transmission or information storage and retrieval, electronic adaptation, computer software, or by similar or dissimilar methodology now known or hereafter developed.
The use of general descriptive names, registered names, trademarks, service marks, etc. in this publication does not imply, even in the absence of a specific statement, that such names are exempt from the relevant protective laws and regulations and therefore free for general use.
The publisher, the authors and the editors are safe to assume that the advice and information in this book are believed to be true and accurate at the date of publication. Neither the publisher nor the authors or the editors give a warranty, expressed or implied, with respect to the material contained herein or for any errors or omissions that may have been made. The publisher remains neutral with regard to jurisdictional claims in published maps and institutional affiliations.

This Springer imprint is published by the registered company Springer Nature Switzerland AG
The registered company address is: Gewerbestrasse 11, 6330 Cham, Switzerland

If disposing of this product, please recycle the paper.

RIME 2025 Preface

The **1st International Workshop on Reconstruction and Imaging Motion Estimation (RIME 2025)** was proudly held in conjunction with the 28th International Conference on Medical Image Computing and Computer-Assisted Intervention (MICCAI 2025) in Daejeon, South Korea.

This inaugural edition of RIME established a new forum for researchers at the forefront of image reconstruction and motion estimation in medical imaging. These two synergistic fields are rapidly reshaping diagnostic and interventional workflows, especially in dynamic and real-time imaging settings. RIME 2025 welcomed innovative contributions—from variational and model-based methods to physics-informed machine learning and state-of-the-art deep learning approaches—that push the boundaries of what is possible in image reconstruction and motion correction.

The workshop showcased a scientific program, highlighted by two inspiring keynote lectures from world-leading experts: Jong Chul Ye (KAIST) and Julia Schnabel (TUMHelmholtz Munich). Alongside the invited talks, the workshop featured a diverse selection of peer-reviewed papers and concluded with a lively open panel discussion. The contributions covered a wide spectrum of timely topics, including motion-aware reconstruction, dynamic and sparse-view imaging, neural operators for spatiotemporal modeling, and advanced methods for self-supervision and domain generalization.

From a total of 15 submissions, 12 full papers were accepted after a rigorous double-blind review process in which each manuscript received at least two independent reviews. No invited contributions appear in these proceedings, and all papers co-authored by members of the organizing committee were subject to strict conflict-of-interest handling and independent review.

We are deeply grateful to all authors, reviewers, and keynote speakers for their outstanding contributions and enthusiastic participation. Their support made RIME 2025 not only a scientifically rewarding experience but also a truly inspiring start to what we hope will become a long-running and impactful workshop series.

Lina Felsner
Thomas Küstner
Andreas Maier
Chen Qin

Organization

Workshop Chairs/Volume Editors

Lina Felsner   Technical University of Munich, Germany
Thomas Küstner   University Hospital of Tübingen, Germany
Andreas Maier   University of Erlangen-Nuremberg, Germany
Chen Qin   Imperial College London, UK

GRAIL-TGI-HGMIA 2025 Preface

The 7th International Workshop on **Graphs in BiomedicAl Image anaLysis (GRAIL 2025)**, satellite track of MICCAI 2025 in Daejeon, South Korea, represented a major step in the evolution of our workshop format within the broader community. This year, GRAIL joined forces with two closely aligned initiatives as special tracks: the **Topology- and Graph-Informed Imaging Informatics (TGI)** workshop and the **Hypergraph Computation for Medical Image Analysis (HGMIA)** workshop. By uniting our communities, we aim to provide a single, integrated forum for advancing graph and topological methods in biomedical image analysis and related domains.

Graphs and topological representations have become powerful tools for modeling complex biomedical data. They underpin geometric deep learning, hypergraph signal processing, probabilistic graphical models, and topological deep learning, enabling interpretable, data-efficient solutions across diverse biomedical tasks. Over the past years, these methods have shown remarkable progress in brain connectomics, computational pathology, population studies, shape modeling, multi-omics data integration, and drug discovery. Their increasing visibility at leading machine learning and computer vision conferences reflects a maturing field whose potential extends well beyond medical imaging.

Our **joint GRAIL-TGI-HGMIA 2025 workshop** reflects this growing scope. Alongside traditional topics such as graph neural networks, spectral methods, and statistical graph analysis, this year's workshop embraced emerging directions, including hypergraph computation, integration of topological priors, foundation models for graph-structured data, and the unification of graphs with large language models and knowledge graphs. Our program featured invited keynotes from leading researchers in these areas, peer-reviewed paper presentations, and a new Spotlight Track for selected MICCAI main-conference papers, fostering cross-pollination between the main conference and the workshop. Towards proceedings, our workshop had 15 submissions and 10 accepted papers (67% acceptance rate), following double-blind review with an average of 2.3 reviews per paper.

By merging with TGI and HGMIA, we strengthen our mission to bridge theory and clinical application. We seek to create a collaborative environment where method developers, biomedical scientists, and clinicians can engage in meaningful dialogue, share datasets and benchmarks, and identify the next generation of research challenges at the interface of imaging, computation, and medicine.

We warmly thank the authors, reviewers, and organizing committee members for their commitment and contributions. Their efforts have made this joint edition of GRAIL, TGI, and HGMIA a unique platform for driving forward the science and application of graph-based and topological methods at MICCAI.

July 2025

Seyed-Ahmad Ahmadi
Anees Kazi
Chao Chen
Xiaoling Hu
Xiangmin Han

Organization

GRAIL Chairs

Program Committee Chairs

Seyed-Ahmad Ahmadi (0000-0002-7082-0739)	NVIDIA, Germany
Anees Kazi (0000-0003-4528-1670)	Harvard Medical School, USA

Co-chairs

Yousef Yegane (0000-0002-3114-6729)	Technical University of Munich, Germany
Kamilia Zaripova (0000-0002-8200-0818)	Technical University of Munich, Germany
Lucy Godson (0000-0002-3419-7628)	National Pathology Imaging Co-operative, UK
Niharika S. D'Souza (0009-0004-7657-3312)	IBM, USA
Seong Tae Kim	Kyung Hee University, South Korea
Johannes C. Paetzold	Weill Cornell Medicine, USA
Mustafa Hajij	University of San Francisco, USA

TGI Chairs

Program Committee Chairs

Chao Chen	Stony Brook University, USA
Xiaoling Hu	Harvard Medical School, USA

Co-chairs

Moo K. Chung	University of Wisconsin-Madison, USA
Johannes C. Paetzold	Weill Cornell Medicine, USA
Martin Andreas Styner	University of North Carolina at Chapel Hill, USA
Tom Fletcher	University of Virginia, USA
Jae-Hun Jung	Pohang University of Science and Technology, South Korea
Anuj Srivastava	Florida State University, USA
Anqi Qiu	Hong Kong Polytechnic University, China
Ilwoo Lyu	Pohang University of Science and Technology, South Korea

HGMIA Chairs

Program Committee Chair

Xiangmin Han	Tsinghua University, China

Co-chairs

Yue Gao	Tsinghua University, China
Angelica I. Aviles-Rivero	Tsinghua University, China
Mingxia Liu	University of North Carolina at Chapel Hill, USA
Pietro Liò	University of Cambridge, UK

Conference Logo

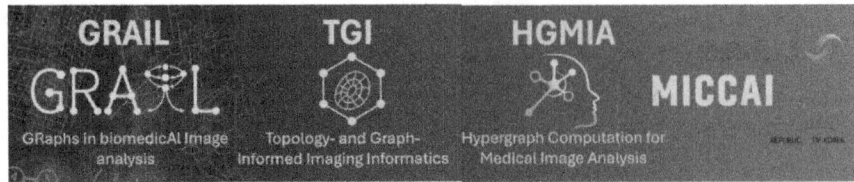

Contents

Proceedings of the 1st International Workshop on Reconstruction and Imaging Motion Estimation (RIME 2025)

Robustness Evaluation of Multi-visit Magnetic Resonance Image Reconstruction 3
  Youssef Beauferris, Abbas Omidi, Amirmohammad Shamaei, Jacob Idoko, Gouri Deshpande, Mariana Bento, and Roberto Souza

NerT-CA: Efficient Dynamic Reconstruction from Sparse-View X-ray Coronary Angiography 13
  Kirsten W. H. Maas, Danny Ruijters, Nicola Pezzotti, and Anna Vilanova

INR Meets Multi-contrast MRI Reconstruction 23
  Natascha Niessen, Carolin M. Pirkl, Ana Beatriz Solana, Hannah Eichhorn, Veronika Spieker, Wenqi Huang, Tim Sprenger, Marion I. Menzel, and Julia A. Schnabel

Markerless Tracking-Based Registration for Medical Image Motion Correction 34
  Luisa Neubig, Deirdre Larsen, Takeshi Ikuma, Markus Kopp, Melda Kunduk, and Andreas M. Kist

MCM: Mamba-Based Cardiac Motion Tracking Using Sequential Images in MRI 44
  Jiahui Yin, Xinxing Cheng, Jinming Duan, Yan Pang, Declan O'Regan, Hadrien Reynaud, and Qingjie Meng

Evaluating Deep Learning Based Domain Generalization for Motion Mitigation in Multi-center Brain MRI 55
  Saad Ashraf, Md Afif Al Mamun, Mumu Aktar, Roberto Souza, and Mariana Bento

Localized FNO for Spatiotemporal Hemodynamic Upsampling in Aneurysm MRI 65
  Kyriakos Flouris, Moritz Halter, Yolanne Y. R. Lee, Samuel Castonguay, Luuk Jacobs, Pietro Dirix, Jonathan Nestmann, Sebastian Kozerke, and Ender Konukoglu

LSTT: Latent Spatio-Temporal Transformer for Non-rigid Motion
Compensation in CBCT 76
    Yipeng Sun, Linda-Sophie Schneider, Annette Schwarz, Mingxuan Gu,
    Siyuan Mei, Chengze Ye, Siming Bayer, and Andreas Maier

cIDIR: Conditioned Implicit Neural Representation for Regularized
Deformable Image Registration 87
    Sidaty El Hadramy, Oumeymah Cherkaoui, and Philippe C. Cattin

Self-supervised Motion-Compensated Reconstruction for Cardiac Cine
MRI 97
    Siying Xu, Aya Ghoul, Kerstin Hammernik, Jens Kuebler,
    Patrick Krumm, Andreas Lingg, Daniel Rueckert, Sergios Gatidis,
    and Thomas Küstner

Generating Realistic Synthetic Motion Curves for MRI Retrospective
Motion Correction 108
    Hristo Georgiev, Jakob Sheye, Madeleine Wyburd, Jens Petersen,
    Vincent Beliveau, and Melanie Ganz

Neural Space-Time Modeling for Motion-Corrected MR Reconstruction 118
    Aizada Nurdinova, Wenqi Huang, Daniel Raz Abraham, Jaehyeok Bae,
    Yimeng Lin, Kawin Setsompop, and Brian Andrew Hargreaves

**Proceedings of Graphs in BiomedicAl Image anaLysis (GRAIL),
Topology- and Graph-Informed Imaging Informatics (TGI), and
Hypergraph Computation for Medical Image Analysis (HGMIA)**

Skip Priors and Add Graph-Based Anatomical Information,
for Point-Based Couinaud Segmentation 131
    Xiaotong Zhang, Alexander Broersen, Gonnie C. M. van Erp,
    Silvia L. Pintea, and Jouke Dijkstra

Prompt-Driven Multi-view Representation Learning for Clinical
Progression Prediction of Significant Memory Concern 141
    Cui Wang, Yongheng Sun, Minhui Yu, Yuzhen Gao, and Mingxia Liu

Improving Late-Life Depression Analysis with Collaborative Domain
Adaptation: Learning from Heterogeneous Structural MRI 151
    Yuzhen Gao, Mengqi Wu, Li Wang, Lihong Wang, David C. Stephens,
    Guy G. Potter, and Mingxia Liu

Decoder-Free Supervoxel GNN for Accurate Brain-Tumor Localization
in Multi-modal MRI 162
   *Andrea Protani, Marc Molina Van De Bosch, Lorenzo Giusti,
Heloisa Barbosa Da Silva, Paolo Cacace, Albert Sund Aillet,
Friedhelm Hummel, and Luigi Serio*

Graph Conditioned Diffusion for Controllable Histopathology Image
Generation 172
   *Sarah Cechnicka, Matthew Baugh, Weitong Zhang,
Mischa Dombrowski, Zhe Li, Johannes C. Paetzold, Candice Roufosse,
and Bernhard Kainz*

X-Node: Self-explanation is All We Need 184
   Prajit Sengupta and Islem Rekik

Population-Graph Post-hoc Correction of Survival Predictions
for Improved Risk Stratification 195
   *Oriane Thiery, Mira Rizkallah, Hakima Laribi, Martin Vallières,
Thomas Carlier, and Diana Mateus*

Spectral Graph Autoregressive Modeling for Conditional Brain Network
Augmentation 209
   *Hayoung Ahn, Seungjoo Lee, Jaeyoon Sim, Yechan Hwang, Hyuna Cho,
Guorong Wu, and Won Hwa Kim*

HFR: Hemodynamic Feature Regression for Physically Constrained
Pressure Drop Estimation 219
   Jakub Chojnacki, Szymon Kopeć, Konrad Duraj, and Maciej Zamorski

WANCDR: Wasserstein Adversarial Network for Cancer Drug Response 229
   Hanjun Choi and Mansu Kim

Author Index 239

Proceedings of the 1st International Workshop on Reconstruction and Imaging Motion Estimation (RIME 2025)

Robustness Evaluation of Multi-visit Magnetic Resonance Image Reconstruction

Youssef Beauferris[1,2], Abbas Omidi[2,3], Amirmohammad Shamaei[2,3], Jacob Idoko[3], Gouri Deshpande[3], Mariana Bento[2,3], and Roberto Souza[2,3(✉)]

[1] Department of Biomedical Engineering, University of Calgary, Calgary, AB, Canada
[2] Department of Electrical and Software Engineering, University of Calgary, Calgary, AB, Canada
roberto.souza2@ucalgary.ca
[3] Hotchkiss Brain Institute, University of Calgary, Calgary, AB, Canada

Abstract. Multi-visit magnetic resonance (MR) image reconstruction models utilize information from a previous MR exam to enhance the efficiency of the current exam, thereby speeding up the process while maintaining the image reconstruction quality. One concern with multi-visit reconstruction models is the potential for biases, such as not properly reflecting anatomical changes, that may appear in the reconstructed images when leveraging information from a previous scan. Moreover, the quality of the reconstructed images may be influenced by the time gap between scans. To address these concerns, a series of challenging test scenarios was devised and performed to evaluate the robustness of the multi-visit integration module (MIM). Our analysis shows that despite variations in the quality of the single-visit reconstruction, the multi-visit reconstruction exhibits significant enhancement ($p < 0.001$) by leveraging prior data available from previous scans. Our findings demonstrate that the MIM is robust to the time interval between scans and that synthetic lesions can be used to further enhance its robustness. Our study offers valuable insights for future advancements in the field.

Keywords: MRI · reconstruction · deep learning · longitudinal imaging

1 Introduction

Brain magnetic resonance (MR) imaging is a radiation-free and non-invasive technique used to identify and monitor the progress of neurodiseases. MR imaging sessions are conducted repeatedly to evaluate the advancement of a disease by monitoring alterations in the brain across time. Nevertheless, the utilization of MR imaging is hindered by its long duration and high costs, leading to significant economic and societal ramifications. The wait times and costs for MRI scans are quickly escalating [15].

Supplementary Information The online version contains supplementary material available at https://doi.org/10.1007/978-3-032-06103-4_1.

There is a significant degree of mutual information between images acquired in longitudinal studies, even for diseased subjects whose brains are often expected to change considerably between time points. Thus, this information can be leveraged to speed up multi-visit MR scans. Souza et al. [14] proposed the multi-visit integration model (MIM), a deep learning approach that aims to incorporate a previous scan during the reconstruction process of an aggressively under-sampled follow-up scan to produce a multi-visit reconstruction. The MIM was validated using a dataset of presumed normal adult subjects that were scanned twice. However, the main limitations of this work are that the brains of presumed healthy adult subjects are not expected to differ significantly over a short time span, and potential biases in the multi-visit reconstruction results, such as the reconstruction not accurately reflecting brain changes occurring over time, were not investigated. A follow-up study by Beauferris et al. [2] investigated the MIM in the context of a cohort of glioblastoma subjects. Glioblastoma is a highly aggressive form of brain cancer known for its rapid and pronounced brain changes [8]. In [14], the MIM for an acceleration factor (R) of 15 obtained better peak signal-to-noise ratio (PSNR) and structural similarity index measure (SSIM) [17] than the single-visit reconstruction models for $R = 5$.

The main limitation of past multi-visit MR image reconstruction studies is that they did not investigate the robustness of the MIM model in terms of potentially biasing the MR image reconstructions towards the anatomy of the previous exam, the time interval between the past scan and the scan presently being reconstructed. This research evaluates the robustness of multi-visit MR image reconstruction in the context of a challenging dataset of glioblastoma patients, where significant anatomical changes are expected to be present between MR scans. Our main contributions are: **1.** We evaluate the sensitivity of the MIM to two different single-visit MR reconstruction algorithms and show that the multi-visit reconstruction improves overall image quality and it inherits the visual characteristics of the single-visit image reconstructions; **2.** We demonstrate that the multi-visit reconstruction can be biased towards the previous MR image; however, this bias can be mitigated by incorporating synthetic image artifacts during model training; **3.** Our results suggest that the multi-visit reconstruction is robust to the time gap between scans.

2 Methods

2.1 Dataset

This work utilizes a dataset courtesy of [2]. It comprises 57 fully sampled, T1-weighted, sagittal MR image acquisitions collected on a 3 T MR scanner (Discovery MR750, General Electric (GE) Healthcare, Waukesha, WI). The 57 scans were collected from 27 glioblastoma subjects, 18 of which had undergone multiple imaging sessions, 11 subjects had two, 6 subjects had three, and 1 subject had eight MR scans (see Supplementary Figure 1). The mean ± standard deviation age of the population is 58.8 ± 9.8 years, with a sex distribution of 55% male and

45% female. A field of view of 256 mm × 256 mm was used, and 150 to 170 contiguous 1.0-mm thick images were collected. The inverse Fourier Transform was taken along the frequency encode axis, and the data was processed in the plane defined by the phase and slice encode directions. The median interval between consecutive scans was 67 d. Upon visual inspection, the dataset reveals structural brain changes across time points, resulting from surgical intervention, therapy, or disease progression. This makes it suitable for evaluating the robustness of multi-visit MR image reconstruction.

2.2 Methodology

The underlying processing model is adopted from Souza et al. [14]. The methodology consists of a three-step model: 1) Obtain a single-visit reconstruction from under-sampled k-space measurements using a deep-learning model; 2) Non-linear registration of the previous scan to the single-visit reconstruction; and 3) Enhancement of the single-visit reconstruction by leveraging the previous scan (i.e., multi-visit reconstruction). These steps collectively will be referred to as the MIM approach. Step 1 utilizes the dual-domain cascade of U-nets model [13] or the Recurrent Variational Network (RecVarNet) [16] to generate the single-visit reconstruction. Cascaded models are commonly used in the literature [5,11], and RecVarNet was the winner of a recent MR reconstruction challenge [3], making them suitable candidates for assessing the robustness of multi-visit MR reconstruction. Step 2 involves accounting for changes in subject orientation between imaging sessions and compensating for non-topological brain changes (e.g., brain atrophy). This step utilizes Elastix [7] for non-linear registration. Step 3 implements the U-Net [10] model for multi-visit reconstruction. Independent models are trained for each acceleration factor, R = {5, 10}, which are simulated using retrospective under-sampling. In some experiments, only the results for R=10 are computed for the sake of simplicity in analysis. A Poisson disc distribution sampling scheme [4] was used. The center of the k-space was fully sampled within a circle of radius 50 and 40 pixels, corresponding to each acceleration factor in ascending order (R = {5, 10}).

2.3 Experiments

Three experiments were designed to evaluate the robustness of the multi-visit MR image reconstructions. These experiments are described in detail below.

(E1) Comparing the Impact of Using Different Single-Visit Reconstruction Models: This experiment investigates the impact of single-visit reconstruction quality on its multi-visit reconstructions counterparts. This experiment involves a comparative analysis between two distinct multi-visit reconstructions generated by leveraging different single-visit reconstruction models. The two single-visit reconstruction models are: 1) the WW-net IKI network [13], and 2) RecVarNet [16]. The single-visit reconstruction models used 11 scans in training and 9 in validation. The multi-visit reconstruction network used six pairs of scans for training and five for validation. The test set had seven pairs of scans. The single-visit reconstruction model only used the present scan for reconstructing the images, while the multi-visit used a pair of scans, i.e., present and past scans.

(E2) Assessing Reconstruction Bias by Leveraging Synthetic Lesions: Accounting for bias is crucial when leveraging a previous scan to facilitate the reconstruction of a follow-up scan (multi-visit reconstruction). The multi-visit reconstruction should accurately reflect anatomical changes occurring between scans. To assess potential bias in the MIM, a Graphical User Interface (GUI) was developed using Python's Visual Toolkit (VTK) library [12]. This GUI facilitates the augmentation of synthetic white matter lesions exclusively on the previous scans. The appearance and location of these lesions were inserted with the input of medical experts and medical physicists. These synthetic lesions are excluded from the follow-up scan to simulate scenarios where lesions suddenly disappear between visits. The presence of these lesions in the multi-visit reconstruction would indicate bias towards the previous scan, resulting in incorrect reconstruction of these regions. This experiment randomly added nine hand-crafted lesions of varying sizes to each previous scan at different slices. These lesions ranged in size between 67 voxels (small) and 680 voxels (large). Two multi-visit reconstruction models were trained, one with lesions added to the previous scan and the other without. Multi-visit reconstructions are generated using both models. Independent predictions were made on the test set using these two models when synthetic lesions were added to the previous scan. The data split for the multi-visit reconstruction module, used for training and validation, is identical to E1. The single-visit reconstructions inputted into the MIM are generated using the RecVarNet for an acceleration factor of $R = 10$.

(E3) Analysis of the Time-Dependence of the Multi-visit Reconstruction: The time interval between successive brain MR scans significantly influences their degree of similarity, particularly in cases where patients are suffering from a progressive disease. In this experiment, a single glioblastoma patient from the same dataset used in the previous experiments was examined. This patient underwent the greatest number of repeated scans, comprising a total of eight scans over a two-year period. The MIM used in this experiment is the same as in experiments 1 and 2. It is noteworthy that none of the scans pertaining to this patient were included in the MIM's training or validation phases. The first seven scans were registered to the final scan acquired on May 23, 2019, as depicted in Supplementary Fig. 1. This visualization highlights the transformation of the diseased anatomical region resulting from disease progression, surgical intervention, or therapy. The RecVarNet was employed to generate the single-visit reconstruction for the final scan collected on May 23, 2019, using an acceleration factor of $R = 10$. Subsequently, a multi-visit reconstruction was generated using each of the previous scans available.

2.4 Quality Assessment Metrics

The single-visit and multi-visit models were evaluated using two quantitative performance measures: pSNR and SSIM [17]. These two quantitative metrics are commonly used to assess MR image reconstruction quality [9]. They were computed by comparing the previous scan registered (PS_{reg}), the single-visit and multi-visit MR reconstructions to the corresponding fully sampled reference. Statistical significance between the single-visit and multi-visit reconstructions was assessed using the Wilcoxon signed-rank test with an α of 0.05.

3 Results

(E1) Comparing the impact of using different single-visit reconstruction models: Two sets of single-visit reconstructions were produced using the WW-net IKI network and the RecVarNet. Figure 1 summarizes all the single-visit and multi-visit reconstruction metrics for both reconstruction networks and acceleration factors. The RecVarNet outperformed the WW-net IKI network based on the SSIM and PSNR metrics for both the single and multi-visit reconstructions. For the RecVarNet, the average SSIM and PSNR of the single-visit reconstructions for an acceleration factor of 10× were 0.939 ± 0.020 and 35.0 ± 1.77, respectively. For the same acceleration factor, the SSIM and PSNR of the single-visit reconstructions produced from the WW-net IKI network were 0.865±0.024 and 34.2±1.77, respectively. The multi-visit reconstruction metrics, when using the RecVarNet as the single-visit reconstruction module, were 0.958 ± 0.0161 and 36.4 ± 1.78 for SSIM and PSNR, respectively. The multi-visit reconstruction metrics, when using the WW-net IKI network as the single-visit reconstruction module, were 0.952 ± 0.0173 and 35.6 ± 2.13 for SSIM and PSNR, respectively.

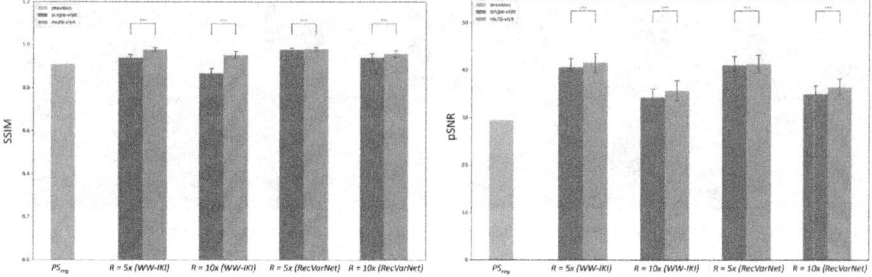

Fig. 1. (E1) SSIM and PSNR computed against the fully sampled reference for the previous scan registered, the single-visit, and the multi-visit reconstructions. Leveraging multi-visit information improved the single-visit reconstruction in all experiments (***$p < 0.001$, Wilcoxon signed-rank test). Notice that the multi-visit reconstruction for $R = 10$ results in better metrics than the single-visit reconstruction for $R = 5$.

Figure 2 allows for a visual comparison of the single and multi-visit reconstructions for both when using the WW-net IKI and RecVarNet as the single-visit reconstruction modules.

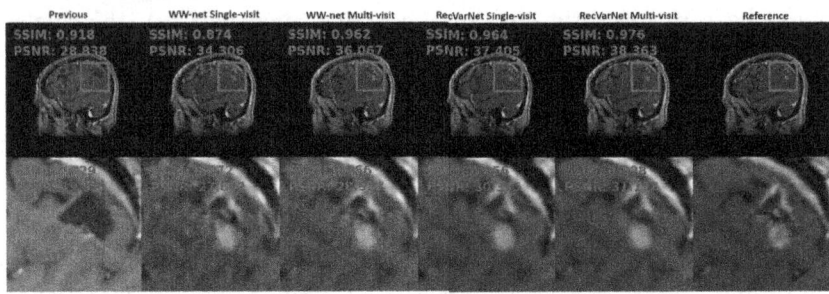

Fig. 2. (E1) Pairs of Single and Multi-visit reconstructions shown for an acceleration factor of R = 10, along-side the previous scan and the fully sampled reference (i.e. target image). SSIM and PSNR are calculated by comparing the previous scan and reconstructions to the reference.

(E2) Assessing reconstruction bias by leveraging synthetic lesions In E2, synthetic white matter lesions added to the previous scan but omitted from the follow-up single-visit reconstruction incorrectly biased the multi-visit reconstruction. However, the addition of synthetic lesions to the previous scans during training improved the model's ability to correctly reconstruct regions where the lesions disappeared in the follow-up study, compared to the model which did not include synthetic lesions during training, as shown in Fig. 3. This observation was consistent for all sizes of lesions.

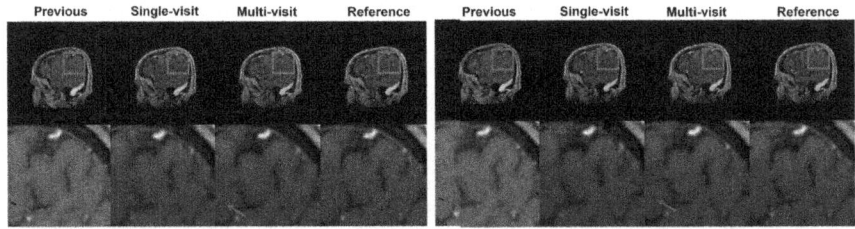

(a) Trained **without** synthetic lesions   (b) Trained **with** synthetic lesions

Fig. 3. (E2) Visual comparison of multi-visit reconstruction performance when the synthetic white matter lesion is added to the previous scan and are no longer present in follow-up. a) When the MIM is trained without synthetic lesions added to the previous scan, the multi-visit reconstruction becomes incorrectly biased towards this previous time point. b) Training with synthetic lesions improves the MIM's ability to detect and correctly reconstruct these lesions when they are present in the previous scan.

(E3) Time-dependence of the multi-visit reconstruction: The multi-visit reconstruction, which leverages the most recently acquired previous scan (2019-03-28), has the second-highest SSIM and highest PSNR of 0.96 and 36.2, respectively, compared to the other multi-visit reconstructions, which leverage older previous scans. Comparing the previous scans to the follow-up reference, the most recently acquired previous scan has only the second-highest SSIM and lowest PSNR as shown in Fig. 4. The multi-visit reconstruction produced using the oldest previous scan (2017-03-09) had the lowest SSIM and PSNR of 0.952 and 34.3, respectively. This degradation in quality is also apparent visually (see Supplementary Fig. 2). The magnified region in row 2 for the multi-visit reconstruction is less sharp than those generated from the previous scans collected in 2018 and 2019. However, across all seven time points, within the magnified regions, bias towards the previous scan is not apparent despite significant differences in the structural information used to enhance this region. Also, when comparing all seven multi-visit reconstructions, the metrics appear to be fairly stable, ranging from 0.952 to 0.961 for SSIM and from 34.3 to 36.2 for PSNR.

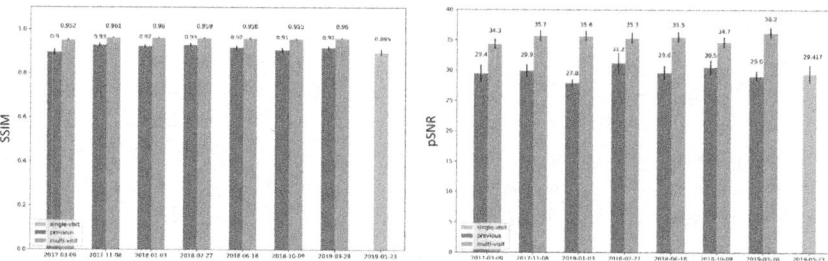

Fig. 4. (E3) Quantitative comparison of multi-visit reconstruction quality when different previous scans are leveraged. Seven different previous scans were independently leveraged to produce a corresponding multi-visit reconstruction from the same single-visit reconstruction (2019-05-23 shown in blue). When considering both SSIM and PSNR leveraging the most recently acquired previous scan (2019-03-28) resulted in the highest quality multi-visit reconstruction. (Color figure online)

4 Discussion

The findings from our experiments shed light on the robustness of multi-visit information in enhancing the quality of an undersampled MR reconstruction, despite the varying degrees of similarity between the previous and follow-up scans. E1 demonstrated that comparable levels of multi-visit reconstruction quality can be achieved, even with a lower quality single-visit reconstruction. However, it is important to note that leveraging a higher quality single-visit reconstruction, particularly with the RecVarNet, yields a significant improvement ($p < 0.001$) in the evaluated metrics. The first experiment successfully validated the robustness of the MIM by demonstrating consistent improvements regardless of the specific single-visit reconstruction model used. This outcome holds significance because, as the field of single-visit MR image reconstruction continuously evolves, the MIM must be able to work with different single-visit reconstruction models to facilitate widespread clinical adoption of this technology. E2 revealed a limitation of the MIM regarding the adaptation of augmented synthetic white matter lesions. Despite the absence of these white matter lesions in a follow-up scan, they still appeared in the multi-visit reconstruction. This suggests that the MIM is not entirely reliable against all degrees of change between scans. However, a potential solution was identified during the investigation to mitigate this concern. By incorporating lesions added to the previous scan into the training of the multi-visit reconstruction module, the bias towards the previous scan is alleviated, and the MIM no longer incorrectly reconstructs lesions. Based on this observation, it is recommended to include more examples of synthetic structural changes to the brain during training to enhance the overall robustness of the MIM. There are a number of works devoted to simulating not just white matter lesions [1] but stroke and cancer lesions as well [6].

In E3, multi-visit reconstructions were generated using seven different previous subject scans from the same patient, repeatedly imaged over two years. Surprisingly, leveraging the highest-quality previous scan based on SSIM and PSNR (SSIM = 0.93, PSNR = 31.2) did not result in the highest-quality multi-visit reconstruction (SSIM = 0.959, PSNR = 35.3). Rather, leveraging the most recently acquired scan, which interestingly did not have the highest degree of similarity to the follow-up according to the quantitative metrics, resulted in the highest quality multi-visit reconstruction (SSIM = 0.96, PSNR = 36.2). Negligible differences in both quantitative and qualitative quality were observed when utilizing previous scans taken within the same year or one year prior, however, the multi-visit reconstruction incorporating information from two years prior (2017-03-09) exhibited the lowest quantitative metrics. This finding indicates that the deciding factor for the MIM in producing high-quality multi-visit reconstructions is not solely reliant on the SSIM and PSNR values computed between the previous scan and the fully sampled follow-up. These results suggest that the superior performance of the lower-quality previous scan is attributed to better agreement in structural information across scans. This, therefore, indicates that the temporal gap between scans should be taken into consideration when implementing this methodology for subjects with progressive diseases in order to optimize the effectiveness of the MIM in producing high-quality multi-visit reconstructions. Also, it is noticeable that the MIM improved the results in cases with anatomical changes between scans. The main limitation of the present work is the relatively small test set used (7 volumetric images per experiment). However, we believe our results can help guide future developments in the field.

5 Conclusion

Our work highlights the untapped potential for advancing the medical imaging field by exploring the limitations that arise with the integration of multi-visit information to enhance MR acquisition for patients with evolving brain diseases. Further work should explore validation of the MIM model using a larger dataset and include expert radiologist evaluations of the reconstructed images. Due to data scarcity, another future research avenue would be to make the MIM self-supervised by leveraging recent advances in generative models and creating synthetic pathologies to simulate longitudinal data [1,6].

Acknowledgments. This work was supported by research funding from the Natural Sciences and Engineering Research Council of Canada (NSERC) Discovery Grant program and the NSERC-Alberta Innovates Advance grant.

Disclosure of Interests. The authors have no competing interests to declare that are relevant to the content of this article.

References

1. Basaran, B.D., et al.: Lesionmix: a lesion-level data augmentation method for medical image segmentation. In: International Conference on Medical Image Computing and Computer-Assisted Intervention. pp. 73–83. Springer (2023). https://doi.org/10.1007/978-3-031-58171-7_8
2. Beauferris, Y., et al.: Leveraging multi-visit information for magnetic resonance image reconstruction: Pilot study on a cohort of glioblastoma subjects. In: 2022 IEEE 19th International Symposium on Biomedical Imaging (ISBI), pp. 1–5. IEEE (2022)
3. Beauferris, Y., et al.: Multi-coil MRI reconstruction challenge-assessing brain MRI reconstruction models and their generalizability to varying coil configurations. Front. Neurosci. **16**, 919186 (2022)
4. Cook, R.L.: Stochastic sampling in computer graphics. ACM Trans. Graph. (TOG) **5**(1), 51–72 (1986)
5. Eo, T., et al.: Kiki-net: cross-domain convolutional neural networks for reconstructing undersampled magnetic resonance images. Magn. Reson. Med. **80**(5), 2188–2201 (2018)
6. Fernandez, V., et al.: Generating multi-pathological and multi-modal images and labels for brain mri. Med. Image Anal. **97**, 103278 (2024)
7. Klein, S., et al.: Elastix: a toolbox for intensity-based medical image registration. IEEE Trans. Med. Imaging **29**(1), 196–205 (2009)
8. Mineo, J.F., et al.: Prognosis factors of survival time in patients with glioblastoma multiforme: a multivariate analysis of 340 patients. Acta Neurochir. **149**, 245–253 (2007)
9. Muckley, M.J., et al.: Results of the 2020 fastMRI challenge for machine learning mr image reconstruction. IEEE Trans. Med. Imaging **40**(9), 2306–2317 (2021)
10. Ronneberger, O., Fischer, P., Brox, T.: U-Net: convolutional networks for biomedical image segmentation. In: Navab, N., Hornegger, J., Wells, W.M., Frangi, A.F. (eds.) MICCAI 2015. LNCS, vol. 9351, pp. 234–241. Springer, Cham (2015). https://doi.org/10.1007/978-3-319-24574-4_28
11. Schlemper, J., et al.: A deep cascade of convolutional neural networks for dynamic mr image reconstruction. IEEE Trans. Med. Imaging **37**(2), 491–503 (2017)
12. Schroeder, W., et al.: The visualization toolkit an object-oriented approach to 3D graphics. Prentice-Hall Inc. (1998)

13. Souza, R., et al.: Dual-domain cascade of U-nets for multi-channel magnetic resonance image reconstruction. Magn. Reson. Imaging **71**, 140–153 (2020)
14. Souza, R., et al.: Enhanced deep-learning-based magnetic resonance image reconstruction by leveraging prior subject-specific brain imaging: Proof-of-concept using a cohort of presumed normal subjects. IEEE J. Selected Topics Signal Process. **14**(6), 1126–1136 (2020)
15. Sutherland, G., et al.: The value of radiology, part ii (2019)
16. Yiasemis, G., et al.: Recurrent variational network: a deep learning inverse problem solver applied to the task of accelerated MRI reconstruction. In: Proceedings of the IEEE/CVF Conference on Computer Vision and Pattern Recognition, pp. 732–741 (2022)
17. Wang, Z., et al.: Image quality assessment: from error visibility to structural similarity. IEEE Trans. Image Process. **13**(4), 600–612 (2004). April

NerT-CA: Efficient Dynamic Reconstruction from Sparse-View X-ray Coronary Angiography

Kirsten W. H. Maas[1]($\boxtimes$), Danny Ruijters[2,3], Nicola Pezzotti[1,2], and Anna Vilanova[1]

[1] Department of Mathematics and Computer Science, Eindhoven University of Technology, Eindhoven, The Netherlands
{k.w.h.maas,a.vilanova}@tue.nl, nicola.pezzotti@proton.me
[2] Philips Healthcare, Best, The Netherlands
danny.ruijters@philips.com
[3] Department of Electrical Engineering, Eindhoven University of Technology, Eindhoven, The Netherlands

Abstract. Three-dimensional (3D) and dynamic 3D+time (4D) reconstruction of coronary arteries from X-ray coronary angiography (CA) has the potential to improve clinical procedures. However, there are multiple challenges to be addressed, most notably, blood-vessel structure sparsity, poor background and blood vessel distinction, sparse-views, and intra-scan motion. State-of-the-art reconstruction approaches rely on time-consuming manual or error-prone automatic segmentations, limiting clinical usability. Recently, approaches based on Neural Radiance Fields (NeRF) have shown promise for automatic reconstructions in the sparse-view setting. However, they suffer from long training times due to their dependence on MLP-based representations. We propose NerT-CA, a hybrid approach of Neural and Tensorial representations for accelerated 4D reconstructions with sparse-view CA. Building on top of the previous NeRF-based work, we model the CA scene as a decomposition of low-rank and sparse components, utilizing fast tensorial fields for low-rank static reconstruction and neural fields for dynamic sparse reconstruction. Our approach outperforms previous works in both training time and reconstruction accuracy, yielding reasonable reconstructions from as few as three angiogram views. We validate our approach quantitatively and qualitatively on representative 4D phantom datasets.

Keywords: 4D reconstruction · X-ray coronary angiography · sparse-views · neural fields · tensorial fields

1 Introduction

X-ray coronary angiography (CA) remains the gold standard for diagnosing and treating coronary artery disease [6]. It captures the coronary arteries as a dynamic sequence of two-dimensional (2D) X-ray projections, referred to as coronary angiograms. These angiograms inherently represent four-dimensional

(4D) information, as they capture the dynamic 3D motion of the coronary arteries largely induced by cardiac and respiratory motion [6]. However, since these X-rays provide only 2D projections of 3D structure, depth foreshortening occurs, which can negatively impact the perception of the vessel structure. To mitigate this, multiple coronary angiograms are acquired from different viewpoints during interventions [10]. Yet, to avoid X-ray exposure, typically no more than four views are captured [7]. A 4D reconstruction could improve CA procedures by providing a comprehensive view of the coronary arteries, aiding, for example, in catheter navigation for roadmap overlays while limiting the X-ray exposure [22].

Even though clinically relevant, the problem of 4D reconstruction from CA data remains challenging. CA data exhibits several complex properties, namely the extremely sparse-view setting, intra-scan motion, and the intricate characteristics of the blood vessel, such as structure sparsity, overlapping background structures, and contrast inhomogeneity [6]. Traditional successful reconstruction methods rely on sequences acquired from many viewpoints or time-consuming manual segmentations, not fitting clinical workflows [6,10,12]. Recent approaches utilize automatic segmentation models to reconstruct from two angiogram sequences, but suffer from noise propagated from the segmentation errors [1,11,32].

Approaches building on Neural Radiance Fields (NeRFs) [19] have been proposed, which directly reconstruct from the x-ray images, avoiding reliance on segmentation models [16,17]. NeRFs are powerful for 3D and 4D reconstruction by learning individual continuous scenes from a limited number of projections [19,30]. Most notably, NeRF-CA [17] recently demonstrated reasonable reconstructions from four synthetic coronary angiograms exhibiting cardiac motion. They decompose the CA scene into a static and dynamic component, each modeled by a multi-layer perceptron (MLP). While significantly outperforming state-of-the-art X-ray NeRF-based techniques in reconstruction quality,

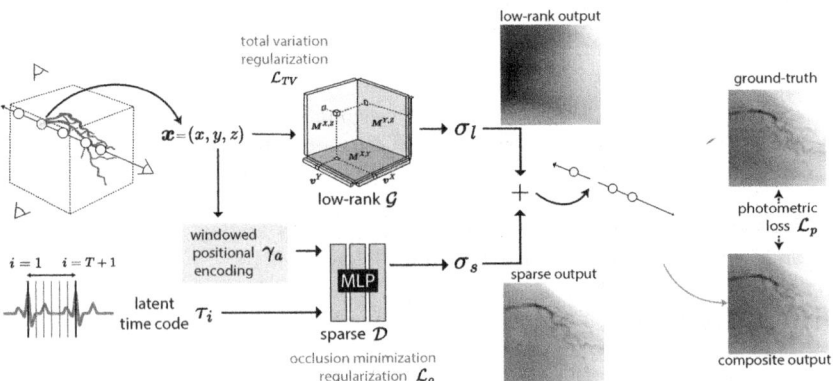

Fig. 1. NerT-CA models the dynamic CA scene as a self-supervised decomposition of low-rank and sparse components. The low-rank component is modeled by a tensorial field and the sparse component by a dynamic neural field.

their method suffers from hour-long training times. Meanwhile, many acceleration methods for general NeRFs have been proposed [5,31], which could potentially be leveraged to accelerate NeRF-CA. For example, Tensorial Radiance Field (TensoRF) [5] models scenes through fast tensorial decompositions [15], which has been shown beneficial to replace MLP-based representations in medical settings [29], but are not directly applicable to the CA setting.

Inspired by these developments, we present NerT-CA, a hybrid approach of Neural and Tensorial representations that efficiently reconstructs dynamic CA scenes from as few as three coronary angiogram views. Building on the foundation of NeRF-CA, we decompose the CA scene into static and dynamic components. In contrast, we extend this decomposition by applying low-rank and sparse decomposition constraints [9], utilizing a combination of neural and tensorial representations with tailored regularizations. Specifically, the low-rank component is modeled as an efficient tensorial field with smoothness regularization. Concurrently, the sparse component is represented as a dynamic neural field, where we impose specific regularization for the sparse-view setting. Not only does this approach gain a significant speed-up, but we also show that it benefits reconstruction quality in the sparse-view setting, outperforming NeRF-CA and state-of-the-art X-ray reconstruction techniques. We evaluate our work quantitatively and qualitatively using representative 4D phantoms [24,26].

2 Method

NerT-CA is built on top of NeRF-CA [17], a method that utilizes static and dynamic decomposition to reconstruct synthetic CA scenes with four coronary angiograms that exhibit cardiac motion. We propose a hybrid approach, combining fast tensorial fields [5] with dynamic neural fields [19,21], to accelerate this method while improving reconstruction quality. Figure 1 provides an overview of NerT-CA. We will detail each element in the next sections. The total loss function is given by

$$\mathcal{L} = \mathcal{L}_p + \lambda_{TV}\mathcal{L}_{TV} + \lambda_o\mathcal{L}_o,$$

where $\mathcal{L}_p$, $\mathcal{L}_{TV}$, and $\mathcal{L}_o$ are the photometric, total variation (TV), and occlusion minimization losses, respectively, with their according weights λ_{TV} and λ_o. Compared to NeRF-CA, this loss term is simplified by removing several regularization techniques, but introducing the $\mathcal{L}_{TV}$ term enabled by the tensorial field.

2.1 Neural and Tensorial Representations

Neural Radiance Fields (NeRFs) encode a continuous scene through a multilayer perceptron (MLP) [19]. For the X-ray setting [25], NeRF-CA learns a function $\mathcal{F}_\Theta(\boldsymbol{x}) = \sigma$, where $\boldsymbol{x} \in \mathbb{R}^3$ is a 3D spatial coordinate, σ is the attenuation coefficient, and Θ represents the weights of the MLP. To model high-frequency details, NeRF-CA applies positional encoding γ to map inputs to a higher-dimensional space, a common approach in NeRFs [19,27]. A dynamic scene is modeled as $\mathcal{D}_\Theta(\boldsymbol{x}, \tau_i) = \sigma$, where a learnable time latent-code τ_i represents each unique time

i in the scene [21]. In NeRF-CA and our method, τ_i represents discrete cardiac phases $i \in 1,\dots,T$ extracted from an ECG signal, as illustrated in Fig. 1. While powerful for high-fidelity scene reconstruction, NeRF-CA, like other NeRF architectures [20], suffers from high computational costs due to the dense sampling of 3D points needed to be evaluated by the MLP.

Tensorial Radiance Field (TensoRF) addresses the computational costs by introducing an explicit volumetric representation of the attenuation field [5]. Instead of utilizing MLPs, TensoRF decomposes the volumetric space using low-rank tensor decomposition [15]. The 3D volume is represented as a tensor $\mathcal{G} \in \mathbb{R}^{D \times D \times D}$, where D represents the resolution of the grid in the dimensions X, Y, and Z. TensoRF specifically proposes vector-matrix (VM) decomposition, where $\mathcal{G}$ is decomposed with a component that consists of orthogonal pairs of learnable vectors $\boldsymbol{v}_r^X, \boldsymbol{v}_r^Y, \boldsymbol{v}_r^Z \in \mathbb{R}^D$ and learnable matrices $\boldsymbol{M}_r^{Y,Z}, \boldsymbol{M}_r^{X,Z}, \boldsymbol{M}_r^{X,Y} \in \mathbb{R}^{D \times D}$. A number of these components or ranks $r \in \{0,\dots,R\}$, where R represents the maximum rank, are then used to approximate the tensor $\mathcal{G}$, where a higher R allows for more detailed scenes to be modeled. Specifically, $\mathcal{G}$ is represented as

$$\mathcal{G} = \sum_{r=1}^{R} \boldsymbol{v}_r^X \circ \boldsymbol{M}_r^{Y,Z} + \boldsymbol{v}_r^Y \circ \boldsymbol{M}_r^{X,Z} + \boldsymbol{v}_r^Z \circ \boldsymbol{M}_r^{X,Y},$$

where $\circ$ represents the outer product. Evaluating $\mathcal{G}(\boldsymbol{x}) = \sigma$ is performed by retrieving its corresponding values from the vectors and matrices, as schematically shown in Fig. 1. The values of these vectors and matrices are optimized with gradient descent. Trilinear interpolation within these vectors and matrices is applied to achieve continuous outputs. The advantage of this tensorial representation, in comparison to the neural representation, is that fewer parameters need to be learned for the same value [5].

2.2 Low-Rank and Sparse Decomposition

In this work, we model the dynamic CA scene as a mixture of low-rank and sparse (L+S) components [9]. Low-rank components are assumed to be stationary and low-frequency, whereas sparse components are assumed to be dynamic with minimal non-zero attenuation values. Given a CA scene as $\mathcal{H}(\boldsymbol{x},\tau_i)$, we model it as a decomposition of low-rank stationary field $\mathcal{G}(\boldsymbol{x}) = \sigma_l$ and sparse dynamic field $\mathcal{D}_\Theta(\boldsymbol{x},\tau_i) = \sigma_s$ as

$$\mathcal{H}(\boldsymbol{x},\tau_i) = \mathcal{G}(\boldsymbol{x}) + \mathcal{D}_\Theta(\gamma_a(\boldsymbol{x}),\tau_i) = \sigma_l + \sigma_s.$$

Prior work on 4D medical scenes has shown that L+S decomposition enforces temporal coherence by sparsifying dynamic components, enabling accurate separation of background and motion [9]. NeRF-CA also decomposed scenes into static and dynamic parts, but via a hard factorization that forces the points to be static or dynamic [17]. Given the sparsity of coronary structures, this can lead to missing vessels in reconstructions. We propose an L+S decomposition as it better reflects CA scenes, where contrast injection naturally produces a composition of background and blood vessel structure.

To enforce the L+S properties, we model the low-rank component as a static tensorial field $\mathcal{G}(\boldsymbol{x})$ and the sparse component as a dynamic neural field $\mathcal{D}_\Theta(\gamma_a(\boldsymbol{x}), \tau_i)$. We use a tensorial field as it is computationally fast and can be regularized to be smooth globally by using a small maximum rank R. Meanwhile, the sparse component is modeled through a dynamic neural $\mathcal{D}_\Theta(\gamma_a(\boldsymbol{x}), \tau_i)$ to model the coronary artery, similarly to NeRF-CA.

Given $\mathcal{H}(\boldsymbol{x}, \tau_i) = \sigma_l + \sigma_s$, we compute the predicted pixel intensities per cardiac phase i using the discretized Beer-Lambert law [18]. The photometric loss $\mathcal{L}_p$ is then defined as the mean-squared error (MSE) between predicted and ground-truth frame intensities, following NeRF-CA.

2.3 Low-Rank and Sparse-View Regularizations

To enable L+S decomposition in the sparse-view setting, we apply regularizations to both the tensorial and neural fields, as shown in Fig. 1. For the smooth low-rank tensorial field, we utilize total variation (TV) regularization $\mathcal{L}_{TV}$ on the learned matrices and vectors directly, similarly to TensoRF [5].

For the neural field, we apply the same two regularization techniques as in NeRF-CA, windowed positional encoding and occlusion minimization, to prevent degenerate solutions in the sparse-view setting. Windowed positional encoding γ_a introduces a coarse-to-fine training schedule to suppress high-frequency artifacts, which is implemented through an annealing schedule on the frequency encoding [28,30]. Occlusion minimization $\mathcal{L}_o$ penalizes density near the X-ray source to prevent floaters near the camera as a masked L1 regularization on the attenuation coefficients [30]. We refer to NeRF-CA [17] for further details.

3 Experimental Setup

Datasets. We evaluate on 4D phantom datasets exhibiting cardiac motion, allowing for quantitative evaluation [17,23]. We use the XCAT [26] and MAGIX [24] phantoms to generate realistic CA data of the left coronary artery (LCA). XCAT is a parameterized 4D phantom sampled at 0.5 mm^3 isotropic resolution [26] as $T = 10$ volumes. MAGIX is a 4D CCTA dataset at $0.4 \times 0.4 \times 2$ mm^3 resolution with $T = 10$ volumes. Like NeRF-CA [17], we utilize the tomographic toolbox TIGRE [2] to generate realistic CA projections of 200×200 pixels for the single-plane C-arm setting with standard clinical angles [7]. NeRF-CA defines this training setting for a minimum of four projections with four validation projections. We use the same setting, where for our three-view training setting, we exclude the projection with the most blood vessel overlap.

Evaluation Metrics. We perform quantitative and qualitative evaluations. The qualitative approach is necessary to overcome challenges in evaluating medical imaging with standard quantitative metrics [13]. Quantitatively, we focus on 2D novel view synthesis for clinical roadmap overlays [22]. Our evaluation focuses on the masked blood vessel Dice score (DSC), evaluating the overlap between the segmented blood vessel structure and the ground-truth, following NeRF-CA. This DSC score is computed by thresholding the maximum intensity projection of the predicted attenuation values. Since our focus is blood vessel reconstruction, DSC is a more suitable metric, as also demonstrated in NeRF-CA. For completeness, we additionally report common image quality metrics, namely peak signal-to-noise ratio (PSNR) and structural similarity (SSIM). We also discuss training times in minutes and inference times in frames per second (FPS). The image quality metrics are reported as the mean of 40 views, which are the 4 validation views across all time steps $T = 10$.

Implementation Details. The models were implemented in PyTorch and trained on an RTX A5000 GPU for 30,000 iterations using a ray batch size of 2048 and 256 samples per ray. Hyperparameters were selected via grid search. The tensorial field uses a grid size of $D = 48$ and maximum rank $R = 3$, while the neural field has 4 layers of 128 ReLU neurons. A softplus activation ensures positive attenuation outputs. We use the Adam optimizer with a linear learning rate schedule: 10^{-2} to 10^{-3} for the tensorial field and 10^{-1} to 10^{-2} for the neural field. To let the tensorial field first capture coarse background, neural field optimization is delayed by 1500 iterations. Hyperparameters are consistent across datasets, except for the total variation weight λ_{TV}, with $\lambda_{TV} = 10^{-3}$ for the MAGIX dataset and $\lambda_{TV} = 10^{-2}$ for the XCAT dataset, adjusted for the finer resolution of the XCAT dataset. We impose a maximum frequency band for the windowed positional encoding of 10 over 15,000 iterations. The occlusion minimization weight is $\lambda_o = 10^{-8}$, with the same distance threshold as NeRF-CA. The code repository of NerT-CA is publicly available[1].

Baselines. State-of-the-art CA reconstruction methods require input segmentations for reconstructions [8,32]. As our method is fully automatic, we compare it to state-of-the-art sparse-view reconstruction methods for X-ray data, besides the dynamic NeRF-CA. These baselines include SAX-NeRF [4], X-Gaussian [3], and R2-Gaussian [31] with default parameters, which all model static X-ray data. The latter two were introduced after NeRF-CA, demonstrating reconstructions from a minimum of 25 views. For these methods, DSC is computed from the directly predicted attenuation values, whereas for our method and NeRF-CA, they are computed from the dynamic attenuation values.

4 Results and Discussion

We report reconstruction quality across different training view settings: 3, 4, and 9 projections. Table 1 shows DSC, PSNR, and SSIM scores for XCAT and MAGIX. Our method consistently outperforms all baselines in DSC, particularly in the 3-view setting. Except for NeRF-CA, baselines fail to reconstruct vessel structures, reflected in low DSC scores. Our method generally achieves better PSNR and SSIM, with occasional exceptions where NeRF-CA scores higher

---

[1] https://github.com/kirstenmaas/NerT-CA.

Table 1. Quantitative results for the XCAT and MAGIX datasets for 3, 4, and 9 training projections. Bolding is used to indicate the best score.

Dataset	Method	3-view			4-view			9-view		
		DSC	PSNR	SSIM	DSC	PSNR	SSIM	DSC	PSNR	SSIM
XCAT	SAX-NeRF	0.00	13.49	0.58	0.00	12.89	0.58	0.00	11.44	0.55
	X-Gaussian	0.05	12.58	0.49	0.22	12.53	0.51	0.37	9.30	0.49
	R2-Gaussian	0.01	13.02	0.57	0.00	12.87	0.58	0.49	15.77	0.71
	NeRF-CA	0.41	11.69	0.65	0.76	11.66	0.72	0.74	**16.25**	0.83
	Ours	**0.75**	**15.34**	**0.79**	**0.84**	**15.16**	**0.80**	**0.87**	14.12	**0.84**
MAGIX	SAX-NeRF	0.00	13.70	0.71	0.00	14.17	0.72	0.00	13.72	0.74
	X-Gaussian	0.26	13.73	0.65	0.20	14.54	0.68	0.47	14.20	0.67
	R2-Gaussian	0.00	13.09	0.55	0.00	13.90	0.58	0.00	16.24	0.70
	NeRF-CA	0.82	**18.46**	0.80	0.81	13.34	0.72	0.90	18.49	0.85
	Ours	**0.88**	14.97	**0.82**	**0.90**	**16.70**	**0.86**	**0.90**	**18.92**	**0.88**

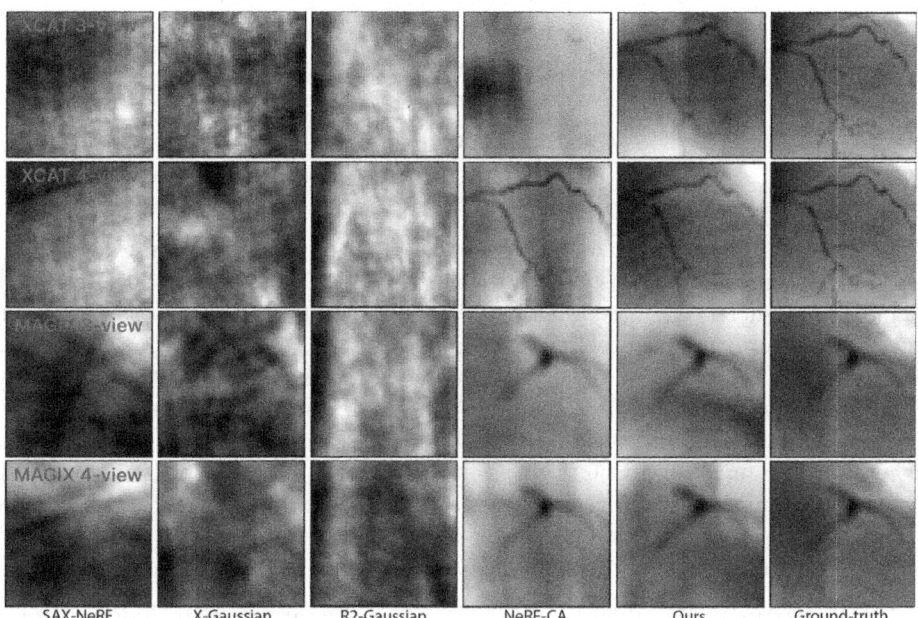

Fig. 2. Qualitative results for XCAT (3 and 4-view) and MAGIX (3 and 4-view).

PSNR, likely due to background oversmoothing from TV regularization. Figure 2 shows qualitative results at time point $i = 1$. The baselines yield degenerate reconstructions due to background distractions, except NeRF-CA. Compared to NeRF-CA, our method recovers finer vessels for 4 projections and robustly reconstructs in the 3-view setting on XCAT, with comparable quality on MAGIX.

We also analyze the evolution of DSC score during training for the 3-view XCAT setting compared to the baselines, as shown in Fig. 3. SAX-NeRF [4] fails to capture a blood vessel structure within one hour, leading to a zero DSC score.

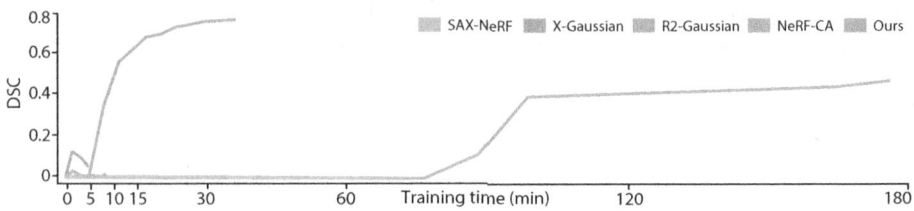

Fig. 3. Dice (DSC) scores over training time in minutes for the XCAT 3 training view setting for our method and the baselines.

Table 2. Ablation of the regularizations in three-view training setting: total variation $\mathcal{L}_{TV}$, windowed positional encoding γ_a, and occlusion minimization $\mathcal{L}_o$.

			XCAT			MAGIX		
$\mathcal{L}_{TV}$	γ_a	$\mathcal{L}_o$	DSC	PSNR	SSIM	DSC	PSNR	SSIM
×	✓	×	0.09	13.97	0.69	0.84	**15.65**	0.70
✓	✓	×	0.58	11.82	0.74	0.54	15.24	0.80
×	✓	✓	0.58	14.32	0.71	0.86	14.37	0.67
✓	✓	✓	**0.75**	**15.34**	**0.79**	**0.88**	14.97	**0.82**

The Gaussian-based baselines [3,31] converge within minutes, but they fail to produce reconstructions, as reflected by their low DSC scores. NeRF-CA [17] achieves a higher DSC score but requires hours-long training time. Notably, our method outperforms all baselines in DSC score with a running time of 37 min. For the other dataset and projection settings, we observe similar trends.

Finally, we report detailed training and inference times. As we have shown, only NeRF-CA and our method are able to reconstruct blood vessel structures. Therefore, we only report the times for these methods. On average, NeRF-CA requires 361 min for training and runs at 0.5 FPS, while our method trains in 37 min and runs at 2 FPS. Therefore, we achieve an approximate 10× speedup in training and 4× in inference, while also improving reconstruction quality.

Ablation Study. We ablate the effect of the regularization techniques, total variation $\mathcal{L}_{TV}$, windowed positional encoding γ_a, and occlusion minimization $\mathcal{L}_o$, in Table 2. Enforcing all regularizations yields the highest DSC scores, highlighting their importance and robustness across both datasets. For XCAT, the regularizations significantly improve vessel reconstruction, while for the lower-resolution MAGIX dataset, improvements are smaller.

5 Conclusion

We propose NerT-CA, an efficient 4D reconstruction method for synthetic CA scenes from as few as three angiogram views. Our method extends a prior NeRF-based approach by modeling the scene as a low-rank and sparse decomposition using a combination of tensorial and neural fields with dedicated regularizations. Experiments demonstrate that NerT-CA reduces training time by a factor of 10 while significantly improving reconstruction quality in sparse-view settings. We conduct an ablation study to validate the importance of our regularization components. Future work could explore integrating 3D Gaussian Splatting [14] into our decomposition framework for further acceleration.

Acknowledgments. This research was performed within the Medusa project as part of the Eindhoven MedTech Innovation Center research collaboration between Eindhoven University of Technology, Philips Healthcare, and Catharina Ziekenhuis Eindhoven.

Disclosure of Interests. The authors have no competing interests to declare that are relevant to the content of this article.

References

1. Bappy, D., Hong, A., Choi, E., Park, J.O., Kim, C.S.: Automated three-dimensional vessel reconstruction based on deep segmentation and bi-plane angiographic projections. Comput. Med. Imaging Graph. **92**, 101956 (2021)
2. Biguri, A., Dosanjh, M., Hancock, S., Soleimani, M.: TIGRE: a MATLAB-GPU toolbox for CBCT image reconstruction. Biomed. Phys. Eng. Exp. **2**(5), 055010 (2016)
3. Cai, Y., et al.: Radiative gaussian splatting for efficient X-ray novel view synthesis. In: European Conference on Computer Vision, pp. 283–299. Springer (2024)
4. Cai, Y., Wang, J., Yuille, A., Zhou, Z., Wang, A.: Structure-aware sparse-view X-ray 3D reconstruction. In: Proceedings of the IEEE/CVF Conference on Computer Vision and Pattern Recognition, pp. 11174–11183 (2024)
5. Chen, A., Xu, Z., Geiger, A., Yu, J., Su, H.: Tensorf: tensorial radiance fields. In: European Conference on Computer Vision, pp. 333–350. Springer (2022)
6. Çimen, S., Gooya, A., Grass, M., Frangi, A.F.: Reconstruction of coronary arteries from X-ray angiography: a review. Med. Image Anal. **32**, 46–68 (2016)
7. Di Mario, C., Sutaria, N.: Coronary angiography in the angioplasty era: projections with a meaning. Heart **91**(7), 968–976 (2005)
8. Fu, X., et al.: 3DGR-CAR: coronary artery reconstruction from ultra-sparse 2D X-ray views with a 3D gaussians representation. In: International Conference on Medical Image Computing and Computer-Assisted Intervention, pp. 14–24. Springer (2024)
9. Gao, H., Cai, J.F., Shen, Z., Zhao, H.: Robust principal component analysis-based four-dimensional computed tomography. Phys. Med. Biol. **56**(11), 3181 (2011)
10. Green, N.E., Chen, S.Y.J., Messenger, J.C., Groves, B.M., Carroll, J.D.: Three-dimensional vascular angiography. Curr. Probl. Cardiol. **29**(3), 104–142 (2004)
11. Hwang, M., et al.: A simple method for automatic 3D reconstruction of coronary arteries from X-ray angiography. Front. Physiol. **12**, 724216 (2021)
12. Iyer, K., Nallamothu, B.K., Figueroa, C.A., Nadakuditi, R.R.: A multi-stage neural network approach for coronary 3D reconstruction from uncalibrated X-ray angiography images. Sci. Rep. **13**(1), 17603 (2023)
13. Kastryulin, S., Zakirov, J., Pezzotti, N., Dylov, D.V.: Image quality assessment for magnetic resonance imaging. IEEE Access **11**, 14154–14168 (2023)

14. Kerbl, B., Kopanas, G., Leimkühler, T., Drettakis, G.: 3D gaussian splatting for real-time radiance field rendering. ACM Trans. Graph. **42**(4), 139-1 (2023)
15. Kolda, T.G., Bader, B.W.: Tensor decompositions and applications. SIAM Rev. **51**(3), 455–500 (2009)
16. Kshirsagar, J., et al.: Generative AI-assisted novel view synthesis of coronary arteries for angiography. In: 2024 IEEE International Symposium on Medical Measurements and Applications (MeMeA), pp. 1–6. IEEE (2024)
17. Maas, K., Ruijters, D., Vilanova, A., Pezzotti, N.: NeRF-CA: dynamic reconstruction of X-ray coronary angiography with extremely sparse-views. IEEE Trans. Vis. Comput. Graph. 1–14 (2025)
18. Max, N.: Optical models for direct volume rendering. IEEE Trans. Visual Comput. Graphics **1**(2), 99–108 (2002)
19. Mildenhall, B., Srinivasan, P.P., Tancik, M., Barron, J.T., Ramamoorthi, R., Ng, R.: Nerf: representing scenes as neural radiance fields for view synthesis. Commun. ACM **65**(1), 99–106 (2021)
20. Müller, T., Evans, A., Schied, C., Keller, A.: Instant neural graphics primitives with a multiresolution hash encoding. ACM Trans. Graph. (TOG) **41**(4), 1–15 (2022)
21. Park, K., et al.: Nerfies: deformable neural radiance fields. In: Proceedings of the IEEE/CVF International Conference on Computer Vision, pp. 5865–5874 (2021)
22. Piayda, K., et al.: Dynamic coronary roadmapping during percutaneous coronary intervention: a feasibility study. Eur. J. Med. Res. **23**, 1–7 (2018)
23. Rohkohl, C., Lauritsch, G., Keil, A., Hornegger, J.: Cavarevan open platform for evaluating 3D and 4D cardiac vasculature reconstruction. Phys. Med. Biol. **55**(10), 2905 (2010)
24. Rosset, A., Spadola, L., Ratib, O.: Osirix: an open-source software for navigating in multidimensional DICOM images. J. Digit. Imaging **17**, 205–216 (2004)
25. Rückert, D., Wang, Y., Li, R., Idoughi, R., Heidrich, W.: Neat: neural adaptive tomography. ACM Trans. Graph. (TOG) **41**(4), 1–13 (2022)
26. Segars, W.P., Sturgeon, G., Mendonca, S., Grimes, J., Tsui, B.M.: 4D XCAT phantom for multimodality imaging research. Med. Phys. **37**(9), 4902–4915 (2010)
27. Tancik, M., et al.: Fourier features let networks learn high frequency functions in low dimensional domains. Adv. Neural. Inf. Process. Syst. **33**, 7537–7547 (2020)
28. Wu, T., Zhong, F., Tagliasacchi, A., Cole, F., Oztireli, C.: D2nerf: self-supervised decoupling of dynamic and static objects from a monocular video. Adv. Neural. Inf. Process. Syst. **35**, 32653–32666 (2022)
29. Yang, C., Wang, K., Wang, Y., Dou, Q., Yang, X., Shen, W.: Efficient deformable tissue reconstruction via orthogonal neural plane. IEEE Trans. Med. Imaging (2024)
30. Yang, J., Pavone, M., Wang, Y.: Freenerf: improving few-shot neural rendering with free frequency regularization. In: Proceedings of the IEEE/CVF Conference on Computer Vision and Pattern Recognition, pp. 8254–8263 (2023)
31. Zha, R., Lin, T.J., Cai, Y., Cao, J., Zhang, Y., Li, H.: R2-gaussian: rectifying radiative gaussian splatting for tomographic reconstruction. Adv. Neural. Inf. Process. Syst. **37**, 44907–44934 (2025)
32. Zhu, Y., Wang, Y., Di, C., Liu, H., Liao, F., Ma, S.: Sparse and transferable three-dimensional dynamic vascular reconstruction for instantaneous diagnosis. Nat. Mach. Intell. 1–13 (2025)

INR Meets Multi-contrast MRI Reconstruction

Natascha Niessen[1,2(✉)], Carolin M. Pirkl[2], Ana Beatriz Solana[2], Hannah Eichhorn[1,3], Veronika Spieker[1,3], Wenqi Huang[1], Tim Sprenger[2,4], Marion I. Menzel[2,5,6], Julia A. Schnabel[1,3,7], and on behalf of the PREDICTOM consortium

[1] School of Computation, Information and Technology, Technical University of Munich, Munich, Germany
natascha.niessen@tum.de
[2] GE HealthCare, Munich, Germany
[3] Institute of Machine Learning in Biomedical Imaging, Helmholtz Munich, Neuherberg, Germany
[4] Department of Clinical Neuroscience, Karolinska Institutet, Stockholm, Sweden
[5] Technische Hochschule Ingolstadt, Ingolstadt, Germany
[6] School of Natural Sciences, Technical University of Munich, Munich, Germany
[7] King's College London, London, UK

Abstract. Multi-contrast MRI sequences allow for the acquisition of images with varying tissue contrast within a single scan. The resulting multi-contrast images can be used to extract quantitative information on tissue microstructure. To make such multi-contrast sequences feasible for clinical routine, the usually very long scan times need to be shortened e.g. through undersampling in k-space. However, this comes with challenges for the reconstruction. In general, advanced reconstruction techniques such as compressed sensing or deep learning-based approaches can enable the acquisition of high-quality images despite the acceleration. In this work, we leverage redundant anatomical information of multi-contrast sequences to achieve even higher acceleration rates. We use undersampling patterns that capture the contrast information located at the k-space center, while performing complementary undersampling across contrasts for high frequencies. To reconstruct this highly sparse k-space data, we propose an implicit neural representation (INR) network that is ideal for using the complementary information acquired across contrasts as it jointly reconstructs all contrast images. We demonstrate the benefits of our proposed INR method by applying it to multi-contrast MRI using the MPnRAGE sequence, where it outperforms the state-of-the-art parallel imaging compressed sensing (PICS) reconstruction method, even at higher acceleration factors. Our code is available at https://github.com/compai-lab/2025-miccai-niessen.

Supplementary Information The online version contains supplementary material available at https://doi.org/10.1007/978-3-032-06103-4_3.

© The Author(s), under exclusive license to Springer Nature Switzerland AG 2026
L. Felsner et al. (Eds.): RIME 2025/GRAIL 2025, LNCS 16150, pp. 23–33, 2026.
https://doi.org/10.1007/978-3-032-06103-4_3

Keywords: Implicit Neural Representation · MRI Reconstruction · Multi-Contrast MRI · Quantitative MRI

1 Introduction

Quantitative assessment of brain tissue microstructure is essential for identifying novel biomarkers e.g. associated with neurodegenerative diseases such as multiple sclerosis and Alzheimer's disease. Multi-contrast magnetic resonance imaging (MRI) sequences [1] provide a rich set of comprehensive tissue contrasts and enable extraction of quantitative information within a single scan by acquiring images with varying contrast, through changes in experimental parameters such as inversion time or echo time. However, these sequences are often time-consuming, limiting clinical feasibility and increasing the risk of motion artifacts. To achieve shorter scan times, one approach is to acquire fewer k-space samples. However, undersampling k-space data is known to produce artifacts and low SNR, oftentimes rendering the MR images clinically meaningless.

Many multi-contrast MRI sequences have in common, that the anatomy of the object being scanned remains constant, disregarding potential motion, field inhomogeneity etc., while the contrast changes with the experimental parameters. In this work, we leverage this property to achieve high acceleration rates for multi-contrast MRI sequences in neuro applications through complementary undersampling in k-space along the additional dimension.

Deep learning-based reconstruction show promising results for accelerated MRI [2,3]. However, supervised approaches require large datasets for training and lack generalizability e.g. with respect to varying MRI sequences. In contrast, self-supervised approaches reconstruct the MR image only based on the acquired data itself which is ideal for multi-contrast settings where training data is very scarce. One prominent self-supervised architecture that emerged from the field of computer vision is implicit neural representation (INR) [4] that allows to represent images or objects continuously with the help of a small neural network and an input encoding. The INR architecture enables a patient-specific, compact representation of high-dimensional multi-contrast images with fewer network parameters, capturing shared information and evolving contrast in a compressed form. Our approach is inspired by recent work leveraging INRs for dynamic MRI [5,6]. In applications such as cardiac cine MRI, the heart as an object remains the same, while undergoing temporal deformation due to cardiac motion. Analogously, in the neuroimaging context we target, multi-contrast sequences yield a series of images in which the brain's anatomy remains the same, while the image contrast evolves across acquisitions.

In this work, we propose an INR based reconstruction framework for accelerated multi-contrast MRI acquisitions. We evaluate our method through retrospective variable density Poisson disk undersampling of fully sampled k-space data of the brain, acquired with the MPnRAGE sequence [7], that provides multiple inversion contrast images. Beyond the rich contrast information provided by MPnRAGE, the resulting images can be used for fitting a quantitative T1 map or to investigate tissue nulling, similar to fluid suppression in FLAIR [8].

2 Related Work

INRs for MRI Reconstruction. Recent works use INRs for undersampled MRI reconstruction either directly in k-space [9,10] or, like in our work, in image domain [5,6,11–15]. However, most of these methods focus on dynamic MRI [12] or more specifically cardiac cine MRI with radial acquisitions [5,6,11]. Lao et al. target multi-parametric quantitative MRI [14] and for Shen et al. [15] a prior reconstructed MR image is required.

Deep Learning-Based Multi-contrast MRI Reconstruction. Joint multi-contrast reconstruction with complementary undersampling is explored for a combination of T1w, T2w and T2-FLAIR images with an unrolled neural network by Polak et al. [16]. A method proposed by Lei et al. [17] guides the unrolled reconstruction of T2w images with the help of fully sampled proton density (PD) reference images. Both approaches are supervised methods requiring large training datasets. Moreover, these methods may be susceptible to image registration effects, a limitation that our approach circumvents by acquiring data simultaneously.

3 Method

3.1 Multi-contrast MRI Reconstruction

In multi-contrast MRI reconstruction, the goal is to reconstruct multiple images $\mathbf{d} \in \mathbb{C}^{(V_y \times V_z) \times N}$ of the same object, acquired with a different contrast $n = 1, ..., N$. In this work, we consider 2D slices of the object with image dimensions V_y and V_z as k_x corresponds to the readout direction. In MRI, noisy linear measurements $\mathbf{D}_c \in \mathbb{C}^{(V_y \times V_z) \times N}$ are acquired according to the forward model

$$\mathbf{D}_c = \mathbf{MFS}_c \mathbf{d} + \mathbf{e}_c \tag{1}$$

at the receiver coils $c = 1, .., C$, where $\mathbf{M}$ is a binary undersampling mask, $\mathbf{F}$ is the 2D Fourier matrix and $\mathbf{S}_c$ is a diagonal matrix containing the coil sensitivity maps of the receiver coil c. The measurements are distorted by additive Gaussian noise $\mathbf{e}_c$.

Complementary Undersampling. In our multi-contrast setup, we consider the undersampling mask $\mathbf{M} = [\mathbf{M}_1, ..., \mathbf{M}_N]$ to be different for each of the $n = 1, ..., N$ acquired contrasts. To sufficiently encode the contrast evolution, we densely sample the k-space center, while the outer k-space is sampled sparsely and complementarily across contrasts. More specifically, we use variable density Poisson disk sampling, a strategy that places sampling points in k-space in a pseudo-random yet spatially balanced manner (see Fig. 1A). This introduces incoherent, noise-like aliasing artifacts, which are effectively mitigated by deep learning-based reconstruction methods [3]. By using different random seeds for each contrast, we generate complementary sampling patterns for each multi-contrast data which collectively enhance k-space coverage.

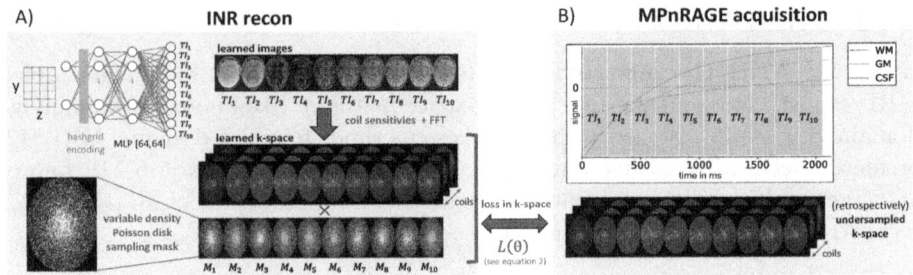

Fig. 1. A) Our proposed INR reconstruction framework for accelerating multi-contrast MRI sequences through complementary undersampling in k-space demonstrated for the MPnRAGE sequence. B) Exemplary inversion recovery signal curves for white matter (WM), gray matter (GM) and cerebrospinal fluid (CSF). The zero-crossings of the different tissues are visible in the TI contrast images in A (learned images). (Color figure online)

3.2 Implicit Neural Representation

For reconstruction, we represent the images with an INR network, $\mathbf{d} = G_\theta(\mathbf{y}, \mathbf{z})$ with model parameters θ, similarly to Huang et al. [5], but jointly embed all multi-contrast images. The INR network learns a continuous mapping $G_\theta(\cdot)$: $\mathbb{R}^2 \to \mathbb{C}^N$ from the image coordinates $(\mathbf{y}, \mathbf{z}) \in \mathbb{R}^2$ to the complex voxel-wise signal evolution with N different contrasts. By jointly reconstructing all contrast images, the INR enforces data consistency for each individual contrast. It leverages the shared anatomical structure present across contrasts, made accessible through the complementary undersampling strategy. Although the inference operates on a voxel-wise basis, the INR captures the entire encoded spatiotemporal information since all contrast images are reconstructed jointly.

Loss Function in k-Space with Distance Weighting. We constructed the loss function such that it promotes each individual contrast image to be in accordance with its acquired k-space data. To calculate the loss at each epoch, we multiply the learned images with the coil sensitivity maps and apply a fast fourier transform (FFT) to the resulting coil images. Hereby, we obtain what we refer to as the "learned k-space" (see Fig. 1A). For data consistency, we apply the complementary undersampling masks, to the learned k-space data. Finally, we compare the resulting masked learned k-space data to the acquired k-space data by calculating the mean-squared error

$$L(\theta) = \sum_{c=1}^{C} \left\| W(\mathbf{k}_y, \mathbf{k}_z) \cdot [\mathbf{MFS}_c G_\theta(\mathbf{y}, \mathbf{z}) - \mathbf{D}_c] \right\|_2^2 \tag{2}$$

using weights proportional to the Euclidean distance from the k-space center

$$W(\mathbf{k}_y, \mathbf{k}_z) = \sqrt{\mathbf{k}_y^2 + \mathbf{k}_z^2} + 1 \tag{3}$$

These weights ensure that all k-space samples contribute to the loss function, despite the high-dynamic range between high intensities at the k-space center and low intensities at higher frequencies.

INR Architecture. The INR network that we use for reconstruction consists of a multi-layer perceptron (MLP) with two hidden linear layers of 64 neurons each, using ReLu as activation function. The relatively low dimensionality of the MLP is possible thanks to the hashgrid encoding [18] of the input coordinates that has proven to contribute to the reconstruction performance of the INR [5,6]. We use the same hyperparameters for all datasets, to show the generalizability of our proposed method that does not require any subject-specific hyperparameter tuning. The INR is optimized per slice, where the final epoch of training simultaneously yields the inference result.

3.3 Data Acquisition Using the MPnRAGE Sequence

While MPRAGE is the state-of-the-art sequence to obtain one single T1-weighted image, MPnRAGE allows to obtain **N** inversion time (TI) images that correspond to different T1 weightings and hence varying tissue contrasts. The scan duration of state of the art MPnRAGE sequences of around 15 min is considered clinically infeasible, but a recent work investigates the sparsity of the sequence through subspace compression and thus proves that there is a potential for acceleration in the TI dimension [19].

The TI images are acquired along the inversion recovery curve, following an inversion preparation module. Figure 1B visualizes the acquisition for $N = 10$, showing that k-space data acquired within a certain TI interval is binned, to form one TI image. The inversion recovery is repeated several times in order to acquire the entire 3D+TI k-space in an iterative way. This segmented k-space readout is performed in 3D with k_x as readout direction. This enables us to apply a one dimensional FFT in k_x direction and reconstruct each axial slice individually. To simulate the acceleration, we retrospectively undersample in the $k_y - k_z$ plane corresponding to the axial view. The fully sampled Cartesian 3D MPnRAGE sequence was implemented in KSFoundation [20]. Healthy volunteer data from three subjects were acquired on a DISCOVERY^{TM} MR750w (GE HealthCare, Waukesha, USA) in the sagittal scan orientation with the following parameters: FOV $24 \times 24 \times 18\,\text{cm}^2$, matrix size $160 \times 160 \times 120$, for 10 TIs ($TI_1 = 26$ ms, $\Delta_{TI} = 249.05$ ms), imaging TR 4.98 ms, imaging flip angle 4°, MPnRAGE TR 2.7 s, acquisition time 13.47 min.

Analysis. For evaluation, we apply a brain mask calculated using the FSL toolbox [21] and perform percentile normalization jointly across all contrasts and slices [22]. The resulting masked and normalized data is then used to compute the structural similarity index measure (SSIM) and the peak signal-to-noise ratio (PSNR) relative to the fully sampled reference.

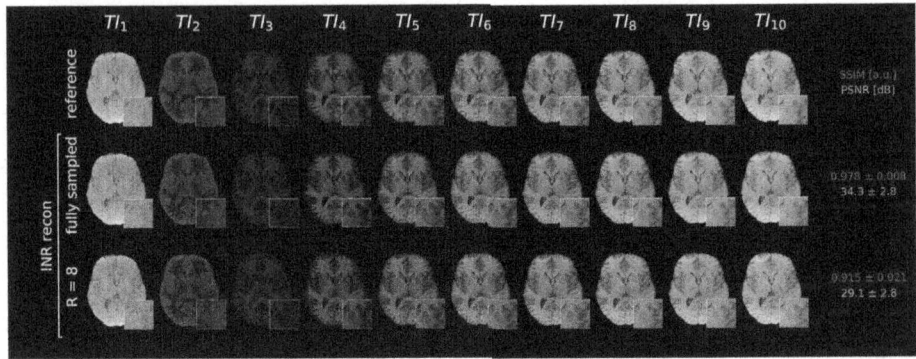

Fig. 2. $N = 10$ inversion time contrast images acquired with the MPnRAGE sequence, reconstructed for fully sampled data with an inverse FFT (reference) and using our joint INR reconstruction for fully sampled and undersampled data (R = 8). The metrics correspond to the shown slice and were averaged over all contrasts.

4 Experiments and Results

Denoising Effect: Fully-Sampled INR Recon vs. Reference Image First, we evaluate the INR framework for fully sampled data. To obtain a reference image, we reconstruct the same fully sampled data with a standard inverse FFT using coil sensitivity maps. In Fig. 2 we compare the reconstructions of the 10 multi-contrast images with the proposed INR framework, based on the fully sampled k-space data, to the reference images. The INR-obtained images preserve fine anatomical details, as evidenced by a high SSIM of 0.978 ± 0.008 and a PSNR of 34.3 ± 2.8 dB, averaged over all slices and contrast images. The denoising effect due to the inherent regularization of the INR becomes clearly visible in the zoomed insets (Fig. 2). The unified training-inference process takes approximately 0.8 s per slice.

Acceleration Through Undersampling: INR Recon vs. PICS. Next, we retrospectively undersample the fully sampled k-space data to simulate the acceleration of the multi-contrast sequence at R = 4, 8 and 12. We compare our method to PICS [23], a classical state-of-the-art approach for mitigating incoherent artifacts, such as those introduced by our random undersampling pattern. The SSIM and PSNR metrics in Table 1 demonstrate that our proposed INR reconstruction framework performs comparably to PICS in the fully sampled case. For accelerations up to R = 12, the SSIM and PSNR of the INR method decline only slightly, indicating strong robustness to undersampling. In contrast, PICS shows a much stronger drop in both metrics, reflecting a significant degradation in image quality which can be seen in the example images in Fig. 3. The proposed INR framework preserves fine structural details even at high acceleration rates.

Table 1. SSIM [a.u.] and PSNR [dB] across different acceleration rates for INR reconstruction and PICS, averaged over all volunteers, contrasts, and slices.

		Fully Sampled	R = 4	R = 8	R = 12
INR recon	SSIM	0.981 ± 0.002	**0.959 ± 0.005**	**0.947 ± 0.006**	**0.935 ± 0.008**
	PSNR	33.9 ± 1.4	31.5 ± 0.7	**30.6 ± 0.9**	**30.0 ± 0.6**
PICS	SSIM	**0.987 ± 0.001**	0.953 ± 0.004	0.924 ± 0.005	0.903 ± 0.006
	PSNR	**36.0 ± 0.9**	**31.8 ± 0.3**	29.4 ± 0.1	28.0 ± 0.2

Cross-Plane Contrast Continuity and Spatial Alignment. As previously mentioned, the reconstruction is performed independently for each axial slice. In Fig. 4 we visualize sagittal and coronal planes extracted from the reconstructed axial slice stack to show the cross-plane contrast continuity achieved with our proposed INR framework. Additionally, the individually reconstructed slices are accurately aligned in space, which demonstrates that our approach maintains the global coordinate space through the data consistency enforced by the loss function in k-space.

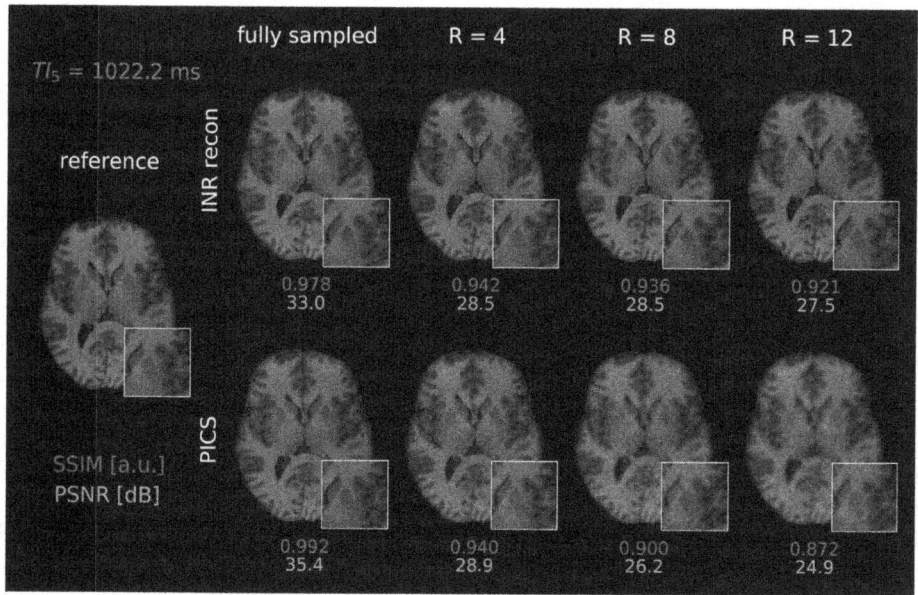

Fig. 3. Exemplary reconstructed contrast images ($TI_5 = 1022.2$ ms) resulting from our joint INR reconstruction and PICS, for fully sampled data and various acceleration factors R. An inverse FFT of the fully sampled data serves as reference image. The metrics correspond to the shown slice and contrast.

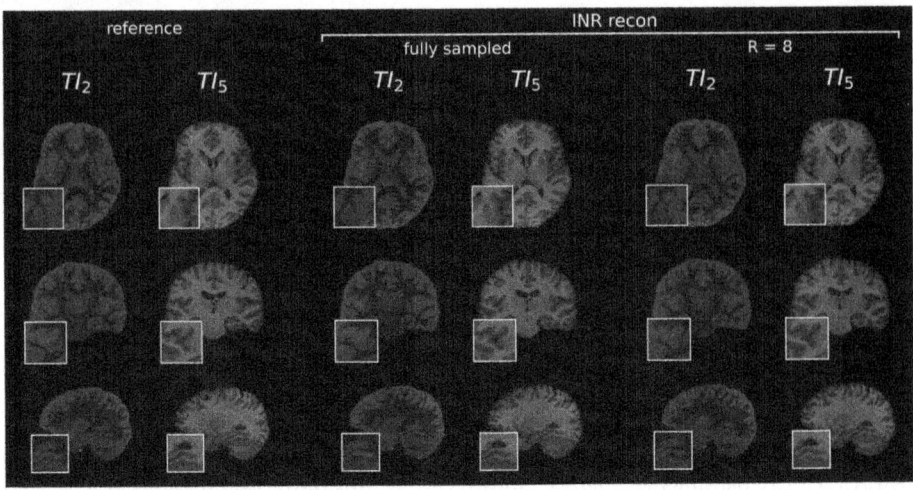

Fig. 4. Reconstructed axial slices, as well as coronal and sagittal views extracted from the stack of axial slices obtained using our joint INR reconstruction with fully sampled and undersampled (R = 8) data, for two exemplary inversion time (TI) contrasts. An inverse FFT of the fully sampled data serves as reference image.

5 Discussion and Conclusion

In this work, we propose a joint INR-based reconstruction framework that leverages shared information across multi-contrast images to produce high-quality reconstructions at high acceleration factors. The proposed method was evaluated using an inversion recovery contrast sequence and based on the shared underlying MR physics, is anticipated to generalize effectively to other multi-contrast sequences, such as T2-weighted imaging. In the fully sampled case, our method performs comparably to PICS. The slight performance gap is likely due to the INR's inherent regularization, which promotes spatial smoothness and therefore acts as a denoiser. This denoising has a negative effect on the metrics, as the reference images themselves contain noise. The reference image for one of the volunteers showed blurring, potentially introduced by slight motion. Our INR framework was able to mitigate parts of this blurring, as observed in the fully sampled case and even at an acceleration factor of R = 4 (Supplementary Material, Fig. 5). This demonstrates the potential of our joint INR-based reconstruction, which leverages data acquired at different time points across all contrasts by embedding them within a shared network.

At high acceleration factors up to R = 12, which corresponds to a shortening of the scan time from initially 13.47 min to 1.12 min, our joint INR framework significantly outperforms PICS both visually and quantitatively. The presented metrics are very promising regarding the robustness of our joint INR framework to increasingly undersampled data, from which we conclude that even higher accelerations could be achieved. However, for Poisson disk sampling, generating even more sparse sampling patterns becomes algorithmically complex, which is why for future work we also consider exploring other undersampling strategies, potentially even in 3D, or optimizing the complementary sampling instead of relying on varying random seeds. The INR reconstructions tend to exhibit increased Gibbs ringing at an acceleration factor of R = 4. A deeper investigation into how the choice of undersampling trajectory interacts with INR-based reconstruction methods to reduce such artifacts is planned as part of our ongoing research efforts. In future work, we aim to implement the most promising sampling strategy in a prospectively accelerated acquisition and compare the reconstruction also to other deep learning-based reconstructions. This will enable high-quality imaging and extremely efficient acquisition of multi-contrast datasets for improved diagnostics and biomarker discovery.

Conclusion. We introduce a joint INR-based reconstruction framework with complementary undersampling for accelerated multi-contrast MRI acquisitions. Our method was validated through retrospective undersampling using variable-density Poisson disk patterns applied to fully sampled in vivo data acquired with the MPnRAGE sequence, providing multiple inversion contrast images. The resulting images demonstrate that our approach enables high-quality reconstructions at substantially reduced scan time.

Acknowledgments. This work is supported by the DAAD programme Konrad Zuse Schools of Excellence in Artificial Intelligence and the Munich Center for Machine Learning, both sponsored by the Federal Ministry of Research, Technology and Space. PREDICTOM is supported by the Innovative Health Initiative Joint Undertaking (IHI JU), under Grant Agreement No 101132356. JU receives support from the European Union's Horizon Europe research and innovation programme, COCIR, EFPIA, EuropaBio, MedTechEurope and Vaccines Europe. The UK participants are supported by UKRI Grant No 10083467 (National Institute for Health and Care Excellence), Grant No 10083181 (King's College London), and Grant No 10091560 (University of Exeter). University of Geneva is supported by the Swiss State Secretariat for Education, Research and Innovation Ref No 113152304. See www.ihi.europa.eu for more details.

Disclosure of Interests. NN, CMP, ABS, TS, and MIM are employees of GE HealthCare, Munich. All other authors declare that they do not have any financial or non-financial conflict of interests.

References

1. Seiberlich, N., et al.: Quantitative Magnetic Resonance Imaging, 1st edn, vol. 1. Elsevier (2020)
2. Hammernik, K., et al.: Physics-driven deep learning for computational magnetic resonance imaging: combining physics and machine learning for improved medical imaging. IEEE Signal Process. Mag. **40**(1), 98–114 (2023)
3. Heckel, R., Jacob, M., Chaudhari, A., Perlman, O., Shimron, E.: Deep learning for accelerated and robust MRI reconstruction. Magn. Reson. Mater. Phy. (2024)

4. Mildenhall, B., Srinivasan, P.P., Tancik, M., Barron, J.T., Ramamoorthi, R., Ng, R.: NeRF: representing scenes as neural radiance fields for view synthesis. Commun. ACM **65**(1), 99–106 (2021)
5. Huang, W., et al.: Subspace Implicit Neural Representations for Real-Time Cardiac Cine MR Imaging, arXiv preprint arXiv:2412.12742 (2024)
6. Feng, J., et al.: Spatiotemporal implicit neural representation for unsupervised dynamic MRI reconstruction. IEEE Trans. Med. Imaging 1 (2025)
7. Kecskemeti, S., Samsonov, A., Hurley, S.A., Dean, D.C., Field, A., Alexander, A.L.: MPnRAGE: a technique to simultaneously acquire hundreds of differently contrasted MPRAGE images with applications to quantitative T1 mapping. Magn. Reson. Med. **75**(3), 1040–1053 (2016)
8. Hajnal, J.V., et al.: Use of fluid attenuated inversion recovery (FLAIR) pulse sequences in MRI of the brain. J. Comput. Assist. Tomogr. **16**(6), 841–844 (1992)
9. Huang, W., Li, H.B., Pan, J., Cruz, G., Rueckert, D., Hammernik, K.: Neural implicit k-space for binning-free non-cartesian cardiac MR imaging. In: Information Processing in Medical Imaging (IPMI 2023). Lecture Notes in Computer Science, vol. 13939, pp. 548–560. Springer (2023)
10. Spieker, V., et al.: ICoNIK: generating respiratory-resolved abdominal MR reconstructions using neural implicit representations in k-space. In: Lecture Notes in Computer Science, vol. 14533, pp. 183–192 (2024)
11. Catalán, T., Courdurier, M., Osses, A., Botnar, R., Sahli Costabal, F., Prieto, C.: Unsupervised reconstruction of accelerated cardiac cine MRI using neural fields. Comput. Biol. Med. **185** (2025)
12. Kunz, J.F., Ruschke, S., Heckel, R.: Implicit neural networks with fourier-feature inputs for free-breathing cardiac MRI reconstruction. IEEE Trans. Comput. Imaging **10**, 1280–1289 (2024)
13. Feng, R., et al.: IMJENSE: scan-specific implicit representation for joint coil sensitivity and image estimation in parallel MRI. IEEE Trans. Med. Imaging **43**(4), 1539–1553 (2024)
14. Lao, G., et al.: Coordinate-based neural representation enabling zero-shot learning for fast 3D multiparametric quantitative MRI. Med. Image Anal. **102**, Article 103530 (2025)
15. Shen, L., Pauly, J., Xing, L.: NeRP: implicit neural representation learning with prior embedding for sparsely sampled image reconstruction. IEEE Trans. Neural Netw. Learn. Syst. (2022)
16. Polak, D., et al.: Joint multi-contrast variational network reconstruction (jVN) with application to rapid 2D and 3D imaging. Magn. Reson. Med. **84**(3), 1456–1469 (2020)
17. Lei, P., Fang, F., Zhang, G., Zeng, T.: Decomposition-based variational network for multi-contrast MRI super-resolution and reconstruction. In: 2023 IEEE/CVF International Conference on Computer Vision (ICCV), 2123949. IEEE (2023)
18. Müller, T., Evans, A., Schied, C., Keller, A.: Instant neural graphics primitives with a multiresolution hash encoding. ACM Trans. Graph. (TOG) **41**(4), 102 (2022)
19. Niessen, N., et al.: Probing the sparsity of the MPnRAGE sequence through subspace compression. In: International Society for Magnetic Resonance in Medicine Conference Proceedings (2025)
20. Skare, S., Avventi, E., Norbeck, O., Ryden, H.: An abstraction layer for simpler EPIC pulse programming on GE MR systems in a clinical environment. In: Proceedings of the 25th Annual Meeting of ISMRM, p. 3813 (2017)
21. Jenkinson, M., Beckmann, C.F., Behrens, T.E.J., Woolrich, M.W., Smith, S.M.: FSL. Neuroimage **62**(2), 782–790 (2012)

22. Marchetto, E., Eichhorn, H., Gallichan, D., et al.: Agreement of image quality metrics with radiological evaluation in the presence of motion artifacts. Magn. Reson. Mater. Phys. Biol. Med. (MAGMA) (2025)
23. Lustig, M., Donoho, D., Pauly, J.M.: Sparse MRI: the application of compressed sensing for rapid MR imaging. Magn. Reson. Med. **58**(6), 1182–1195 (2007). https://doi.org/10.1002/mrm.21391.

Markerless Tracking-Based Registration for Medical Image Motion Correction

Luisa Neubig[1]($\boxtimes$), Deirdre Larsen[2], Takeshi Ikuma[3], Markus Kopp[4], Melda Kunduk[5], and Andreas M. Kist[1]

[1] Department Artificial Intelligence in Biomedical Engineering, FAU Erlangen-Nürnberg, Erlangen, Germany
luisa.e.neubig@fau.de
[2] Department of Communication Sciences and Disorders, East Carolina University, Greenville, NC, USA
[3] Department of Otolaryngology–Head and Neck Surgery, Louisiana State University Health Sciences Center, New Orleans, LA, USA
[4] Department of Radiology, FAU Erlangen-Nürnberg, Erlangen, Germany
[5] Department of Communication Sciences and Disorders, Louisiana State University, Baton Rouge, LA, USA

Abstract. Our study focuses on isolating swallowing dynamics from interfering patient motion in videofluoroscopy, an X-ray technique that records patients swallowing a radiopaque bolus. These recordings capture multiple motion sources, including head movement, anatomical displacements, and bolus transit. To enable precise analysis of swallowing physiology, we aim to eliminate distracting motion, particularly head movement, while preserving essential swallowing-related dynamics. Optical flow methods fail due to artifacts like flickering and instability, making them unreliable for distinguishing different motion groups. We evaluated markerless tracking approaches (CoTracker, PIPs++, TAP-Net) and quantified tracking accuracy in key medical regions of interest. Our findings demonstrate that even sparse tracking points can generate morphing displacement fields that outperform leading registration methods, including ANTs, LDDMM, and VoxelMorph. To compare all approaches, we evaluated their performance using MSE and SSIM metrics after registration. We introduce a novel motion correction algorithm that effectively removes disruptive motion while preserving swallowing dynamics and surpassing competitive registration techniques. Our code is openly available at https://github.com/neuluna/markerless-motion-correction.

Keywords: point tracking · registration · motion correction

1 Introduction

Motion analysis plays a critical role in medical imaging, particularly in the evaluation of dynamic processes such as swallowing [7]. Videofluoroscopy Swallowing Studies (VFSS) capture high-temporal-resolution (up to 30 fps) X-ray recordings of patients swallowing a radiopaque bolus to assess swallowing function [11]. However, these recordings contain multiple overlapping sources of motion, including bolus flow, head movement, and shifts in anatomical structures. This makes it difficult to focus and isolate clinically relevant swallowing dynamics. Accurately correcting for these confounding motions is crucial for precise physiological analysis. To achieve this, we must (i) assess global motion artifacts, (ii) identify distinct moving structures, and (iii) register relevant features to compensate for unwanted motion. In this study, we address these challenges by isolating unwanted motion, such as head movement, while preserving anatomical motions and accurate depiction of the bolus suitable for downstream quantitative analysis. By leveraging advanced tracking algorithms, we compute velocity and displacement fields to effectively suppress large, distracting shifts. We show that our proposed method works on rather still and more challenging data, as well as generalizes to data provided by other hospitals.

2 Related Works

2.1 Optical Flow-Based Methods in Medical Imaging

Optical flow methods are widely used in medical image analysis to estimate motion from pixel displacements and enhance computer vision tasks like segmentation and image registration. Xue et al. [19] propose a novel segmentation method for echocardiography that effectively utilizes motion information by accurately predicting the optical flow. This information and initial segmentation guesses result in a motion-enhanced segmentation module for final segmentation. Suji et al. analyzed how optical flow methods (i.e., Farneback, Horn-Schunck and Lucas-Kanade) can improve motion-based segmentation of lung nodules in thin-sliced CT scans [16]. In [20], motion information is used to discriminate between healthy and affected patients. They use an optical flow-based method to quantitatively analyze the deformation of the right diaphragm in ultrasound imaging to track respiratory motion.

2.2 Image Registration in Medical Imaging

Image registration enables the matching of medical images across time, modalities, or patients, as well as multimodal fusion to improve diagnosis and treatment. Deformable image registration for image-guided adaptive radiotherapy has issues in low-contrast regions. Meng et al. [12] address this problem by integrating a finite element method (FEM) with the 'demon' algorithm and refining a tetrahedral mesh to improve displacement accuracy and reduce registration errors. A common issue in medical image registration is that labeling corresponding features is a very time-consuming task, which is why image registration is preferably performed unsupervised. [6] overcomes the labeling of corresponding areas by using foundation segmentation models, such as Segment Anything (SAM), to define ROI correspondences for registration.

2.3 Tracking in Videofluoroscopies

Previous studies on VFSS have focused on the dynamics of anatomical landmarks, assuming a motionless patient. However, we are not aware of any study that has investigated or corrected for patient motion during VFSS recordings. On rare occasions, the community acknowledges this issue: [10] draws attention to the importance of tracking the movement of the hyoid bone during VFSS. We believe that quantitative measurements, such as those in [10] and [9], or the tracking of cervical vertebrae, such as in [13], can benefit from compensating for the patient's head movement, as we propose in our study.

3 Methods

3.1 Dataset

We created two datasets termed *motion* and *non-motion* containing each of 10 manually selected VFSS recordings with a single swallowing event. The VFSS recordings were acquired at the Lady of the Lake Hospital in Baton Rouge, US, and in accordance with the granted IRB (#IRBAM-21-0625). The videos have different lengths, ranging from approximately 30 to 170 frames. All videos have a temporal resolution of 30 fps and frames were resized to a uniform resolution of 256×256 px. For validation, we rely on example data from the University Hospital in Erlangen, Germany. This data differs from the American dataset, i.e., the temporal resolution is only 7.5 fps, and the aperture is rectangular instead of circular.

3.2 Assessing Optical Flow Using Markerless Tracking

We first analyzed multiple optical flow methods, including classical approaches (Lucas-Kanade Method, Farneback) and Deep Learning-based methods (RAFT, SpyNet, FlowFormer, PWCNet, UnFlow). Next to these methods, we tested three different markerless tracking algorithms (CoTracker, PIPs++ and TAP-Net) to determine correspondences. We conducted all experiments on a NVIDIA GeForce RTX 4090 with 24GB RAM. Additionally, we experimented with different grid sampling strategies to assess their impact on tracking performance and adjusted grid sizes to meet the GPU RAM limitations.

CoTracker utilizes a query-based mechanism, allowing for adaptive point selection and refinement across frames [8]. We employed CoTrackerPredictor3 in its default offline configuration with the provided pre-trained model [1].

PIPs++ uses a sliding window to update predictions iteratively and maintain trajectory consistency [21]. We utilized PIPs++ in its default configuration and leveraged the provided pre-trained model [5].

TAP-Net follows a two-stage approach to estimate trajectories accurately over time. In the matching stage, TAP-Net independently searches for a suitable candidate point match for the query point across all frames, leveraging local and global spatial information. We use the provided pre-trained BootsTAPIR model for our analysis [4].

3.3 Motion Correction Using Velocity Flow Fields

We stabilize the patient during swallowing by correcting motion through image registration using the predicted velocity field. To evaluate our approach, we compare it with cross-domain state-of-the-art (SOTA) registration algorithms, as no existing methods are specifically designed for VFSS. The comparison includes Large Deformation Diffeomorphic Metric Mapping (LDDMM) [3], Advanced Normalization Tools (ANTs) [17], and VoxelMorph [2]. For LDDMM, we empirically set spatial smoothness to $\sigma = 20$ and used the "gdr" optimizer; for ANTs, we utilized the SyNOnly transformation mode; and for VoxelMorph, we evaluated its publicly available pretrained model.

LDDMMs formulate registration as finding a smooth, time-varying velocity field $\{v_t\}_{t \in [0,T]}$ whose flow integrates to a diffeomorphism. Specifically, we define the transformation $\phi_t : \Omega \to \Omega$ by

$$\frac{\partial \phi_t(x)}{\partial t} = v_t(\phi_t(x)), \quad \phi_0(x) = x, \tag{1}$$

so that ϕ_1 is the final deformation at $t = 1$. The objective function typically balances a regularization on the velocity field (encouraging smoothness) and an image similarity term:

$$\min_{v_t} \int_0^1 \mathcal{R}(v_t) \, dt + \mathcal{D}(I_0 \circ \phi_1^{-1}, I_1), \tag{2}$$

where $\mathcal{R}$ denotes a regularization functional (e.g., $\|\nabla v_t\|^2$) and $\mathcal{D}$ is a measure of image dissimilarity (e.g., squared difference or mutual information).

ANTs implements a symmetric diffeomorphic registration (SyN), jointly estimating forward and inverse transformations in a single optimization framework. A velocity field is used to generate diffeomorphic transformations via the exponential map, ensuring invertibility. The energy functional typically combines an image similarity term, $\mathcal{D}$, and a smoothness regularizer, $\mathcal{R}$, on the deformation:

$$\phi^* = \arg\min_\phi \Big[\mathcal{D}(I_0 \circ \phi, I_1) + \lambda \mathcal{R}(\phi)\Big]. \tag{3}$$

Common choices for $\mathcal{D}$ include cross-correlation or mutual information, while $\mathcal{R}$ is often based on the norm of the velocity field or its spatial derivatives.

VoxelMorph is a deep learning-based framework that leverages a convolutional neural network (CNN), often U-Net style, to predict a dense displacement field ϕ_θ directly from the moving and fixed images (I_0, I_1). The network parameters θ are learned in an unsupervised manner by minimizing a joint loss function:

$$\mathcal{L}(\theta) = \mathcal{D}(I_0 \circ \phi_\theta, I_1) + \alpha \mathcal{R}(\phi_\theta), \tag{4}$$

where $\mathcal{D}$ measures similarity between the warped moving image $I_0 \circ \phi_\theta$ using a local cross-correlation and the fixed image I_1, and $\mathcal{R}$ enforces spatial smoothness or regularity of the predicted deformation field, similar to LDDMMs.

Ours Using Markerless Tracking. Let $\Omega \subset \mathbb{R}^n$ be the domain of our images. Each tracking method (see above) provides a discrete velocity field $\mathbf{v}^{(M)}$ with M being any tracking method. We interpret each $\mathbf{v}^{(M)}$ as describing how points in the image move over time, discretized as:

$$\phi_{t+1}^{(M)}(\mathbf{x}) = \phi_t^{(M)}(\mathbf{x}) + \Delta t\, \mathbf{v}^{(M)}\bigl(\phi_t^{(M)}(\mathbf{x})\bigr), \quad \phi_0^{(M)}(\mathbf{x}) = \mathbf{x}. \tag{5}$$

After iterating these updates from $t = 0$ to $t = T$, we obtain the final displacement field $\phi^{(M)}(\mathbf{x})$, which maps any point $\mathbf{x} \in \Omega$ to its new location under the chosen velocity field.

Given an input image $I : \Omega \to \mathbb{R}$, we produce the warped image $I'(\mathbf{x})$ by evaluating

$$I'(\mathbf{x}) = I\bigl(\phi^{(M)}(\mathbf{x})\bigr) \tag{6}$$

using the scipy function `map_coordinates` [14].

3.4 Analysis Methods

To assess the accuracy of the tracking algorithms, we manually labeled three important point structures of the patient's head in each of the videos. We labeled the rightmost point of the first vertebra, the leftmost point of the hard palate, and the leftmost point of the mandible to best describe the motion caused by the head and jaw motion. We determined the mean precision error (MAPE). To quantify the registration efficiency, we measured the mean square error to determine intensity differences and the structural similarity index measure (SSIM [18]) to account for anatomically close relationships.

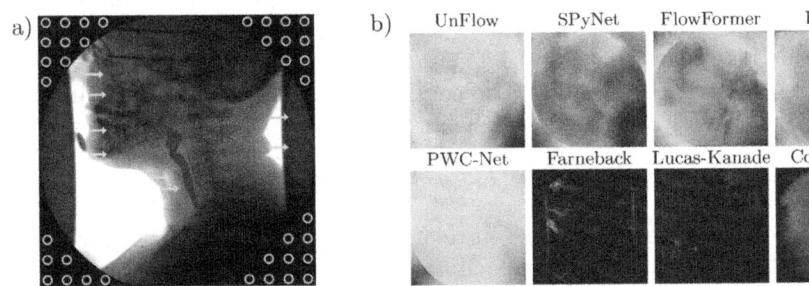

Fig. 1. VFSS recording motion sources and their estimation. a) shows a VFSS frame, where the blue circles mark the static background, yellow arrows indicate the moving head, and the bolus with its moving direction is highlighted in purple. b) presents the normalized summed flow fields throughout the same recording as in a) for Deep Learning-based optical flow methods (consisting of UnFlow, SPyNet, Flowformer, RAFT, and PWC-Net), the Farneback and Lucas-Kanade algorithm as well as one tracking-based algorithm (CoTracker). (Color figure online)

4 Results

The analysis of swallowing physiology presents challenges, including the presence of a static background, bolus transit, and unwanted head movement (Fig. 1a). To address these complexities, we assessed whether optical flow could effectively estimate different motion sources in VFSS data. However, as shown in Fig. 1b), deep learning-based optical flow methods performed poorly, detecting excessive motion throughout the recording. Traditional methods like Farneback and Lucas-Kanade captured only minimal motion aspects. We then explored markerless tracking algorithms for motion pattern detection in medical imaging. Among them, CoTracker demonstrated promising results, despite not being specifically trained on or designed for medical imaging data. We then systematically tested three SOTA markerless tracking algorithms, namely CoTracker, PIPs++, and TAP-Net on manually annotated key landmarks (see Methods). We found only minor differences across tracking algorithms when analyzing the MAPE. In recordings with little motion, the median MAPE is around 1%, whereas in motion-dominated recordings, it increases to approximately 3% with larger deviations observed for PIPs++ (Fig. 2a).

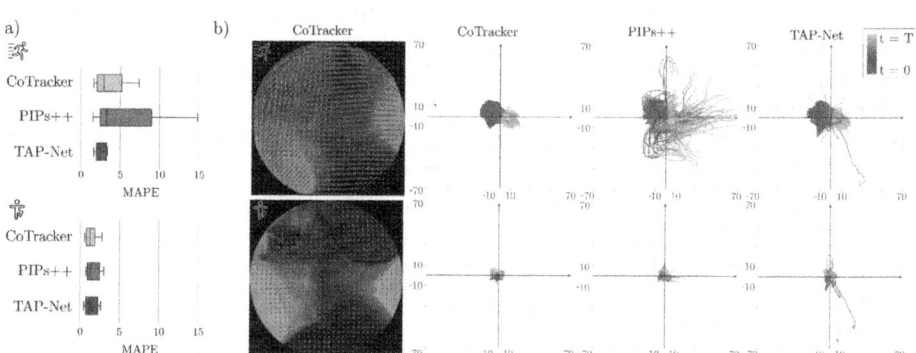

Fig. 2. Markerless tracking reliably recovers moving parts. a) MAPE evaluation of the tracking algorithms for three landmarks. b) Visualizes the characteristics of the point tracks with individual trajectories. The magenta-blue color gradient encodes the transition from $t = 0$ to $t = T$. The upper and lower panels in a) and b) show the results for images with and without unwanted motion respectively. (Color figure online)

Knowing that all trackers can track manually selected landmarks, we were interested in an unsupervised approach to capturing global motion. We investigated several strategies: random point sampling, good features to track [15], and a uniform grid across all three markerless tracking methods. We found that the uniformly distributed points best represented the overall patient motion. Therefore, we focused on analyzing the uniform grid intensively (Suppl. Movie 1).

Figure 2b provides an overview of the grid point trajectories for a patient with hardly any motion (lower panel) and one with significantly higher motion (upper panel). All three trackers showed a similar angle and path distribution of points over the entire video. CoTracker provides a clear, uniform characteristic of point trajectories, whereas PIPs++ and TAP-Net suffer from individual outliers that do not correspond to the general motion present in the video. We consistently observe this behavior for PIPs++, especially in videos with high-motion components. TAP-Net shows this behavior for both high-motion and low-motion videos. In summary, CoTracker is best at capturing the overall motion by showing a very uniform profile across all point trajectories. This confirms our assumption that there is not necessarily the "best point" or ROI to capture the motion; however, the sparse representation of the image with regularly spaced points supports the analysis of head movement.

We hypothesized that we could utilize the regular grids to estimate a displacement field Φ suitable for a non-linear registration step (see Methods). Therefore, we next analyzed the optimal grid size to estimate the full size Φ. Figure 3 shows the MSE and SSIM for each algorithm, averaged across motion and non-motion recordings, against varying grid sizes (see Suppl. Movie 2). CoTracker performs best with the full, dense grid (256×256), reaching an SSIM of 0.875/0.926 and an MSE of 423.6/26.6 for motion/non-motion recordings, respectively. In contrast, TAP-Net performs the worst among tracker-driven motion correction methods (see Table 1, Suppl. Movie 3). Notably, even very small grid sizes (4×4) exhibit stable motion correction capabilities for CoTracker. Finally, we tested whether our approach is competitive with SOTA registration methods. Qualitatively, the preservation of the patient's anatomy and structure is best with our proposed approach using CoTracker, whereas all competitors, namely VoxelMorph, LDDMM, and ANTs, show high deformation of the patient's head, which influences the patient's appearance and consecutively the unequivocal visibility of the bolus (Fig. 4, Suppl. Movie 4). Quantitatively, the CoTracker-based method outperforms LDDMM for motion-dominated recordings but performs comparably for VFSS with no motion, where LDDMM achieves a slightly lower MSE for both types of videos. However, CoTracker surpasses the other two baselines in SSIM and remains competitive in MSE, as shown in Table 1. Further, our

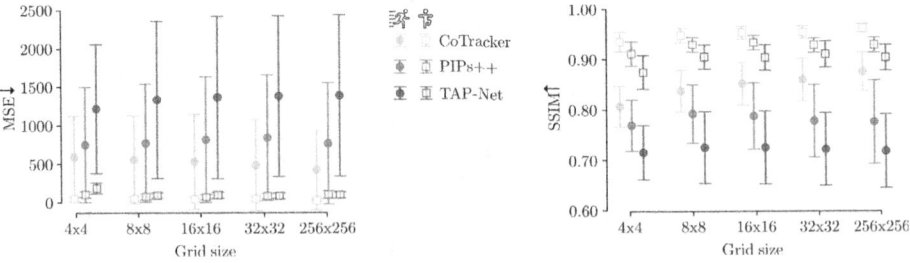

Fig. 3. Analysis of the influence of the grid size on the MSE and SSIM **per video** for the three evaluated tracking algorithms CoTracker, PIPs++, and TAP-Net.

method generalizes to VFSS data from a different hospital, which differs greatly in the acquisition settings (see Suppl. Movie 5).

5 Conclusion

In our study, we first showed that optical flow methods are not sufficiently usable in the presented context, at least in their current form. As an alternative, we investigated markerless tracking algorithms and demonstrated their utility in tracking anatomical structures in VFSS recordings (Fig. 2). Our results demonstrate that small tracker-derived velocity fields are sufficient for motion correction while remaining computationally efficient (Fig. 3). Quantitative (Table 1) and qualitative analyses (Fig. 4, Suppl. Movie 4) show that our approach performs comparably to or better than established baseline methods. This work

Table 1. Comparison of SOTA methods ANTs, VoxelMorph, and LDDMM against our markerless tracker-based registration method with a grid size of 256×256 in terms of SSIM and MSE **per video**. The analysis considers videos with higher patient motion and nearly motionless recordings. (MC = map_coordinates)

	Method	🏃 SSIM ↑	🏃 MSE ↓	🧍 SSIM ↑	🧍 MSE ↓
SOTA	ANTs	0.761 ± 0.072	1020.4 ± 920.5	0.953 ± 0.014	27.3 ± 15.2
SOTA	VoxelMorph	0.822 ± 0.040	875.2 ± 865.2	0.948 ± 0.012	46.8 ± 22.2
SOTA	LDDMM	0.850 ± 0.030	231.8 ± 299.0	**0.962** ± 0.010	16.7 ± 6.1
OURS	PIPs++ → MC	0.775 ± 0.083	767.1 ± 789.6	0.928 ± 0.015	99.2 ± 54.4
OURS	TAP-Net → MC	0.717 ± 0.074	1390.2 ± 1048.3	0.903 ± 0.026	94.8 ± 41.0
OURS	CoTracker → MC	**0.875** ± 0.039	423.6 ± 514.1	**0.962** ± 0.009	26.6 ± 11.4

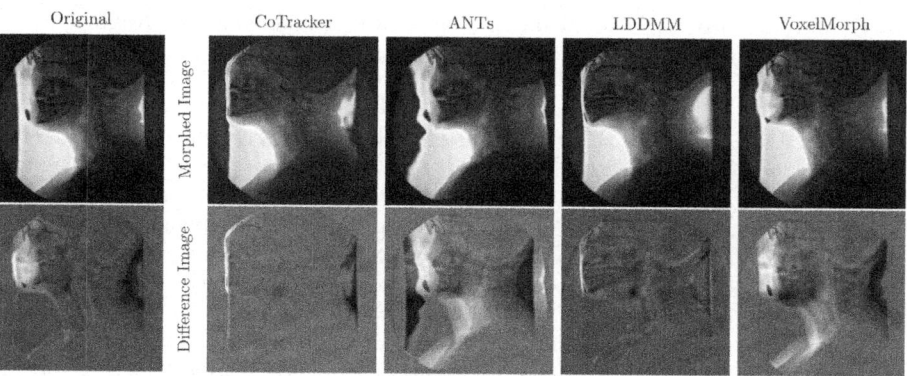

Fig. 4. Visual comparison of the last image of the morphed sequence (top row) and the difference between the last and first images (bottom row) for CoTracker, ANTs, LDDMM, and VoxelMorph. Difference image: $P \in [white > gray = 0 > black]$. (Color figure online)

does not explicitly assess auxiliary strategies for selecting tracking points, such as those based on segmentation of the background, patient, or specific anatomical regions. Nevertheless, the proposed method represents an efficient markerless solution for motion correction, demonstrated here in the context of VFSS. With further development and integration of auxiliary components, it holds promise as a generalizable tool for motion correction in medical imaging applications.

Disclosure of Interests. The authors have no competing interests to declare that are relevant to the content of this article.

References

1. AI, M.: Cotracker3 - pretrained model weights (2024). https://huggingface.co/facebook/cotracker3/resolve/main/scaled_offline.pth. Accessed 20 Feb 2025
2. Balakrishnan, G., Zhao, A., Sabuncu, M., Guttag, J., Dalca, A.V.: VoxelMorph: a learning framework for deformable medical image registration. IEEE TMI: Trans. Med. Imaging **38**, 1788–1800 (2019). https://doi.org/10.1109/TMI.2019.2897538
3. Beg, M.F., Miller, M.I., Trouvé, A., Younes, L.: Computing large deformation metric mappings via geodesic flows of diffeomorphisms. Int. J. Comput. Vis. **61**(2), 139–157 (2005). https://doi.org/10.1023/B:VISI.0000043755.93987.aa
4. Doersch, C., et al.: BootsTAP: Bootstrapped training for tracking-any-point (2024). https://github.com/google-deepmind/tapnet, gitHub repository. Accessed 20 Feb 2025
5. Harley, A.W.: PIPs2: predicting interactions for pixel streams (2024). https://github.com/aharley/pips2, gitHub repository. Accessed 20 Feb 2025
6. Huang, S., et al.: SAMReg: SAM-enabled image registration with ROI-based correspondence (2024). https://arxiv.org/abs/2410.14083
7. Kang, B.S., Oh, B.M., Kim, I.S., Chung, S.G., Kim, S.J., Han, T.R.: Influence of aging on movement of the hyoid bone and epiglottis during normal swallowing: a motion analysis. Gerontology **56**(5), 474–482 (2010). https://doi.org/10.1159/000274517
8. Karaev, N., Rocco, I., Graham, B., Neverova, N., Vedaldi, A., Rupprecht, C.: CoTracker: it is better to track together (2024). https://arxiv.org/abs/2307.07635
9. Kim, H.I., Kim, Y., Kim, B., Shin, D.Y., Lee, S.J., Choi, S.I.: Hyoid bone tracking in a videofluoroscopic swallowing study using a deep-learning-based segmentation network. Diagnostics **11**(7) (2021). https://doi.org/10.3390/diagnostics11071147, https://www.mdpi.com/2075-4418/11/7/1147
10. Kim, W.S., Zeng, P., Shi, J.Q., Lee, Y., Paik, N.J.: Semi-automatic tracking, smoothing and segmentation of hyoid bone motion from videofluoroscopic swallowing study. PLoS ONE **12**(11), e0188684 (2017). https://doi.org/10.1371/journal.pone.0188684, http://dx.doi.org/10.1371/journal.pone.0188684
11. Martin-Harris, B., Canon, C.L., Bonilha, H.S., Murray, J., Davidson, K., Lefton-Greif, M.A.: Best practices in modified barium swallow studies. Am. J. Speech Lang. Pathol. **29**(2S), 1078–1093 (2020). https://doi.org/10.1044/2020_AJSLP-19-00189
12. Meng, M., Bi, L., Fulham, M., Feng, D.D., Kim, J.: Enhancing medical image registration via appearance adjustment networks. Neuroimage **259**, 119444 (2022). https://doi.org/10.1016/j.neuroimage.2022.119444, https://www.sciencedirect.com/science/article/pii/S1053811922005614

13. Reinartz, R., Platel, B., Boselie, T., van Mameren, H., van Santbrink, H., ter Haar Romeny, B.: Cervical vertebrae tracking in video-fluoroscopy using the normalized gradient field. In: Yang, G.Z., Hawkes, D., Rueckert, D., Noble, A., Taylor, C. (eds.) Medical Image Computing and Computer-Assisted Intervention - MICCAI 2009, pp. 524–531. Springer, Berlin, Heidelberg (2009). https://doi.org/10.1007/978-3-642-04268-3_65
14. SciPy Community: scipy.ndimage.map_coordinates. https://docs.scipy.org/doc/scipy/reference/generated/scipy.ndimage.map_coordinates.html. Accessed 22 Feb 2025
15. Shi, J., Tomasi: Good features to track. In: 1994 Proceedings of IEEE Conference on Computer Vision and Pattern Recognition, pp. 593–600 (1994). https://doi.org/10.1109/CVPR.1994.323794
16. Suji, R.J., Bhadouria, S.S., Dhar, J., Godfrey, W.W.: Optical flow methods for lung nodule segmentation on LIDC-IDRI images. J. Digit. Imaging **33**(5), 1306–1324 (2020). https://doi.org/10.1007/s10278-020-00346-w, http://dx.doi.org/10.1007/s10278-020-00346-w
17. Tustison, N.J., et al.: The ANTsX ecosystem for quantitative biological and medical imaging. Sci. Rep. **11**(1), 9068 (2021). https://doi.org/10.1038/s41598-021-87564-6, https://doi.org/10.1038/s41598-021-87564-6
18. Wang, Z., Bovik, A., Sheikh, H., Simoncelli, E.: Image quality assessment: from error visibility to structural similarity. IEEE Trans. Image Process. **13**(4), 600–612 (2004). https://doi.org/10.1109/TIP.2003.819861
19. Xue, W., Cao, H., Ma, J., Bai, T., Wang, T., Ni, D.: Improved segmentation of echocardiography with orientation-congruency of optical flow and motion-enhanced segmentation. IEEE J. Biomed. Health Inform. **26**(12), 6105–6115 (2022). https://doi.org/10.1109/JBHI.2022.3221429
20. Zhang, Q., Yang, D., Zhu, Y., Liu, Y., Ye, X.: An optimized optical-flow-based method for quantitative tracking of ultrasound-guided right diaphragm deformation. BMC Med. Imaging **23**(1) (2023). https://doi.org/10.1186/s12880-023-01066-7, http://dx.doi.org/10.1186/s12880-023-01066-7
21. Zheng, Y., Harley, A.W., Shen, B., Wetzstein, G., Guibas, L.J.: PointOdyssey: a large-scale synthetic dataset for long-term point tracking (2023). https://arxiv.org/abs/2307.15055

MCM: Mamba-Based Cardiac Motion Tracking Using Sequential Images in MRI

Jiahui Yin[1], Xinxing Cheng[1], Jinming Duan[1,2,3], Yan Pang[4], Declan O'Regan[5], Hadrien Reynaud[6,7], and Qingjie Meng[1,7(✉)]

[1] School of Computer Science, University of Birmingham, Birmingham, UK
{jxy427,m.qingjie}@bham.ac.uk
[2] Division of Informatics, Imaging and Data Sciences, University of Manchester, Manchester, UK
[3] Centre for Computational Imaging and Modelling in Medicine, University of Manchester, Manchester, UK
[4] Guangdong Provincial Key Laboratory of Computer Vision and Virtual Reality Technology, Shenzhen Institute of Advanced Technology, Chinese Academy of Sciences, Shenzhen, China
[5] MRC Laboratory of Medical Sciences, Imperial College London, London, UK
[6] UKRI CDT in AI for Healthcare, Imperial College London, London, UK
[7] Department of Computing, Imperial College London, London, UK

Abstract. Myocardial motion tracking is important for assessing cardiac function and diagnosing cardiovascular diseases, for which cine cardiac magnetic resonance (CMR) has been established as the gold standard imaging modality. Many existing methods learn motion from single image pairs consisting of a reference frame and a randomly selected target frame from the cardiac cycle. However, these methods overlook the continuous nature of cardiac motion and often yield inconsistent and non-smooth motion estimations. In this work, we propose a novel Mamba-based cardiac motion tracking network (MCM) that explicitly incorporates target image sequence from the cardiac cycle to achieve smooth and temporally consistent motion tracking. By developing a bi-directional Mamba block equipped with a bi-directional scanning mechanism, our method facilitates the estimation of plausible deformation fields. With our proposed motion decoder that integrates motion information from frames adjacent to the target frame, our method further enhances temporal coherence. Moreover, by taking advantage of Mamba's structured state-space formulation, the proposed method learns the continuous dynamics of the myocardium from sequential images without increasing computational complexity. We evaluate the proposed method on two public datasets. The experimental results demonstrate that the proposed method quantitatively and qualitatively outperforms both conventional and state-of-the-art learning-based cardiac motion tracking methods. The code is available at https://github.com/yjh-0104/MCM.

Keywords: Heart motion tracking · Mamba · Sequential images · MRI

1 Introduction

Left ventricular (LV) myocardial motion tracking enables the assessment of LV function spatially and temporally [13,19]. This facilitates the early and accurate detection of LV dysfunction and myocardial diseases [4,8,12]. Cine cardiac magnetic resonance (CMR) imaging is widely employed in myocardial motion tracking, as it provides high-resolution 2D image sequences that capture detailed structural and functional information of the heart. Recent advancements in deep learning have been leveraged for cardiac motion estimation in CMR images [3,16–18,20,29,30]. Many methods train neural networks to learn the motion between a reference frame and a randomly selected target frame from the cardiac cycle. However, by focusing on isolated target frame, these approaches overlook the continuous nature of cardiac motion. This often results in motion estimations that lack consistency and smoothness. Although incorporating the entire sequence of images could address these issues, it would introduce significant memory and computational challenges.

In this work, we propose a novel Mamba-based network that utilizes a sequence of target frames for improved myocardial motion tracking. Our method explicitly incorporates neighboring frames around the target frame to estimate the motion between the reference and the target frame, which enables the estimation of consistent and smooth motion fields. By leveraging Mamba's structured state-space formulation, the proposed approach effectively learns the continuous dynamics of the myocardium from the target image sequences with no significant increase in computational complexity. Moreover, our method integrates spatiotemporal information from both forward and backward directions, facilitating the estimation of plausible deformation fields during motion tracking.

Contributions: (1) We propose an end-to-end trainable Mamba-based cardiac motion tracking network (MCM) that leverages sequential images to achieve smooth and consistent myocardial motion estimation without incurring significant computational overhead. (2) We introduce bi-directional Mamba blocks to extract deformation features at multiple scales. Each block incorporates a novel bi-directional scanning mechanism that captures spatiotemporal information in both forward and backward directions, facilitating the estimation of plausible deformation fields. (3) We develop a motion decoder that estimates motion fields by fusing deformation features across multiple scales, incorporating a novel dual-path fusion head to enhance the temporal consistency of motion estimation.

Related Works: Deformable image registration methods have been widely applied to cardiac motion tracking, where traditional techniques have demonstrated their efficacy [9,21,23]. For instance, Vercauteren et al. [23] introduced the non-parametric diffeomorphic approach based on the demons algorithm [22], which has been effectively used for cardiac motion tracking [20]. More recently, deep learning-based image registration methods have gained increased attention. Balakrishnan et al. [1] proposed VoxelMorph, which employs a U-Net architecture for registration and has been extended to cardiac motion estimation [18]. Chen et al. [7] developed TransMorph, utilizing vision transformers

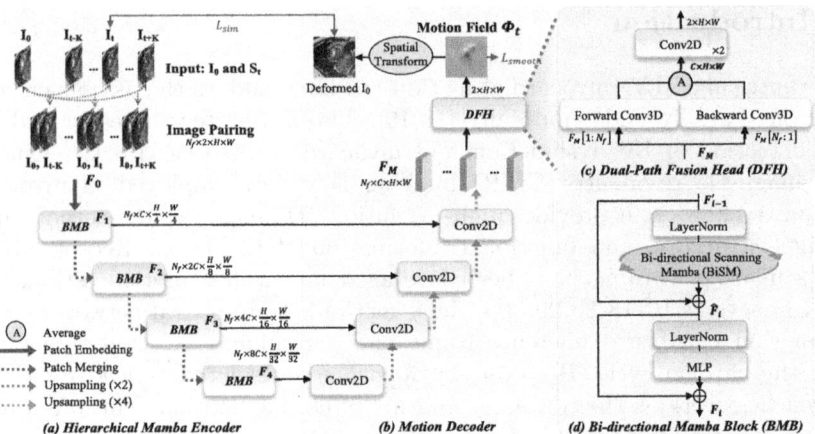

Fig. 1. An overview of the proposed method: (a) The Hierarchical Mamba encoder pairs the reference image with the target image sequence and learns deformation features at different scales; (b) The motion decoder combines the learned deformation features at various scales and predicts the motion field Φ_t via a Dual-Path Fusion Head (DFH); (c) The detailed network architecture of DFH; (d) The detailed network architecture of Bi-directional Mamba Block (BMB).

to capture long-range spatial relationships. Building on neighborhood attention, Wang et al. [24] introduced ModeT to further improve the interpretability and consistency of deformation estimation. Lately, inspired by State Space Models (SSM) [2], Mamba [10] has been developed to address the limitations of modeling lengthy sequences, and it has been explored in various medical image analysis tasks [11,14,15,25–28]. In contrast to existing cardiac motion estimation methods that rely on isolated frame pairs, our method leverages Mamba to process sequences of target images, enhancing the temporal consistency of the estimated motion fields. Our proposed bi-directional scanning mechanism is tailored to sequential cardiac image frames, going beyond prior methods such as Vision Mamba [31], which apply bidirectional scanning only to single 2D images.

2 Method

Our goal is to estimate LV myocardial motion from 2D short-axis (SAX) CMR images. Our task is formulated as follows: Let I_0 be the SAX image of the end-diastole (ED) frame, *i.e.*, reference frame, and I_t be the image of the t-th frame ($0 \leq t \leq T-1$), *i.e.*, target frame. T is the number of frames in the cardiac cycle. We aim to estimate a motion field Φ_t between ED and t-th frame.

The schematic architecture of the proposed method is shown in Fig. 1. Our method leverages sequences of target images for motion estimation, using the ED frame and K neighboring frames around the t-th frame to estimate Φ_t. We denote the sequence of target frames $S_t = \{I_{t-K}, \ldots, I_t, \ldots, I_{t+K}\}$. The method

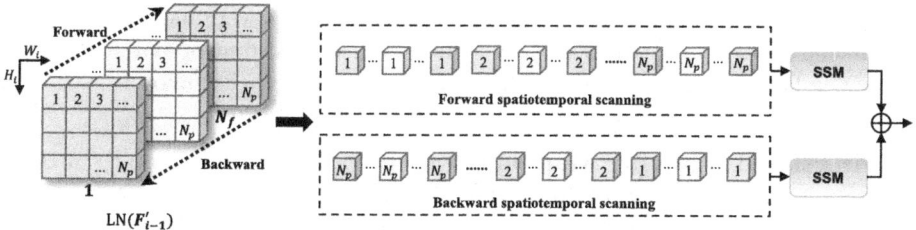

Fig. 2. The proposed bi-directional scanning Mamba (BiSM), including forward and backward spatiotemporal scanning. $\text{LN}(F'_{i-1})$ is F'_{i-1} after layer normalization. $N_p = H_i \times W_i$ is the total number of spatial positions.

comprises two main components. First, a hierarchical Mamba encoder pairs the input images (reference and targets) and learns deformation features at multiple scales via bi-directional Mamba blocks. Within each Mamba block, the proposed bi-directional scanning mechanism is used to integrate spatiotemporal information from both forward and backward directions, facilitating the estimation of plausible deformation. Second, a motion decoder combines the learned deformation features across all scales to predict the motion field Φ_t. Particularly, a dual-path fusion head is developed to strengthen the temporal consistency of Φ_t.

2.1 Hierarchical Mamba Encoder

The hierarchical Mamba encoder learns multi-scale deformation features F_i from the input images. Specifically, the input images I_0 and $\mathcal{S}_t$ are paired into an input sequence, which is then forwarded to the hierarchical Mamba blocks to learn F_i. Within each Mamba block, a bi-directional scanning mechanism is developed to learn spatiotemporal information from the input sequence.

Image Pairing: In this part, we pair I_0 with $\mathcal{S}_t$ to form the input sequence F_0. In detail, each frame from $\mathcal{S}_t$ is paired with the same input image I_0 and $F_0 = \{(I_0, I_{t-K}), \ldots, (I_0, I_t), \ldots, (I_0, I_{t+K})\}$. Each pair in F_0 has a shape of $[2, H, W]$, where H and W are the height and width of the input images, and F_0 has a length of $N_f = 2K + 1$. Note that if $t < K$ or $t > T - K - 1$, we use the nearest available frame for padding e.g. for $t = 1, K = 2, \mathcal{S}_1 = \{I_0, I_0, I_1, I_2, I_3\}$.

Hierarchical Mamba Blocks: From F_0, hierarchical bi-directional Mamba blocks (BMBs) are utilized to learn deformation features $F_i \in \mathbb{R}^{N_f \times C_i \times H_i \times W_i}$ at multiple scales. Here, C_i, H_i, and W_i represent the number of channels, height, and width of F_i at the i-th level. Patch embedding or patch merging [7] are used to downsample F_i between two BMBs. As shown in Fig. 1(d), the i-th BMB process the deformation features as:

$$\hat{F}_i = \text{BiSM}(\text{LN}(F'_{i-1})) + F'_{i-1}, \quad (1)$$
$$F_i = \text{MLP}(\text{LN}(\hat{F}_i)) + \hat{F}_i, \quad i \in [1, 4]. \quad (2)$$

Here, BiSM(·) represents the bi-directional scanning Mamba, which will be discussed next. F'_{i-1} is the F_{i-1} after the patch embedding or patch merging. LN is the layer normalization and MLP is a multi-layer perceptron.

Bi-directional Scanning Mamba (BiSM): Each BMB incorporates a BiSM to integrate spatiotemporal information at level i, as illustrated in Fig. 2. To prepare for temporal modeling, $\text{LN}(F'_{i-1})$ is split into N_p spatial positions per frame, where $N_p = H_i \times W_i$ is the number of positions. These positions are then temporally ordered in both forward and backward directions and fed into two parallel SSMs. Each SSM captures temporal dynamics by recursively updating hidden states through learned linear recurrence. The outputs from both directions are summed to form a unified spatiotemporal representation, enabling smooth and consistent deformation estimation.

2.2 Motion Decoder

The proposed motion decoder estimates the motion field Φ_t by progressively integrating multi-scale deformation features F_i. It consists of a progressive upsampling pathway and a dual-path fusion head (shown in Fig. 1(b)). The progressive upsampling pathway PUP(·) fuses deformation features $F_i, i \in [1, 4]$ via multiple upsampling and convolutional layers and estimates the motion feature $F_M \in \mathbb{R}^{N_f \times C \times H \times W}$ that represents the deformation of the sequential images:

$$F_M = \text{PUP}(\{F_i \mid i \in [1, 4]\}). \tag{3}$$

To further enforce temporal coherence across frames, a dual-path fusion head DFH(·) is introduced to estimate $\Phi_t \in \mathbb{R}^{2 \times H \times W}$ from F_M. The architecture of DFH(·) is shown in Fig. 1(c). Specifically, F_M is simultaneously passed in the forward direction (from 1 to N_f) and the backward direction (from N_f to 1) via 3D convolutional layers operating along N_f, H and W. Subsequently, the results from both paths are averaged, and then passed to 2D convolutional layers to estimate Φ_t:

$$\overline{F}_M = \frac{1}{2}\left(\text{Conv3D}_{\text{fwd}}(F_M[1:N_f]) + \text{Conv3D}_{\text{bwd}}(F_M[N_f:1])\right), \tag{4}$$

$$\Phi_t = \text{DFH}(F_M) = \text{Conv2Ds}\left(\overline{F}_M\right). \tag{5}$$

2.3 Optimization

Our model is an end-to-end trainable framework, and the overall objective is a linear combination of two loss terms:

$$\mathcal{L} = \underbrace{\frac{1}{|\Omega|} \sum_{p \in \Omega} \|I_t(p) - I_0 \circ \Phi_t(p)\|^2}_{\mathcal{L}_{sim}} + \lambda \underbrace{\sum_{p \in \Omega} \|\nabla \Phi_t(p)\|^2}_{\mathcal{L}_{smooth}}, \tag{6}$$

Table 1. Quantitative comparison of other cardiac motion tracking methods. The results are reported as "mean (standard deviation)". ↑ indicates the higher value the better while ↓ indicates the lower value the better. Best results in bold.

		Basal			Mid-ventricle			Apical		
		Dice%↑	$\|J\|_{<0}$%↓	$\|\|J\|-1\|$↓	Dice%↑	$\|J\|_{<0}$%↓	$\|\|J\|-1\|$↓	Dice%↑	$\|J\|_{<0}$%↓	$\|\|J\|-1\|$↓
ACDC	dD [23]	78.9(10.7)	0.35(0.30)	0.29(0.05)	80.9(7.2)	0.36(0.24)	0.30(0.05)	78.6(8.7)	0.28(0.19)	0.29(0.05)
	VM [1]	81.5(6.9)	0.27(0.42)	0.25(0.16)	81.0(7.1)	0.08(0.14)	0.27(0.13)	79.1(8.5)	0.03(0.09)	0.28(0.13)
	TM [7]	82.6(7.3)	0.28(0.40)	0.19(0.07)	83.7(4.9)	0.05(0.09)	0.19(0.07)	82.4(5.9)	0.02(0.05)	0.19(0.09)
	MM [11]	82.2(6.8)	0.33(0.48)	0.19(0.07)	83.7(5.3)	0.05(0.09)	0.19(0.07)	82.3(5.8)	0.05(0.10)	0.20(0.08)
	Ours	**83.4(7.1)**	**0.14(0.31)**	**0.17(0.06)**	**84.6(4.9)**	**0.02(0.04)**	**0.18(0.06)**	**82.8(5.5)**	**0.01(0.02)**	**0.17(0.06)**
M&Ms	dD [23]	75.7(11.3)	0.26(0.21)	0.30(0.06)	78.1(8.9)	0.29(0.22)	0.27(0.05)	73.4(13.0)	0.24(0.20)	0.30(0.07)
	VM [1]	74.6(12.5)	0.09(0.17)	0.30(0.14)	79.5(9.8)	0.21(0.37)	0.29(0.14)	74.6(12.3)	0.29(0.38)	0.27(0.12)
	TM [7]	79.1(8.5)	0.11(0.24)	0.20(0.08)	82.0(6.0)	0.23(0.40)	0.20(0.07)	76.4(11.7)	0.26(0.51)	0.20(0.09)
	MM [11]	78.7(8.9)	0.08(0.17)	**0.19(0.07)**	82.2(6.2)	0.20(0.35)	0.19(0.07)	76.1(12.0)	0.22(0.46)	0.20(0.09)
	Ours	**79.9(8.4)**	**0.03(0.09)**	0.19(0.07)	**83.6(6.2)**	**0.12(0.29)**	**0.18(0.06)**	**77.6(11.5)**	**0.15(0.40)**	**0.19(0.08)**

where λ is the weight of the regularization term, p is a pixel in the image domain Ω and $|\Omega|$ is the total number of pixels. The similarity loss $\mathcal{L}_{sim}$ is defined by the mean squared error while $\mathcal{L}_{smooth}$ is the smoothness regularization.

3 Experiments

Dataset: We evaluate the proposed method on two publicly available cine CMR datasets: ACDC [5] and M&Ms [6]. Both datasets provide a series of short-axis (SAX) slices covering the left ventricle (LV) from the base to the apex. All image slices are resampled to a resolution of 1.5 × 1.5 mm, center-cropped to 128 × 128 pixels and normalized to [0, 1]. The ACDC dataset is divided into 80/20/50 for training, validation, and testing, respectively, while the M&Ms dataset follows a 150/34/136 split.

Evaluation Metrics: Quantitative evaluation is performed using three commonly used metrics: the Dice coefficient to assess motion tracking accuracy, the percentage of negative Jacobian determinant values ($|J|_{<0}$%) to evaluate diffeomorphism, and the mean absolute difference between $|J|$ and 1 (i.e., $||J|-1|$) to measure volume preservation. A higher Dice score indicates better tracking performance, while lower $|J|_{<0}$% and $||J|-1|$ values indicate improved diffeomorphic properties and volume consistency, respectively.

Implementation: The proposed model is implemented in PyTorch and trained on an NVIDIA A100-SXM4 GPU with 40GB of memory. Network optimization is performed using the Adam optimizer with a learning rate of 10^{-4}. The model is trained for 200 epochs on both datasets with a batch size of 32. The hyperparameter in Eq. 6 is set to $\lambda = 0.05$ for both datasets. We estimate the motion fields for all frames in the cardiac cycle.

Comparison Study: The proposed method is compared to one conventional cardiac motion tracking method, dDemons (dD) [23] and three the state-of-the-art deep learning-based methods, including VoxelMorph (VM) [1], TransMorph (TM) [7] and MambaMorph (MM) [11]. All methods are implemented using their officially released code, with optimal parameters tuned on the validation sets. Quantitative comparisons were performed on three representative short-axis slices: basal, mid-ventricular and apical slices, corresponding to 25%, 50% and 75% of the LV length, respectively. We choose $K = 2$, and thus have the input sequential images with $N_f = 5$ frames. In this experiment, we estimate the motion field between the ED frame and the end-systolic (ES) frame and warp the ED frame segmentation to the ES frame, and compute evaluation metrics by comparing the wrapped segmentation with the ground truth ES segmentation. Table 1 shows that the proposed method outperforms all baseline methods, demonstrating the effectiveness of the proposed method for estimating motion fields. In addition, the proposed method achieves the best performance on $|J|_{<0}\%$ and $||J|-1|$ for all three slices, indicating that the proposed method is more capable of estimating smooth motion fields and preserving the volume of the myocardial wall during cardiac motion tracking. We further qualitatively compare the proposed method with baselines. Figure 3 shows that the motion field generated by the proposed method performs best in warping the ED segmentation to the ES frame, and it is the smoothest. This demonstrates that our method is able to estimate smooth and consistent motion fields.

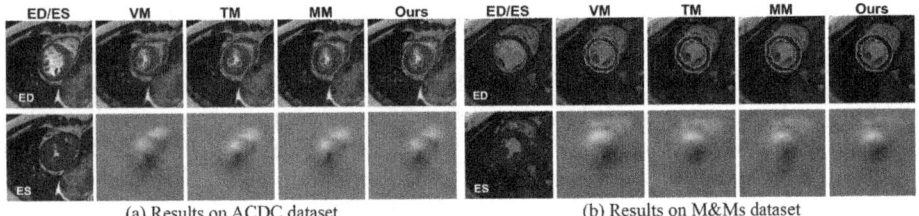

Fig. 3. Motion tracking results using proposed method and baselines. We warp the ED segmentation to the ES frame. The top row shows the deformed ED myocardial contour (green) vs. the ground truth ES myocardial contour (red). The bottom row shows the estimated motion fields. (Color figure online)

Table 2. Motion estimation without BMBs and with BMBs using different scanning strategies.

		Basal		Mid-ventricle		Apical	
		Dice%↑	$|J|_{<0}\%$ ↓	Dice%↑	$|J|_{<0}\%$ ↓	Dice%↑	$|J|_{<0}\%$ ↓
ACDC	Without BMBs	81.8(7.8)	0.15(0.29)	82.4(4.7)	**0.01(0.02)**	81.4(6.1)	**0.01(0.02)**
	BMBs+forward scanning	83.1(6.6)	0.18(0.31)	84.0(4.7)	0.02(0.04)	82.2(5.4)	**0.01(0.02)**
	BMBs+backward scanning	82.8(6.8)	0.15(0.28)	83.4(4.8)	**0.01(0.02)**	82.0(5.3)	**0.01(0.02)**
	BMBs+BiSM (ours)	**83.4(7.1)**	**0.14(0.31)**	**84.6(4.9)**	0.02(0.04)	**82.8(5.5)**	**0.01(0.02)**

Ablation Study: On the ACDC dataset, we explore the importance of BMBs, BiSM and DFH, as well as the effects of hyper-parameters. Table 2 shows that our method, incorporating both BMBs and BiSM, achieves the best performance, while removing BMBs results in the poorest performance. This indicates that the performance gain stems from our proposed approach rather than merely from an increased number of input frames. Figure 4 (b) shows that the motion estimation with DFH achieves better temporal consistency across the cardiac cycle, supporting the importance of the proposed DFH. Figure 4 (c) shows the temporal consistency variations when using different target sequence lengths. We observe that using more neighboring frames achieves better temporal smoothness. Figure 5 presents the quantitative metrics with various λ in Eq. 6. We observe that a strong constraint on motion field smoothness may sacrifice motion estimation accuracy.

Computational Cost: We evaluate model efficiency using GPU training memory (i.e., VRAM) and inference time. As shown in Table 3, while VRAM usage increases with larger N_f due to buffering multiple frames, the inference time remains comparable to baselines, indicating efficient use of sequential images without significant overhead.

Discussion: We quantitatively evaluated the performance of our model for ED to ES motion estimation. This is because ground truth segmentation are only available for the ED and ES frames in both datasets. Motion fields were estimated on three representative SAX slices across the LV, consistent with existing motion estimation studies [20]. Our method also facilitates motion estimation using all slices, at the cost of increased GPU memory usage and longer training time. As our bi-directional Mamba is designed to improve temporal consistency, increasing N_f from 1 to 5 yields only modest gains in quantitative metrics (*e.g.*, +0.5% in Dice) but results in visibly smoother motion fields, as shown in Fig. 4. Our experiments focus on 2D motion tracking due to the use of publicly available 2D datasets. Future work may extend our framework to 3D by integrating 3D convolutions.

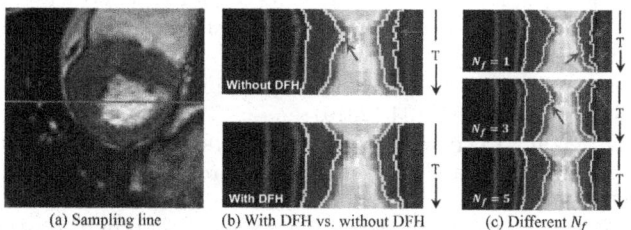

Fig. 4. Temporal consistency across the cardiac cycle. The red line in (a) denotes the temporal axis for (b) and (c). (Color figure online)

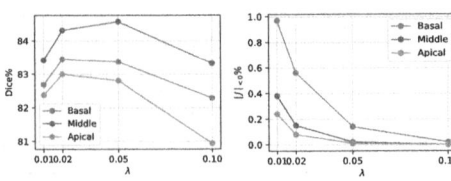

Fig. 5. Motion estimation with different values of λ.

Table 3. GPU VRAM and inference time of comparison methods.

	VRAM (GB)	time (ms)
VM [23]	1.5	7.9
TM [1]	3.6	14.9
MM [11]	2.7	22.9
Ours ($N_f=1$)	3.2	16.3
Ours ($N_f=3$)	7.8	16.5
Ours ($N_f=5$)	12.4	17.1

4 Conclusion

In this paper, we propose an end-to-end trainable, Mamba-based network for myocardial motion tracking. Our method leverages sequential images to achieve smooth and temporally consistent motion estimation while maintaining computational efficiency. Experimental results on two datasets demonstrate that the proposed method outperforms competing methods.

Acknowledgement. The computations described in this research were performed using the Baskerville Tier 2 HPC service. Baskerville was funded by the EPSRC and UKRI through the World Class Labs scheme (EP/T022221/1) and the Digital Research Infrastructure programme (EP/W032244/1) and is operated by Advanced Research Computing at the University of Birmingham.

References

1. Balakrishnan, G., Zhao, A., Sabuncu, M.R., Guttag, J.V., Dalca, A.V.: Voxelmorph: a learning framework for deformable medical image registration. IEEE Trans. Med. Imaging 38(8), 1788–1800 (2019)
2. Basar, T.: A New Approach to Linear Filtering and Prediction Problems, pp. 167–179 (2001)
3. Beetz, M., Banerjee, A., Grau, V.: Modeling 3d cardiac contraction and relaxation with point cloud deformation networks. IEEE J. Biomed. Health Inform. 28(8), 4810–4819 (2024)
4. Bello, G., et al.: Deep learning cardiac motion analysis for human survival prediction. Nat Mach Intell 1, 95–104 (2019)
5. Bernard, O., et al.: Deep learning techniques for automatic mri cardiac multi-structures segmentation and diagnosis: is the problem solved? IEEE Trans. Med. Imaging 37(11), 2514–2525 (2018)
6. Campello, V.M., et al.: Multi-centre, multi-vendor and multi-disease cardiac segmentation: the m&ms challenge. IEEE Trans. Med. Imaging 40(12), 3543–3554 (2021)
7. Chen, J., Frey, E.C., He, Y., Segars, W.P., Li, Y., Du, Y.: Transmorph: transformer for unsupervised medical image registration. Med. Imag. Anal. 82, 102615 (2022)
8. Claus, P., Omar, A.M.S., Pedrizzetti, G., Sengupta, P.P., Nagel, E.: Tissue tracking technology for assessing cardiac mechanics: principles, normal values, and clinical applications. JACC Cardiovasc. Imaging 8(12), 1444–1460 (2015)

9. Craene, M.D., et al.: Temporal diffeomorphic free-form deformation: application to motion and strain estimation from 3D echocardiography. Med. Imag. Anal. **16**(2), 427–450 (2012)
10. Gu, A., Dao, T.: Mamba: linear-time sequence modeling with selective state spaces. In: First Conference on Language Modeling (2024)
11. Guo, T., et al.: Mambamorph: a mamba-based framework for medical mr-ct deformable registration. arXiv preprint arXiv:2401.13934 (2024)
12. Ibrahim, E.S.H.: Myocardial tagging by cardiovascular magnetic resonance: evolution of techniques–pulse sequences, analysis algorithms, and applications. J. Cardiovasc. Magn. Reson. **13**(36) (2011)
13. Inácio, M.H.d.A., et al.: Cardiac age prediction using graph neural networks. medRxiv (2023)
14. Liu, A., et al.: LM-UNet: whole-body PET-CT lesion segmentation with dual-modality-based annotations driven by latent mamba U-Net. In: MICCAI (2024). https://doi.org/10.1007/978-3-031-72114-4_39
15. Ma, J., Li, F., Wang, B.: U-mamba: Enhancing long-range dependency for biomedical image segmentation. arXiv preprint arXiv:2401.04722 (2024)
16. Meng, Q., Bai, W., Liu, T., O'Regan, D.P., Rueckert, D.: Mesh-based 3d motion tracking in cardiac mri using deep learning. In: MICCAI (2022). https://doi.org/10.1007/978-3-031-16446-0_24
17. Meng, Q., Bai, W., O'Regan, D.P., Rueckert, D.: Deepmesh: mesh-based cardiac motion tracking using deep learning. IEEE Trans. Med. Imaging **43**(4), 1489–1500 (2024)
18. Meng, Q., et al.: MulViMotion: shape-aware 3D myocardial motion tracking from multi-view cardiac MRI. IEEE Trans Med Imaging (2022)
19. Puyol-Antón, E., et al.: Regional multi-view learning for cardiac motion analysis: application to identification of dilated cardiomyopathy patients. IEEE Trans. Biomed. Eng. **66**(4), 956–966 (2019)
20. Qin, C., Wang, S., Chen, C., Bai, W., Rueckert, D.: Generative myocardial motion tracking via latent space exploration with biomechanics-informed prior. Med. Imag. Anal. **83**, 102682 (2023)
21. Rueckert, D., Sonoda, L., Hayes, C., Hill, D., Leach, M., Hawkes, D.: Nonrigid registration using free-form deformations: application to breast MR images. IEEE Trans. Med. Imaging **18**(8), 712–721 (1999)
22. Thirion, J.P.: Image matching as a diffusion process: an analogy with Maxwell's demons. Med. Imag. Anal. **2**(3), 243–260 (1998)
23. Vercauteren, T., Pennec, X., Perchant, A., Ayache, N.: Non-parametric diffeomorphic image registration with the demons algorithm. In: Ayache, N., Ourselin, S., Maeder, A. (eds.) MICCAI 2007. LNCS, vol. 4792, pp. 319–326. Springer, Heidelberg (2007). https://doi.org/10.1007/978-3-540-75759-7_39
24. Wang, H., Ni, D., Wang, Y.: Modet: Learning deformable image registration viaămotion decomposition transformer. In: MICCAI (2023). https://doi.org/10.1007/978-3-031-43999-5_70
25. Wang, H., Lin, Y., Ding, X., Li, X.: Tri-plane mamba: efficiently adapting segment anything model for 3D medical images. In: MICCAI (2024). https://doi.org/10.1007/978-3-031-72114-4_61
26. Wang, Z., Zheng, J., Ma, C., Guo, T.: Vmambamorph: a multi-modality deformable image registration framework based on visual state space model with cross-scan module. arXiv preprint arXiv:2402.05105 (2024)

27. Yang, S., Wang, Y., Chen, H.: MambaMIL: enhancing long sequence modeling with sequence reordering in computational pathology. In: MICCAI (2024). https://doi.org/10.1007/978-3-031-72083-3_28
28. Yang, Z., Zhang, J., Wang, G., Kalra, M.K., Yan, P.: cardiovascular disease detection from multi-view chest X-rays with BI-Mamba . In: MICCAI (2024). https://doi.org/10.1007/978-3-031-72086-4_13
29. Ye, M., et al.: Deeptag: an unsupervised deep learning method for motion tracking on cardiac tagging magnetic resonance images. In: CVPR (2021)
30. Yu, H., Chen, X., Shi, H., Chen, T., Huang, T.S., Sun, S.: Motion pyramid networks for accurate and efficient cardiac motion estimation. In: Martel, A.L., et al. (eds.) MICCAI 2020. LNCS, vol. 12266, pp. 436–446. Springer, Cham (2020). https://doi.org/10.1007/978-3-030-59725-2_42
31. Zhu, L., Liao, B., Zhang, Q., Wang, X., Liu, W., Wang, X.: Vision mamba: efficient visual representation learning with bidirectional state space model. In: ICML (2024)

Evaluating Deep Learning Based Domain Generalization for Motion Mitigation in Multi-center Brain MRI

Saad Ashraf[✉], Md Afif Al Mamun, Mumu Aktar, Roberto Souza, and Mariana Bento

University of Calgary, Calgary, Canada
{saad.ashraf,afif.mamun,mumu.aktar,roberto.souza2,
mariana.pinheirobent}@ucalgary.ca

Abstract. Magnetic Resonance Imaging (MRI) is an essential tool for diagnosing brain conditions. The scan procedure is time-consuming, during which the patient must remain still. Any movement during the scan can cause motion artifacts, appearing as noise artifacts in the reconstructed image, complicating diagnosis. Recent Deep Learning (DL) models are effective in tasks such as skull stripping, tissue segmentation, and motion mitigation. However, DL models struggle with distribution shifts occurring due to MRI scans collected in different centers, making it harder to adapt to different datasets. Additionally, unlike other tasks, motion mitigation works with noisy MRI scans, which are harder to denoise since the original scan is distorted. Most of the motion mitigation models have been trained on single datasets; however, for usage in real life, it is crucial to explore the domain adaptability of such models. In our study, we have used three open datasets, collected from 11 different centers. Our chosen baseline 3D UNet model was trained on individual datasets and also in different combinations of these datasets. The model trained on a large, diverse dataset could preserve its knowledge compared to the current literature, while models trained on smaller datasets performed better on datasets with similar properties to the training dataset. These insights can be used to drive further data preprocessing techniques for domain adaptation research concerning motion-mitigation. The source code is available at https://github.com/afifaniks/tiny_brains.

Keywords: Brain MRI · Motion Mitigation and Deep Learning · Heterogeneous Multi-center Datasets · Generalization

1 Introduction

Magnetic Resonance Imaging (MRI) scans produce high-contrast images of the scanned area. We can have excellent soft tissue contrast with brain MRI scans. However, the scan procedure can take around 40 min [9]. This makes it challenging for the patient to stay still for the duration of the scan. It is even harder for non-compliant patients, such as pediatric or neonatal populations. Movements during the scan procedure cause motion artifacts, resulting in blurring and ghosting in the final image [27]. Intense motion artifacts can lead to non-diagnosable images. Prospective techniques to deal with motion include respiratory gating [7], optical tracking system [4] for adult patients, feed-and-wrap for pediatric patients [27]. These methods employ additional steps or equipment in the scan setup, which can increase the total MRI exam duration.

In terms of retrospective motion mitigation, Deep Learning (DL) models are effective in handling motion artifacts by taking the corrupted scan as an input [12,14,16–18]. These models are trained on single-center datasets. However, DL models are susceptible to data distribution shifts, which can significantly degrade their performance. Different MRI datasets are acquired in different centers, having different properties such as scanner configuration, acquisition parameters and pre-processing software. The variations affect the shape, contrast, and image quality of the scan, resulting in distribution shifts. As a result, adapting the learned knowledge from one dataset to another makes motion mitigation tasks harder. These challenges highlight the necessity of analyzing the effect of training a model on large, diverse, open datasets to study the adaptability of the model. This can lead to develop domain adaptation approaches for effective adaptation without compromising performance.

We have studied the adaptability of a baseline 3D UNet model [28] trained on a large, combined, multi-center adult dataset made up of IXI Dataset [1], Calgary-Campinas (CC) Public Brain MR Dataset [24], and OASIS-3 Dataset [13]. Specifically, we have explored:

1. The performance of the baseline model trained on a combined, multi-center (11 centers) dataset. The model could produce results that are similar to the current motion-mitigation literature.
2. The performance of the baseline model trained on each of the datasets and different combinations of the datasets.
3. Analyzing the properties of the datasets to explain the model's behaviour.

2 Background

2.1 Deep Learning Models for Motion Mitigation

Motion mitigation using DL models can be categorized into image-based and k-space-based motion mitigation methods, where image-based methods take the scan as input and k-space-based methods leverage additional information from the raw k-space data [25]. Image-based motion mitigation is solved mostly using Convolutional Neural Networks (CNN) [23,26] and Convolutional encoder-decoder structures, such as U-nets [21]. There are a few variations used for U-nets, such as using residual blocks for encoder-decoder [11], stacking multiple encoder-decoder structures consecutively [3], 3D UNet [19]. Apart from basic CNN and encoder-decoder structures, there are convolutional long/short-term memory models [2] and self-attention mechanisms to address long-distance spatial dependencies of motion artifacts [8]. Many of these models were developed and tested on single and private datasets [3,15,19,23], which makes it hard to reproduce and generalize for researchers. We have addressed this gap by choosing multiple open datasets and selecting a basic 3D UNet [28] model as our baseline to assess generalizability, since UNets are the most common architectures for motion mitigation and 3D models are recommended for processing 3D voxels [25].

2.2 Motion Simulation

Most DL motion mitigation methods are supervised and require paired motion-corrupted and motion-free data, which is costly to acquire. As a result, simulated motion is often used to create training datasets. Current literature has devised ways to simulate motion artifacts that can be seen in real life. These motion artifact simulations follow a common pattern [15,16,22]. First, a random movement is generated by sampling a probability distribution for translation and rotational movement along each of the x, y, and z axes. Thus, we have six degrees of freedom: Three for translation (Tx, Ty, Tz) and three for rotation (Rx, Ry, Rz). Second, we pick a small portion of k-space lines from the moved image and create a composite image by superimposing these lines onto the original image. These steps can be repeated to increase the amount of motion artifacts seen in the brain scan. We have used the approach proposed by Shaw et al. [22] for our study.

3 Methods

3.1 Dataset

Selection and Sampling: We have aggregated T1 weighted MRI scans from cognitively normal adults from 3 multi-center datasets - IXI Dataset [1], Calgary-Campinas (CC) Public Brain MR Dataset [24] and OASIS-3 Dataset [13].

IXI dataset aggregates data from three different centers, from two vendors (Philips and GE) and two magnetic field strengths (1.5 T and 3T). CC data is a heterogeneous dataset acquired in six different centers from three different vendors (Siemens, Philips and GE) and two magnetic field strengths (1.5T and 3T). OASIS aggregates samples acquired from three different Siemens scanners with two different magnetic field strengths (1.5T and 3T).

We utilized the final reconstructed T1-weighted (T1w) scans, resulting in 490, 359, and 576 scans from the OASIS, CC, and IXI datasets, respectively. Due to computational resource constraints, we have randomly chosen 50% of the data from each of the datasets. Then, the scans from each dataset were randomly split into train-test-validation sets (75%-15%-10%, Table 1).

Preprocessing and Simulating Motion Artifacts: The scans from the three datasets were skull-stripped and normalized (min-max). The CC dataset was skull-stripped with the masks provided with the dataset. A skull-stripped version of the IXI dataset was made available as part of a study by Chen et al. [6]. The

Table 1. Training, validation and testing set splits.

Dataset	Number of samples	Train Samples	Validation Samples	Test Samples
IXI Dataset	288	216	43	29
CC Dataset	178	134	26	18
OASIS Dataset	244	183	36	25
Combined	710	533	105	72

OASIS dataset was skull-stripped using the preprocessing method proposed by He et al. [10].

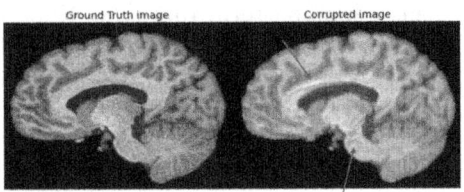

Fig. 1. Sample of simulated motion artifact. The corrupted image is blurry and has ghosting artifacts (marked with red arrows) simulated from the ground truth image (left). (Color figure online)

Motion artifacts were simulated using the TorchIO library in Python [20], following Shaw et al. [22]. The rotation and translation were set at 5°C and 10 pixels, respectively. This means at each step, we had selected a random value from 0 to 5°C of rotation and 0 to 10 pixels of translation along each of the x, y and z axes. This step was done twice to get the level of motion simulation generated for this experiment. This simulation resulted in blurring and ghosting artifacts (Fig. 1).

3.2 Experiment Design

We have used a 3D UNet [28] model for our experiments, since we are processing 3D voxels for motion mitigation. To analyze the effect and knowledge retention of the model trained on a large dataset, we have trained this model on the combined dataset containing images from the CC, OASIS and IXI datasets (Table 1). Apart from the model trained on the combined (CC-OASIS-IXI) dataset, we also analyzed the performance of models trained on the individual (CC, OASIS, IXI) and different combinations of the datasets (CC-OASIS, OASIS-IXI, CC-IXI). These models trained on 7 different datasets were evaluated individually on each of the datasets, providing us with rich insights into the UNet's behavior.

3.3 Evaluation Metrics

We have chosen Mean Squared Error (MSE) (Eq. 1), Structural Similarity Index Measure (SSIM) (Eq. 2), and Peak Signal-to-Noise Ratio (PSNR) (Eq. 3) for our model training and evaluation, since they are the common choices in motion mitigation literature [25]. Specifically, a hybrid of MSE and SSIM was used as the loss function, while SSIM and PSNR were used for evaluating the model's performance.

$$MSE(x, \hat{x}) = \frac{1}{HWD} \sum_{i=1, j=1, k=1}^{H,W,D} (x_{ijk} - \hat{x}_{ijk})^2 \qquad (1)$$

$$SSIM(x, \hat{x}) = \frac{(2\mu_x \mu_{\hat{x}} + C_1)(2\sigma_{x\hat{x}} + C_2)}{(\mu_x^2 + \mu_{\hat{x}}^2 + C_1)(\sigma_x^2 + \sigma_{\hat{x}}^2 + C_2)} \qquad (2)$$

$$PSNR(x, \hat{x}) = 10 \cdot \log_{10} \left(\frac{MAX_{\hat{x}}^2}{MSE(x, \hat{x})} \right) \qquad (3)$$

In the equations, x and $\hat{x}$ denote the ground truth and model prediction, respectively. MSE measures the mean squared pixel-wise difference, SSIM captures the difference in intensity and contrast, and PSNR compares the amount of noise and clarity between the model output and the ground truth.

4 Results and Discussion

4.1 Performance Analysis

Comparison with the Literature: We have achieved an SSIM of 0.98 and a PSNR of 29.74 while testing our model on the combined test set. This is on par with the literature for motion mitigation (Table 2). Despite training on a large, diverse dataset, the 3D UNet could preserve its performance across the test set. One thing to note here is that the performance metrics for the motion mitigation literature in Table 2 are reported by the authors on their experiment setup. The datasets used were private (except [5]). Most of these models are 2D [3,5,15,17], and the 3D model [19] utilizes multi parametric MRI scans, whereas we are investigating the impact of a baseline 3D UNet performance on open T1w datasets. Hence, the metric values are used solely to provide an idea about the 3D UNet model's performance compared to recent motion mitigation approaches.

Comparison Among Models Trained on Different Combination of Training Sets: We have observed a few interesting patterns in models trained on single datasets, shown in Fig. 2 (these models will be referred to by their respective training datasets). All three models performed similarly in terms of the SSIM metric (Fig. 2a), except for the OASIS model lagging slightly behind when tested on the IXI dataset (0.966), which is still a high SSIM value. However, the PSNR metric fluctuates more in these tests (Fig. 2b). All the models lose performance on the IXI dataset compared to the other datasets. The

Table 2. Performance comparison of different models. (*) The datasets are private and were collected by the original authors.

Literature	Dataset	SSIM (↑)	PSNR (↑)
Chatterjee et al. [5] (2D)	IXI (100 T2w subjects), 3 centers, Philips(1.5T, 3T) and GE(1.5T)	0.97	–
DRN-DCMB* [17] (2D)	217 T1w subjects, 2 centers, GE (1.5T) scanners	0.965 ± 0.02	–
MC^2-Net* [15] (2D)	41 subjects, 1 center (Siemens 3T) (T1w, T2w, T2-FLAIR)	0.9905	–
Stacked U-Nets* [3] (2D)	83 T1w subjects, 1 center (GE 3T)	0.950	25.76
Pirkl et al.* [19] (3D)	8 subjects (T1w, T2w, PD) two GE scanners (1.5T and 3.0T)		32.76
3D U-Nets (Ours)	710 subjects (T1w) 3 datasets (11 centers combined) Siemens, Philips and GE scanners (1.5T and 3T)	0.98 ± 0.01	29.74 ± 3.62

OASIS model also achieves lower performance when tested on the CC dataset (27.58). Meanwhile, the IXI model outperforms all the other models on each of the datasets.

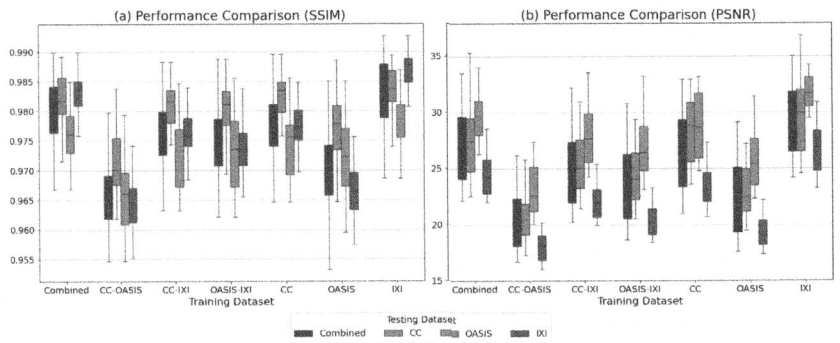

Fig. 2. Baseline model performance trained using different datasets (x-axis), with results stratified by testing datasets (identified by color). (a) SSIM and (b) PSNR metrics are scaled for better visualization. (Color figure online)

These patterns can be attributed to the contrast and intensity differences in the images. A model trained on a distribution of intensities might fall behind while predicting a different intensity distribution, since PSNR is calculated using the Mean Squared Error (MSE)(Eq. 3). Figure 3 shows the histograms from all the scanners in each of the datasets. The CC dataset distribution covers a diverse set of intensities and contrasts (Fig. 3a, Fig. 3c) – hence it can perform similarly in the OASIS dataset (Fig. 3b), given the model has seen the same properties. On the other hand, the OASIS dataset consisted of only two scanners covering a portion of the whole spectrum of the CC dataset, causing the drop in PSNR testing on the CC dataset. In terms of the IXI dataset, the intensity distribution is stretched; it also covers a bigger area compared to the other datasets (Fig. 3d). This results in a drop in performance for the CC and OASIS models, while the IXI model achieves better PSNR across all the test sets.

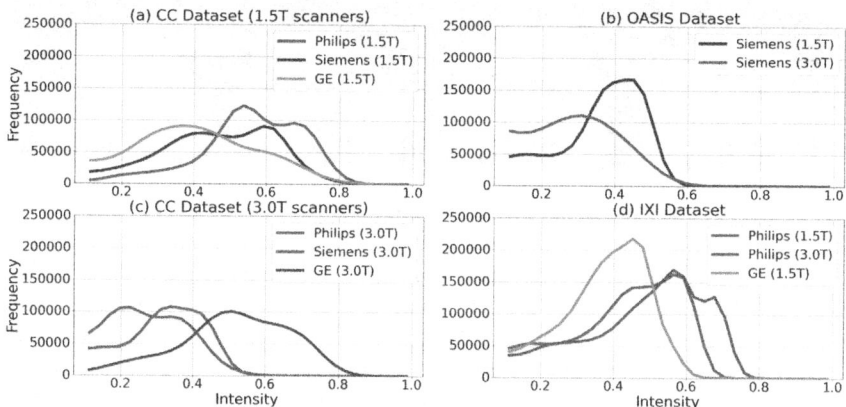

Fig. 3. Histogram distribution for the CC dataset (a, c), combining six scanners, is spread across the spectrum uniformly. (b) The OASIS dataset covers a portion of the intensity spectrum with the two scanners. (d) The IXI dataset with the three scanners is spread across the intensity spectrum, as well as having the most frequency across intensities.

Another interesting phenomenon is the performance drop when trained on multiple datasets (Combined, CC-OASIS, CC-IXI, OASIS-IXI), compared to single datasets. For example, the IXI model performs better than the Combined, CC-IXI, OASIS-IXI models; however, it was expected that the model would preserve the learned knowledge, since it is looking at more diverse data. We aim to study this further by training the model with different combinations of data distributions and assessing their effects on knowledge retention.

4.2 Qualitative Analysis

We have picked the best (IXI) and the worst (CC-OASIS) performing models for analyzing the visual outputs (Fig. 4). Looking at the model outputs, there is not a lot of difference between the best and the worst performing models, as seen in the evaluation metrics. However, despite mitigating the intense artifacts, the IXI model failed to address some of the subtle motion artifacts (for example, the motion in the corpus callosum region) for the CC and the OASIS datasets, while

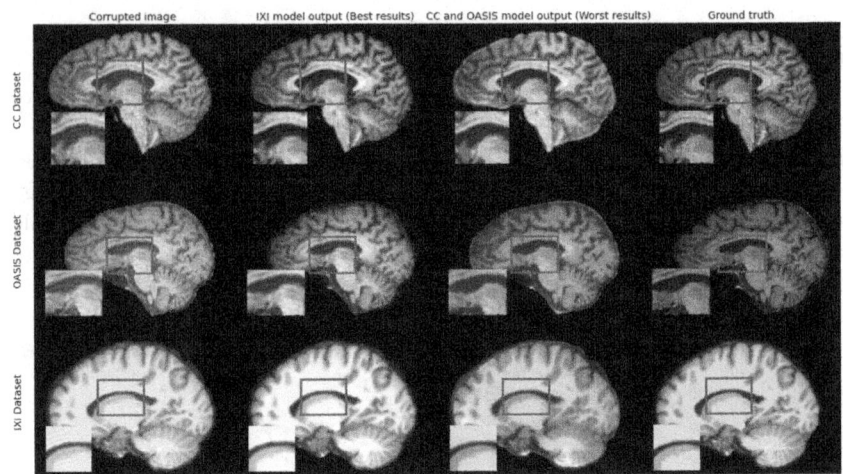

Fig. 4. Example outputs from the model.

the CC-OASIS model could clean them. The presence of these subtle motion artifacts in the output images is inconsistent with the metric values, which generally focus on the structure and noisiness of the brain scans.

4.3 Limitations and Future Work

We have experienced relatively blurry outputs alongside subtle motion artifacts while testing our model, lowering the PSNR values. We aim to address this by incorporating a blur detection loss in our training, with a downstream task such as tissue segmentation as a metric for detecting subtle motion artifacts. Downstream tasks can also focus on the usability of the scans.

We also want to investigate how the intensity range affects the training (Fig. 3), targeting knowledge retention of a baseline model trained on diverse datasets. Additionally, we want to study if histograms can be used as a preprocessing step to adapt our models to different scanners, modalities and age groups.

5 Conclusion

Our study looked into the potential of training several baseline 3D UNet models to test and explain their generalizability across sufficiently diverse, multi-center datasets, without incorporating a domain adaptation component or data augmentation for generalization. The model trained with the large combined adult datasets could retain its performance, on par with the current literature. Models trained on individual datasets could also perform motion mitigation on unseen adult datasets, which implies that, given enough data, the baseline model develops a general idea of motion mitigation. The output image quality from the models trained on individual datasets in our experiments correlated with the breadth of the data distribution covered by the training dataset. This approach can be used for data augmentation for addressing generalizable motion mitigation research.

Acknowledgments. This work was supported by research funding from the Natural Sciences and Engineering Research Council of Canada (NSERC) Discovery Grant program and the Alberta Innovates LevMax-Health grant.

Disclosure of Interests. The authors have no competing interests to declare that are relevant to the content of this article.

References

1. IXI dataset. https://brain-development.org/ixi-dataset/
2. Abdi, M., Bilchick, K.C., Epstein, F.H.: Compensation for respiratory motion-induced signal loss and phase corruption in free-breathing self-navigated cine DENSE using deep learning. Magn. Reson. Med. **89**(5), 1975–1989 (2023)
3. Al-masni, M.A., et al.: Stacked U-Nets with self-assisted priors towards robust correction of rigid motion artifact in brain MRI. Neuroimage **259**, 119411 (2022)
4. Callaghan, M.F., et al.: An evaluation of prospective motion correction (pmc) for high resolution quantitative mri. Front. Neurosci. **9** (2015)
5. Chatterjee, S., et al.: Retrospective motion correction of mr images using prior-assisted deep learning. arXiv preprint arXiv:2011.14134 (2020)
6. Chen, J., et al.: TransMorph: transformer for unsupervised medical image registration. Med. Image Anal. **82**, 102615 (2022)
7. Ehman, R., et al.: Magnetic resonance imaging with respiratory gating: techniques and advantages. Am. J. Roentgenol. **143**(6), 1175–1182 (1984)
8. der Goten, V., Alexander, L., Smith, K.: Wide range MRI artifact removal with transformers. arXiv preprint arXiv:2210.07976 (2022)
9. Hayatghaibi, S.E., et al.: Turnaround time and efficiency of pediatric outpatient brain magnetic resonance imaging: a multi-institutional cross-sectional study. Pediatr. Radiol. **53**(6), 1144–1152 (2023)
10. He, X., Wang, A.Q., Sabuncu, M.R.: Neural pre-processing: a learning framework for end-to-end brain mri pre-processing. In: Greenspan, H., et al. (eds.) Medical Image Computing and Computer Assisted Intervention - MICCAI 2023, pp. 258–267. Springer Nature Switzerland, Cham (2023). https://doi.org/10.1007/978-3-031-43993-3_25
11. Kerfoot, E., Clough, J., Oksuz, I., Lee, J., King, A.P., Schnabel, J.A.: Left-ventricle quantification using residual U-Net. In: Pop, M., et al. (eds.) STACOM 2018. LNCS, vol. 11395, pp. 371–380. Springer, Cham (2019). https://doi.org/10.1007/978-3-030-12029-0_40
12. Küstner, T., et al.: Retrospective correction of motion-affected MR images using deep learning frameworks. Magn. Reson. Med. **82**(4), 1527–1540 (2019)
13. LaMontagne, P.J., et al.: OASIS-3: Longitudinal Neuroimaging, Clinical, and Cognitive Dataset for Normal Aging and Alzheimer Disease (Dec 2019)
14. Lee, J., Kim, B., Park, H.: Mc2-net: motion correction network for multi-contrast brain MRI. Magn. Reson. Med. **86**(2), 1077–1092 (2021)

15. Lee, S., et al.: Deep learning in MR motion correction: a brief review and a new motion simulation tool (view2Dmotion). Investigative Magnetic Resonance Imaging **24**(4), 196–206 (2020)
16. Levac, B., et al.: FSE compensated motion correction for MRI using data driven methods. In: Wang, L., et al. (eds.) Medical Image Computing and Computer Assisted Intervention – MICCAI 2022, pp. 707–716. Springer Nature Switzerland, Cham (2022). https://doi.org/10.1007/978-3-031-16446-0_67
17. Liu, J., et al.: Motion artifacts reduction in brain MRI by means of a deep residual network with densely connected multi-resolution blocks (DRN-DCMB). Magn. Reson. Imaging **71**, 69–79 (2020)
18. Pawar, K., et al.: Clinical utility of deep learning motion correction for T1 weighted MPRAGE MR images. Eur. J. Radiol. **133**, 109384 (2020)
19. Pirkl, C.M., et al.: Learning residual motion correction for fast and robust 3D multiparametric MRI. Med. Image Anal. **77**, 102387 (2022). https://doi.org/10.1016/j.media.2022.102387
20. Pérez-García, F., Sparks, R., Ourselin, S.: TorchIO: a Python library for efficient loading, preprocessing, augmentation and patch-based sampling of medical images in deep learning. Comput. Methods Programs Biomed. **208**, 106236 (2021)
21. Ronneberger, O., Fischer, P., Brox, T.: U-Net: convolutional networks for biomedical image segmentation. In: Navab, N., Hornegger, J., Wells, W.M., Frangi, A.F. (eds.) MICCAI 2015. LNCS, vol. 9351, pp. 234–241. Springer, Cham (2015). https://doi.org/10.1007/978-3-319-24574-4_28
22. Shaw, R., Sudre, C., Ourselin, S., Cardoso, M.J.: MRI k-space motion artefact augmentation: model robustness and task-specific uncertainty. In: Cardoso, M.J., et al. (eds.) Proceedings of The 2nd International Conference on Medical Imaging with Deep Learning. Proceedings of Machine Learning Research, 08–10 Jul, vol. 102, pp. 427–436. PMLR (2019)
23. Sommer, K., et al.: Correction of motion artifacts using a multiscale fully convolutional neural network. Am. J. Neuroradiol. **41**(3), 416–423 (2020)
24. Souza, R., et al.: An open, multi-vendor, multi-field-strength brain mr dataset and analysis of publicly available skull stripping methods agreement. Neuroimage **170**, 482–494 (2018)
25. Spieker, V., et al.: Deep learning for retrospective motion correction in MRI: a comprehensive review. IEEE Trans. Med. Imaging **43**(2), 846–859 (2024)
26. Tamada, D., et al.: Motion artifact reduction using a convolutional neural network for dynamic contrast enhanced MR imaging of the liver. Magn. Reson. Med. Sci. **19**(1), 64–76 (2020)
27. Zaitsev, M., Maclaren, J., Herbst, M.: Motion artifacts in MRI: a complex problem with many partial solutions. J. Magn. Reson. Imaging **42**(4), 887–901 (2015)
28. Çiçek, ., et al.: 3D U-Net: learning dense volumetric segmentation from sparse annotation. In: Ourselin, S., et al. (eds.) Medical Image Computing and Computer-Assisted Intervention – MICCAI 2016, pp. 424–432. Springer International Publishing, Cham (2016). https://doi.org/10.1007/978-3-319-46723-8_49

Localized FNO for Spatiotemporal Hemodynamic Upsampling in Aneurysm MRI

Kyriakos Flouris[1(✉)] [iD], Moritz Halter[1], Yolanne Y. R. Lee[2], Samuel Castonguay[3,4], Luuk Jacobs[5], Pietro Dirix[5], Jonathan Nestmann[5], Sebastian Kozerke[5], and Ender Konukoglu[1]

[1] Biomedical Imaging Group, Computer Vision Lab, ETH Zurich, Zurich, Switzerland
{kflouris,mhalter,kender}@ethz.ch
[2] Department of Computer Science, University College London, London, UK
yolanne.lee.19@ucl.ac.uk
[3] Institute of Environmental Engineering, Swiss Federal Institute of Technology, ETH Zurich, Zurich, Switzerland
[4] Biodiversity and Conservation Biology Unit, Swiss Federal Institute for Forest, Snow and Landscape Research, WSL, Zurich, Switzerland
[5] Institute for Biomedical Engineering, University and ETH Zurich, Zurich, Switzerland

Abstract. Hemodynamic analysis is essential for predicting aneurysm rupture and guiding treatment. While magnetic resonance flow imaging enables time-resolved volumetric blood velocity measurements, its low spatiotemporal resolution and signal-to-noise ratio limit its diagnostic utility. To address this, we propose the Localized Fourier Neural Operator (LoFNO), a novel 3D architecture that enhances both spatial and temporal resolution with the ability to predict wall shear stress (WSS) directly from clinical imaging data. LoFNO integrates Laplacian eigenvectors as geometric priors for improved structural awareness on irregular, unseen geometries and employs an Enhanced Deep Super-Resolution Network (EDSR) layer for robust upsampling. By combining geometric priors with neural operator frameworks, LoFNO de-noises and spatiotemporally upsamples flow data, achieving superior velocity and WSS predictions compared to interpolation and alternative deep learning methods, enabling more precise cerebrovascular diagnostics. The code, ablations and hyperparameters are available at: https://github.com/moritz-halter/deepflow.

Keywords: 4D flow MRI · Super-resolution · Hemodynamics · Anuerysm

---

K. Flouris and M. Halter—These authors contributed equally to this work.

1 Introduction

Intracranial aneurysms (IAs) and arteriovenous malformations (AVMs) are major causes of hemorrhagic strokes, leading to significant morbidity and mortality [23]. Timely surgical or endovascular interventions are critical to reducing rupture risk and preventing cerebral hemorrhage in high-risk individuals [3,20]. Accurate diagnostic imaging is essential for identifying at-risk patients, assessing disease severity, and guiding treatment planning [8,17].

Current clinical imaging primarily captures morphological data, often overlooking hemodynamic parameters that could enable earlier rupture prediction [7,19]. Hemodynamic analyses based solely on morphology rely on costly computational fluid dynamics (CFD) simulations or simplified models derived from limited imaging slices, which may not fully capture the complexity of cerebral blood flow [16]. Recent advances in 4D flow magnetic resonance (MR) imaging now allow time-resolved volumetric measurements of blood velocity vector fields throughout the brain [18], providing real data that was previously only available through CFD simulations. These measurements enable detailed flow pattern analyses, collateral flow activation, arterial and venous pulsatility, pressure gradients, and wall shear stress (WSS, aiding in the detection and diagnosis of IA [29]. While 4D flow MR imaging eliminates the computational burden of CFD and the labor-intensive construction of geometric models, its low spatiotemporal resolution and signal-to-noise ratio remain major limitations. Interpolation methods can improve resolution, but they require manual selection of methods and parameters for each case, limiting automation and scalability.

Neural networks are widely used for medical image upsampling, with models like Enhanced Deep Super-Resolution Network (EDSR) [13] excelling in single-image super-resolution. However, the anatomical variability of IA complicates both upsampling and clinical tasks like diagnosis and rupture prevention [29]. Beyond image enhancement, neural networks have been explored for predicting hemodynamic parameters directly from imaging data, bypassing the computational bottlenecks of CFD. However, solving temporal dynamics and medical imaging super-resolution remains an open challenge.

We propose a novel, domain-agnostic Fourier Neural Operator (FNO) architecture, Localized Fourier Neural Operator (LoFNO), which incorporates embeddings of Laplacian eigenvectors to enhance geometric awareness [14]. Designed to both upsample spatial resolution and temporal dynamics, enabling the prediction of intermediate time steps from sparse inputs. LoFNO predicts hemodynamic parameters directly from routinely acquired flow from MR imaging data, providing an efficient alternative to traditional CFDs.

Our approach introduces three key contributions: *(1)* integrating *Laplacian eigenvectors* as geometric priors to improve generalization across irregular vascular geometries, *(2)* incorporating an *EDSR-based super-resolution layer* to enhance image quality and robustness to noise, and *(3)* implementing Localized Fourier Neural Operator (LoFNO), on a synthetic 4D flow IA dataset. Real data analysis is not feasible due to the need for high-quality training flow and geometry data, as obtaining very high-resolution 4D flow MRI is highly impractical.

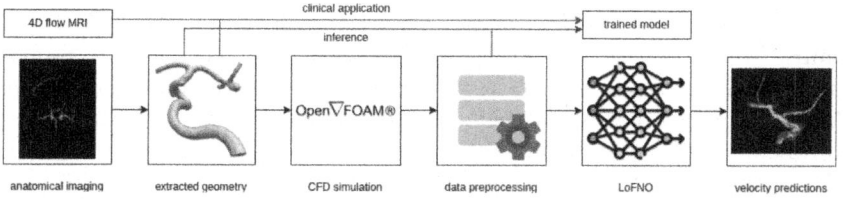

Fig. 1. Method pipeline: Geometry and boundary conditions are extracted from cerebral angiograms. Simulations are run, the relevant section is segmented, Laplacian eigenvectors are computed, and noise is added. Flow and eigenvectors are sampled on a voxel grid for preprocessing. The model is trained to predict noise-free flow on a high-resolution grid and tested on unseen data. In clinical use, it can be applied directly to 4D flow MRI *without* CFD simulations.

However, once trained, the model can be used to upsample low-resolution flow data, making it more applicable in a clinical setting. Figure 1 outlines the complete pipeline, from aneurysm extraction and CFD data preparation to flow parameters prediction.

Relevant Works. Deep neural networks are widely used for upsampling and super-resolution [5,9,25], with Super-Resolution Convolutional Neural Network (SRCNN) [4] and EDSR [13] playing key roles. Physics-informed methods integrate physical priors into training, including Physics-Informed Neural Network (PINN)s [2] and other specialized frameworks [6,15,24]. While effective, these methods struggle with fixed discretization when solving continuous equations and face challenges in enforcing boundary conditions on complex geometries.

Domain Agnostic Fourier Neural Operator (DAFNO) [14] and Geo-FNO [11] aim to alleviate the constraint of operator methods on fixed geometries by enabling applications to general geometries, while Incremental FNO [28] introduces adaptive training schedules to improve efficiency. Despite these advances, generalization to unseen geometries remains challenging, as these methods were not originally designed for intricate biological boundaries such as those in aneurysms.

There are numerous techniques that leverage spectral coordinates for geometric processing [10,22], based on geometric deep learning [1]. In this work, we introduce a novel architecture that builds upon DAFNO, leveraging spectral embeddings [1,10] and incorporating EDSR to efficiently upsample the spatial and temporal resoltuion of 4D flow imaging.

2 Method

The LoFNO consists of an EDSR module that takes low-resolution flow data and Laplacian eigenvectors as input, refining spatiotemporal resolution while integrating geometric features from spectral coordinates. The output, is then processed by a DAFNO layer with implicit FNOs, where a domain characteristic function ensures computations remain localized to the aneurysm geometry.

Operator methods enable learning generalized physical models across diverse domains. We aim to learn a mapping $\tilde{G} : \mathcal{A} \to \mathcal{U}$, where $\mathcal{A}(\mathbb{R}^{d_{in}})$ represents input velocity fields $u_i(x)$ and $\mathcal{U}(\mathbb{R}^{d_{out}})$ represents high-resolution output velocity fields $\hat{u}_i(x)$ over a patient-specific aneurysm domain $\Omega_i \subset \mathbb{R}^3$. This mapping is parameterized by θ and trained to approximate $\tilde{G}[u_i; \theta](x) \approx \hat{u}_i(x), \forall x \in \Omega_i$. Unlike traditional neural networks, the Neural Operator (NO) maintains discretization invariance, allowing evaluation at arbitrary resolutions.

The FNO [12] leverages Fourier transforms for efficient modeling of spatial dependencies in partial differential equations. It applies a convolution-based layer where a tensor kernel $\kappa \in \mathbb{R}^{d_h \times d_h}$ with parameters v^l operates over the computational domain $\bar{\Omega}$, which contains the aneurysm geometry Ω. By utilizing the Fourier transform $\mathcal{F}$ and its inverse $\mathcal{F}^{-1}$, FNO exploits the efficiency of Fourier Transform (FFT)/inverse Fast Fourier Transform (iFFT). However, it requires inputs on a regular grid, making it unsuitable for irregular aneurysm geometries. To address this, a periodic domain grid $\mathbb{O}$ is overlaid onto $\bar{\Omega}$, enabling Fourier-based computations but it has limited effectiveness.

Localizing to the Relevant Domains and Geomteric Prior. Mapping the aneurysm geometry Ω to the grid $\mathbb{O}$ introduces out-of-domain voxels in $\mathbb{O} \setminus \Omega$, increasing complexity and computational inefficiency. These out-of-domain voxels are trivial and detract from the solution, particularly as boundary nodes are crucial for calculating WSS. DAFNOs [14] extend FNO to improve generalization across unseen domains while reducing computational overhead from $\mathbb{O} \setminus \Omega$.

To enforce domain localization, $\chi(x)$ is defined as $\chi(x) = 1$ for $x \in \Omega$ and $\chi(x) = 0$ for $x \in \mathbb{O} \setminus \Omega$. The Fourier layer is modified to ensure interactions remain within the physical domain, Ω, while preserving FFT efficiency. The operator is expressed compactly as:

$$\mathcal{J}^l[h](x) = \sigma\Big(\chi(x)\left(I(\chi(\cdot)h(\cdot); v^l) - h(x)I(\chi(\cdot); v^l)\right) + W^l h(x) + c^l\Big),$$

where $I(\cdot; v^l) = \mathcal{F}^{-1}\left[\mathcal{F}[\kappa(\cdot; v^l)] \cdot \mathcal{F}[\cdot]\right]$ represents the Fourier-based integral computation. $W^l \in \mathbb{R}^{d_h \times d_h}$ and $c^l \in \mathbb{R}^{d_h}$ are learnable parameters. This ensures domain-specific relevance while leveraging FFT for efficient evaluation. Additionally, we adopt implicit FNOs [27], making learnable parameters layer-independent to mitigate overfitting and vanishing gradients.

To map the trained solution to complex and unseen geometries, we use *spectral coordinates*. Laplacian eigenvectors are computed on a graph $\mathcal{G} = (\mathcal{V}, \mathcal{E})$, where $\mathcal{V}$ is the set of vertices and $\mathcal{E}$ the edges. The adjacency matrix $A \in \mathbb{R}^{N \times N}$ encodes connectivity, with $A_{ij} = 1$ if vertices i and j are connected, and the diagonal degree matrix D has D_{ii} as the vertex degree. The normalized graph Laplacian is defined as $L^{\text{sym}} = I - (D^+)^{1/2} A (D^+)^{1/2}$, where I is the identity matrix and D^+ the Moore-Penrose pseudoinverse. Eigenvectors $\mathbf{v}_i$ and eigenvalues λ_i are obtained from $L^{\text{sym}} \mathbf{v}_i = \lambda_i \mathbf{v}_i$, with eigenvalues sorted in descenting order. The first $k = 32$ eigenvectors, corresponding to the largest nonzero eigenvalues, capture key geometric features and are normalized as $\mathbf{v}_i \leftarrow \mathbf{v}_i / |\mathbf{v}_i|_2, \forall i$.

Enhanced Architecture. Inspired from SRNO [26], our approach first increases the resolution of the input data to prepare it for the Fourier layers.

Instead of interpolation, we use an EDSR network, designed for single-image super-resolution, leveraging convolutional layers and residual connections to handle noise in the input flow velocity data $u(x)$ sampled on a $d_x \times d_y \times d_z$ grid. This learnable upsampling improves resolution while reducing noise.

The EDSR network takes as input the low-resolution flow velocity data concatenated with the selected eigenvectors $e(x)$ of the Laplacian operator. It outputs a high-resolution representation, which, along with the domain characteristic function $\chi(x)$, is fed into the DAFNO. Within DAFNO, the input is lifted by an MLP layer P, then processed through Fourier layers, where FFT, matrix multiplications with learnable parameters v^l, and nonlinear activations σ are applied. The characteristic function $\chi(x)$ ensures domain-aware computations by restricting operations to relevant regions.

Finally, the output from the Fourier layers is projected to the target space through another MLP layer Q, reducing dimensionality and producing the noiseless flow velocity prediction $\hat{u}(x)$. See Fig. 2 for a detailed schematic of the LoFNO architecture.

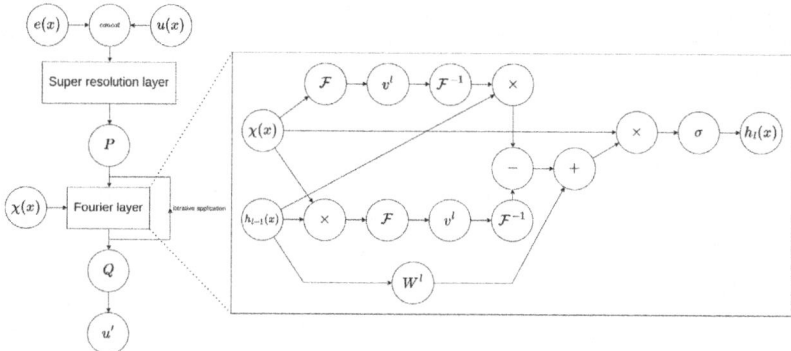

Fig. 2. Computational graph for LoFNO. Both $e(x)$, a selection of eigenvectors, and $u(x)$, the flow velocity with added noise, are sampled on a regular grid, while $\hat{u}(x)$ (prediction for noiseless flow), $\chi(x)$ (domain characteristic function), and $h_l(x)$ (Fourier layer input/output, $l \in \{0, \ldots, L\}$) are sampled on a higher resolution grid. The lifting layer P and projection layer Q are MLPs, mapping $(3 + N_e)$ to d_h and d_h to 3, respectively, where N_e is the number of eigenvectors and d_h the Fourier layer dimensionality. Nodes $\mathcal{F}$ and $\mathcal{F}^{-1}$ perform FFT and iFFT, while v^l applies matrix multiplication selecting the first N_m modes for each dimension. The node W^l performs pointwise multiplication with $W^l \in \mathbb{R}^{d_h \times d_h}$, and $+$, $-$, $\times$, and σ perform pointwise addition, subtraction, multiplication, and activation, respectively. The architecture and hyperparameters were optimized via ablation studies.

3 Experiments and Results

The Blood flow in the vessels was simulated by solving the three-dimensional, unsteady, incompressible Navier–Stokes equations. The blood was modeled as a

Newtonian, incompressible fluid with a density of $1060\,\mathrm{kg/m^3}$ and a kinematic viscosity of $3.5 \times 10^{-3}\,\mathrm{Pa \cdot s}$. In this study, the Navier Stokes equations were solved using a large eddy simulation (LES) approach implemented in Open-FOAM v2212. The wall-adapting local eddy viscosity (WALE) model was chosen as the subgrid-scale model. Second-order central difference and backward Euler schemes were used for spatial and temporal discretization respectively. An adaptive time-stepping strategy was employed to optimize simulation efficiency.

Using these simulations, we have carried out two sets of experiments, the spatial super-resolution and temporal upsampling. We generated a dataset of 95 pulse flows imposed on geometries from the Aneurisk dataset [21]. The dataset contains 3D reconstructions of internal carotid arteries and their associated aneurysms from cerebral angiographies. Simulations provided detailed hemodynamic parameters, including flow velocity, pressure, and WSS, capturing the complex fluid dynamics within patient-specific vascular geometries. Testing was performed on an unseen subset of geometries and their respective eigenvectors.

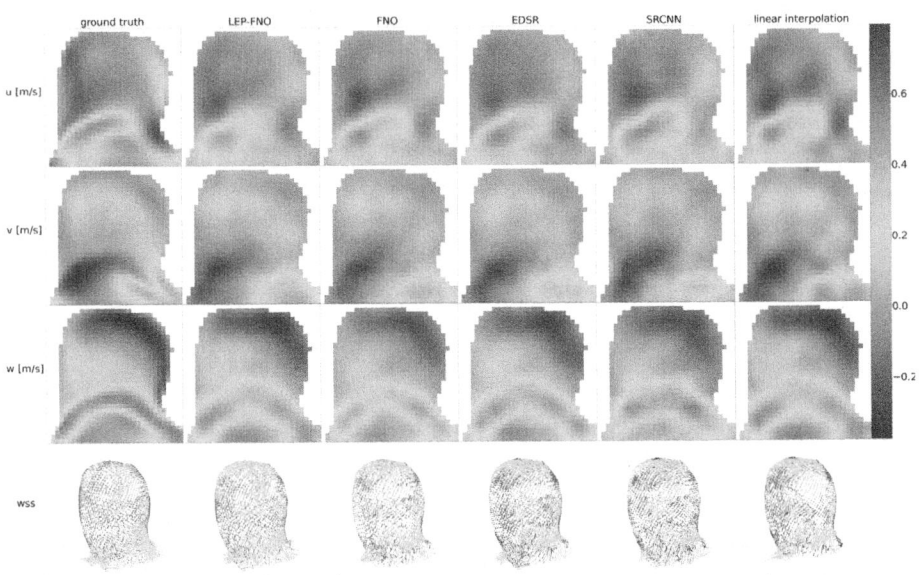

Fig. 3. Time-snapshot of hemodynamics upsampling for Scale ×4: Top: Flow velocity components u, v, and w in the x, y, and z directions for five models compared against the ground truth. Bottom: WSS evaluated on the surface geometry, with the length of vectors colored black for the smallest to yellow for the largest.

We selected 80 aneurysm-containing regions, computed Laplacian eigenvectors on the surface mesh, and resampled them onto a Cartesian grid using VTK's `vtkResampleWithDataSet`. Interior vessel points within subdomain Ω were marked, and white noise with a signal-to-noise ratio of 10 was added. Enforcing divergence-free interpolation was omitted as it caused artifacts, likely due to an overdetermined system from fixed points and the divergence-free constraint. Laplacian eigenvalues were computed on the VTK mesh, defining the graph, and resampled to the grid using point interpolation via vtkPointInterpolator with a Gaussian kernel to preserve boundary information.

We trained several deep learning models, including SRCNN, EDSR, FNO, and the proposed LoFNO, conducting experiments. Additionally, we evaluated classical interpolation methods such as Radial Basis Function (RBF) interpolation, Data-Driven Flow Interpolation (DFI), and linear interpolation. While PINNs performed well with sparse data, they exhibited instability with denser inputs and proved inefficient for large 3D datasets, thus not included.

Models were trained for 500 epochs, reaching near convergence. For spatial upsampling, models reconstructed a noiseless high-resolution flow field of either 32^3 or 24^3 from noisy low-resolution inputs of 16^3 or 8^3, obtained by subsampling the high-resolution data across all 24 timesteps. For temporal upsampling, models reconstructed a full high-resolution $32^3 \times 24$ flow field from noisy, high-spatial-resolution but temporally undersampled inputs, including $32^3 \times 12$, $32^3 \times 6$, and $32^3 \times 1$, where for the latter the entire sequence was predicted from the initial timestep only. For all models, we minimized the relative loss $= \sum_{x \in \Omega} \sum_t |u(x,t) - \hat{u}(x,t)|/|u(x,t)|$, where $u(x,t)$ is the ground truth flow velocity and $\hat{u}(x,t)$ the predicted velocity at spatial point x and time t.

Results. For testing, we use a subset of 10 geometries that were unseen during training. The predicted results for the best-performing methods are shown in Fig. 3. While none of the methods achieved perfect reconstruction, as expected given the complexity of the problem and the unseen geometries, LoFNO qualitatively demonstrated better performance than the other methods in reconstructing velocities and WSS.

To evaluate model performance more concretely, we define the error function as $\text{err} f(x,t) = ||f(x,t) - \hat{f}(x,t)||_2$, which can be applied to flow velocity $u(x,t)$ or WSS. Each model is benchmarked by averaging the error over all spatial and temporal dimensions across the test set. The differences are highlighted in the quantitative comparison in Table 1, where complete LoFNO outperformed other approaches, underscoring the significance of domain and geometric priors.

Combining neural operators with super-resolution networks yielded significant gains, as seen by the $\sim 25\%$ achieved by the LoFNO *wo* LEP. The results are improved with the addition of the Laplacian eigenvectors as geometric priors. Noteworthily, not only does LoFNO improve on the aggregated metrics, but the results consistently outperform the other methods for every individual test case. The localized geometry nature of our framework enhanced robustness and adaptability. While all methods train efficiently on modern hardware, interpolation techniques are slower during evaluation. We have also trained and tested on noiseless data, yielding similar trends, but omitted it due to limited real-world applicability.

For temporal super-resolution, our model showed even greater improvements, as seen in Table 2. Even standalone FNO performed well across all tested scales. However, despite full spatial resolution being available, augmenting FNOs with

Table 1. Test-error, L^{test}, for hemodynamics parameters upsampling.

Model	Scale ×4		Scale ×3		Scale ×2		Training /
	u	wss	u	wss	u	wss	Evaluation
Linear Interp.	0.0684	1.0914	0.04661	0.5680	0.0497	0.8366	$na/<1$ min
RBF Interp.	0.0813	1.4261	0.0563	0.6886	0.0591	0.9387	$na/\sim 15$min
DFI Interp.	0.2054	2.8680	0.1339	1.3554	0.1864	2.5229	$na/\sim 1$h
SRCNN	0.0807	1.1081	0.0574	0.6527	0.0389	0.7969	~ 15min$/\sim 10$s
EDSR	0.0683	1.1483	0.0460	0.5840	0.0300	0.6265	~ 15min$/\sim 10$s
FNO[a]	0.0647	0.9832	0.0388	0.4529	0.0272	0.5288	~ 1h$/\sim 10$s
LoFNO wo LEP[b]	0.0464	0.7699	0.0292	0.3750	0.0201	0.4220	~ 1h$/\sim 10$s
LoFNO	**0.0452**	**0.7625**	**0.0291**	**0.3637**	**0.0198**	**0.4139**	~ 1h$/\sim 10$s

[a] A FNO with a preceding EDSR layer. [b] LoFNO without Laplacian Eigenvalues Prior (LEP).

Table 2. Test-error, L^{test}, for hemodynamics parameters temporal upsampling.

Model	Scale ×2		Scale ×4		Prediction		Training /
	u	wss	u	wss	u	wss	Evaluation
Linear Interp.	0.0632	0.5631	0.0699	0.6107	0.1145	0.9520	$na/<1$ min
SRCNN	0.0461	0.7141	0.0607	0.8298	0.0600	0.8082	~ 15min$/\sim 10$s
EDSR	0.0317	0.4689	0.0459	0.5765	0.0624	0.7671	~ 15min$/\sim 10$s
FNO[a]	0.0384	0.4367	0.0382	0.4247	0.0678	0.7546	~ 1h$/\sim 10$s
LoFNO w/o LEP[b]	0.0245	0.2987	0.0315	0.3684	0.0531	0.6028	~ 1h$/\sim 10$s
LoFNO	**0.0195**	**0.2723**	**0.0280**	**0.3424**	**0.0536**	**0.5941**	~ 1h$/\sim 10$s

[a] A FNO with a preceding EDSR layer. [b] LoFNO without Laplacian Eigenvalues Prior (LEP).

domain or geometric priors, such as the domain characteristic function and Laplacian eigenvectors i.e. the LoFNO, further enhanced performance.

The complete prediction task, $32^3 \times 1$ to $32^3 \times 24$, where 23 future time steps are predicted from a single initial one, is particularly challenging. Despite this, all machine learning models, and especially our method, maintained reasonable accuracy, while interpolation methods suffered a severe performance drop. Even with an extra 25th time step, interpolation remained far less accurate. This disparity arises because the CFD simulation applies the same pressure pulse at the vessel inlet. Machine learning models, trained on slight pulse variations, effectively captured the underlying dynamics, enabling accurate predictions even on unseen geometries. In contrast, interpolation methods, lacking system knowledge, relied only on given data points, limiting their predictive capability.

Limitations. While LoFNO offers efficiency and scalability, it depends on CFD data for training, as no real data exist for this task, which may not capture real-world variability. These methods also lack inherent temporal mechanisms, limiting interpretability for dynamic flow modeling. We evaluated several super-resolution alternatives and found that the chosen methods, including the gold standard FNO, consistently matched or outperformed others. We were also constrained to methods that could be adapted to large 3D datasets.

4 Conclusions

We proposed LoFNO, a novel architecture that outperforms interpolation and alternative methods in test-error for spatial and temporal upsampling of velocity and WSS predictions. Our focus in this work is to address a key methodological challenge: super-resolving fluid dynamics variables in complex boundaries remains unsolved, even with large-scale, high-fidelity simulated data. Generalizing to unseen geometries is not only relevant to aneurysms but to many other problems. The proposed model directly targets this. To demonstrate its effectiveness, we focus on quantitative evaluation using high-resolution simulated data, where ground truth is available. This assessment would not have been possible with clinical data because the ground truth cannot be easily acquired. By enabling precise, non-invasive prediction of hemodynamic parameters, LoFNO can improve clinical assessment, enhance disease progression predictions, and optimize therapeutic outcomes, ultimately advancing patient care and safety.

Acknowledgement. This project was supported by grants #2022-531 and #2022-643 of the Strategic Focus Area "Personalized Health and Related Technologies (PHRT)" of the ETH Domain (Swiss Federal Institutes of Technology).

Disclosure of Interests. The authors declare that they have no competing interests.

References

1. Bronstein, M.M., Bruna, J., Cohen, T., Velickovic, P.: Geometric deep learning: grids, groups, graphs, geodesics, and gauges. CoRR **abs/2104.13478** (2021). https://arxiv.org/abs/2104.13478
2. Cai, S., Mao, Z., Wang, Z., Yin, M., Karniadakis, G.E.: Physics-informed neural networks (PINNs) for fluid mechanics: a review. Acta. Mech. Sin. **37**(12), 1727–1738 (2021)
3. Deshmukh, A.S., et al.: The management of intracranial aneurysms: current trends and future directions. Neurol. Int. **16**(1), 74–94 (2024)
4. Dong, C., Loy, C.C., He, K., Tang, X.: Image super-resolution using deep convolutional networks. IEEE Trans. Pattern Anal. Mach. Intell. **38**(2), 295–307 (2015)
5. Flouris, K., Konukoglu, E.: Canonical normalizing flows for manifold learning. In: Oh, A., Naumann, T., Globerson, A., Saenko, K., Hardt, M., Levine, S. (eds.) Advances in Neural Information Processing Systems, vol. 36, pp. 27294–27314. Curran Associates, Inc. (2023)
6. Flouris, K., Volokitin, A., Bredell, G., Konukoglu, E.: Explicit and data-efficient encoding via gradient flow. In: Proceedings of the NeurIPS 2024 Workshop on Machine Learning and the Physical Sciences (ML4PHY) (2024). https://arxiv.org/abs/2412.00864, arXiv:2412.00864

7. Han, P., et al.: The prognostic effects of hemodynamic parameters on rupture of intracranial aneurysm: a systematic review and meta-analysis. Int. J. Surg. **86**, 15–23 (2021)
8. Hussein, A., Malguria, N.: Imaging of vascular malformations. radiologic. Clinics **58**(4), 815–830 (2020)
9. Ledig, C., et al.: Photo-realistic single image super-resolution using a generative adversarial network. In: 2017 IEEE Conference on Computer Vision and Pattern Recognition (CVPR), pp. 105–114 (2017)
10. Li, X.J., Yang, J., Zhang, F.L.: Laplacian mesh transformer: dual attention and topology aware network for 3D mesh classification and segmentation. In: Avidan, S., Brostow, G., Cissé, M., Farinella, G.M., Hassner, T. (eds.) Computer Vision - ECCV 2022, pp. 541–560. Springer Nature Switzerland, Cham (2022)
11. Li, Z., Huang, D.Z., Liu, B., Anandkumar, A.: Fourier neural operator with learned deformations for PDES on general geometries. J. Mach. Learn. Res. **24**(388), 1–26 (2023)
12. Li, Z., et al.: Fourier neural operator for parametric partial differential equations (2021)
13. Lim, B., Son, S., Kim, H., Nah, S., Mu Lee, K.: Enhanced deep residual networks for single image super-resolution. In: Proceedings of the IEEE Conference on Computer Vision and Pattern Recognition Workshops, pp. 136–144 (2017)
14. Liu, N., Jafarzadeh, S., Yu, Y.: Domain agnostic Fourier neural operators. Adv. Neural Info. Process .Syst. **36** (2024)
15. Liu, X.Y., Zhu, M., Lu, L., Sun, H., Wang, J.X.: Multi-resolution partial differential equations preserved learning framework for spatiotemporal dynamics. Commun. Phys. **7**(1) (Jan 2024)
16. Maramkandam, E.B., et al.: Review of CFD based simulations to study the hemodynamics of cerebral aneurysms. J. Indian Inst. Sci. **104**(1), 1–34 (2024)
17. Maupu, C., Lebas, H., Boulaftali, Y.: Imaging modalities for intracranial aneurysm: more than meets the eye. Front. Cardiov. Med. **9**, 793072 (2022)
18. Morgan, A.G., Thrippleton, M.J., Wardlaw, J.M., Marshall, I.: 4D flow MRI for non-invasive measurement of blood flow in the brain: a systematic review. J. Cereb. Blood Flow Metab. **41**(2), 206–218 (2021)
19. Nico, E., Hossa, J., McGuire, L.S., Alaraj, A.: Rupture-risk stratifying patients with cerebral arteriovenous malformations using quantitative hemodynamic flow measurements. World Neurosurg. **179**, 68–76 (2023)
20. Pan, P., et al.: Review of treatment and therapeutic targets in brain arteriovenous malformation. J. Cerebr. Blood Flow Metab. **41**(12), 3141–3156 (2021)
21. Sangalli, L.M., Secchi, P., Vantini, S.: AneuRisk65: a dataset of three-dimensional cerebral vascular geometries. Electr. J. Stat. **8**(2), 1879–1890 (2014)
22. Torres, L., Chan, K.S., Eliassi-Rad, T.: GLEE: Geometric Laplacian eigenmap embedding. J. Compl. Netw. **8**(2), cnaa007 (2020)
23. Van Gijn, J., Kerr, R.S., Rinkel, G.J.: Subarachnoid haemorrhage. The Lancet **369**(9558), 306–318 (2007)
24. Wang, R., Kashinath, K., Mustafa, M., Albert, A., Yu, R.: Towards physics-informed deep learning for turbulent flow prediction (2020)
25. Wang, Y., Lee, Y.Y.R., Dolfini, A., Reischl, M., Konukoglu, E., Flouris, K.: Energy-based prior latent space diffusion model for reconstruction of lumbar vertebrae from thick slice MRI. In: Mukhopadhyay, A., Oksuz, I., Engelhardt, S., Mehrof, D., Yuan, Y. (eds.) Deep Generative Models, pp. 22–32. Springer Nature Switzerland, Cham (2025)

26. Wei, M., Zhang, X.: Super-resolution neural operator. In: 2023 IEEE/CVF Conference on Computer Vision and Pattern Recognition (CVPR), pp. 18247–18256. IEEE Computer Society, Los Alamitos, CA, USA (2023)
27. You, H., Zhang, Q., Ross, C.J., Lee, C.H., Yu, Y.: Learning deep implicit Fourier neural operators (IFNOs) with applications to heterogeneous material modeling. Comput. Methods Appl. Mech. Eng. **398**, 115296 (2022)
28. Zhao, J., George, R.J., Zhang, Y., Li, Z., Anandkumar, A.: Incremental Fourier neural operator. arXiv preprint arXiv:2211.15188 (2022)
29. Zhou, G., Zhu, Y., Yin, Y., Su, M., Li, M.: Association of wall shear stress with intracranial aneurysm rupture: systematic review and meta-analysis. Sci. Rep. **7**(1), 5331 (2017)

LSTT: Latent Spatio-Temporal Transformer for Non-rigid Motion Compensation in CBCT

Yipeng Sun(✉), Linda-Sophie Schneider, Annette Schwarz, Mingxuan Gu, Siyuan Mei, Chengze Ye, Siming Bayer, and Andreas Maier

Friedrich-Alexander-University Erlangen-Nuremberg, Erlangen, Germany
yipeng.sun@fau.de

Abstract. Non-rigid physiological motion during Cone-Beam Computed Tomography (CBCT) acquisitions remains a significant clinical challenge. To address this, we introduce the Latent Spatio-Temporal Transformer (LSTT), an end-to-end framework designed to directly correct motion artifacts from projection data. Our entirely image-based approach requires only the original CBCT projections and imaging geometry, eliminating the necessity for respiratory or ECG gating and external monitoring devices. The LSTT architecture integrates a VQ-VAE to tokenize projections into a robust latent space, a temporal Transformer to capture global motion dynamics, and a decoder to produce explicit 2D displacement fields. Central to our framework is a differentiable Feldkamp-Davis-Kress (FDK) reconstruction layer, which enables true end-to-end training by optimizing the objective function on the final reconstructed volume. This approach compels the network to learn a physically meaningful policy for non-rigid motion, explicitly tailored for high-fidelity volumetric reconstruction. We validate our framework using a realistic respiratory motion phantom, demonstrating significant improvements over the standard clinical baseline in both artifact suppression and structural preservation.

Keywords: Non-Rigid Motion · Cone-Beam CT · Deep Learning

1 Introduction

Cone-Beam Computed Tomography (CBCT) is a crucial imaging modality in modern medicine, especially in fields such as image-guided radiotherapy and interventional radiology [23], where motion artifacts remain a persistent challenge for various tomographic techniques [11]. The quality of CBCT reconstructions depends heavily on the critical assumption that the patient's anatomy remains stationary throughout data acquisition. However, this static-scene assumption is frequently violated by involuntary, non-rigid physiological movements, most notably respiration [20] and cardiac motion [14,18]. Due to the typically prolonged scan times required for CBCT, these motion-induced inconsistencies in the projection data often result in pronounced image artifacts, such as blurring, ghosting, and streaking. These artifacts can significantly compromise the accuracy of patient positioning and treatment delivery [5,6].

Traditionally, strategies to mitigate motion have spanned procedural solutions like breath-holding, hardware-based gating [17], and computationally expensive model-based iterative reconstruction (MBIR) methods [2,4,8]. While each has its merits, they also come with significant drawbacks. For instance, gating requires external monitoring hardware (e.g., respiratory belts or ECG) and increases scan time [19]. This motivates the development of markerless techniques [12], while MBIR methods have prohibitive reconstruction times.

The rise of deep learning has unlocked powerful new avenues for motion compensation [27,28]. However, many existing strategies have inherent limitations. The most common approach is image-domain post-processing, where a network (e.g., a U-Net [22] or GAN [7]) attempts to "de-artifact" an already reconstructed volume [10,16]. This is fundamentally limited, as it tries to recover information that has been irrecoverably corrupted during the ill-posed reconstruction of inconsistent data. A more physically-principled approach is to correct for motion at its source, in the projection domain [13]. While methods for projection alignment exist, they often rely on optimizing surrogate consistency metrics rather than the final image quality, which can be an unstable objective for complex, non-rigid motion [21].

In this work, we introduce the Latent Spatio-Temporal Transformer (LSTT), an end-to-end learning framework that is entirely image-based. It requires only the raw CBCT projections and the system geometric parameters, eliminating the need for any external gating or monitoring signals. The LSTT operates in the projection domain but is guided by a supervisory signal from the final 3D volume. This is achieved by integrating a fully differentiable CBCT reconstruction operator directly into the training loop, allowing us to optimize the projection warping process based on the quality of the final 3D image. Our key contributions are:

1. **A novel architecture for projection-domain non-rigid motion compensation:** The LSTT synergistically combines a Vector Quantised-Variational AutoEncoder (VQ-VAE) [25] for robust feature tokenization into a latent space with a temporal Transformer [26] that explicitly models the global, long-range spatio-temporal dynamics of physiological motion across the entire scan.
2. **True end-to-end volumetric supervision:** By incorporating a differentiable FDK backprojection operator, we enable the network to directly optimize for the final 3D volume quality, which directly aligns the network's objective with clinical goals and provides a significant advantage over methods that rely on indirect projection-space consistency.
3. **Generation of explicit and interpretable motion fields:** The network explicitly predicts 2D displacement vector fields, providing a physically mean-

ingful and interpretable representation of the learned motion model, a key advantage over black-box image-to-image approaches.

2 Methodology

Mathematically, the challenge of non-rigid motion in CBCT arises from the violation of the static scene assumption. Let the patient's anatomy be represented by a time-varying 3D volume $\mathbf{V}(t)$. This dynamic volume can be modeled as a deformation of a single, static reference volume $\mathbf{V}_c$ (e.g., the end-exhalation phase) by a time-dependent 3D displacement vector field (DVF) $\mathcal{M}_t$. During a CBCT scan, a sequence of 2D projections $\{P_m^{(t)}\}_{t=1}^T$ is acquired at different gantry angles ϕ_t. Each projection is a snapshot of the anatomy at a specific instant, capturing the deformed state: $P_m^{(t)} = \mathcal{P}_{\phi_t}(\mathbf{V}(t)) = \mathcal{P}_{\phi_t}(\mathbf{V}_c \circ \mathcal{M}_t)$, where $\mathcal{P}_{\phi_t}$ is the forward projection operator and $\circ$ denotes warping. The core problem is that the collected sinogram $\mathbf{P}_m$ is physically inconsistent; it does not correspond to any single static object. Applying a standard reconstruction algorithm like FDK to $\mathbf{P}_m$ thus results in a heavily artifacted volume. The objective of motion compensation is therefore to recover an accurate representation of the static reference volume $\mathbf{V}_c$ from the inconsistent projection data $\mathbf{P}_m$.

Our framework is designed to learn a mapping from a sequence of motion-corrupted projections $\mathbf{P}_m = \{P_m^{(t)}\}_{t=1}^T$ to a motion-corrected volume $\hat{\mathbf{V}}_c$ that is maximally similar to a ground-truth motion-free volume $\mathbf{V}_c$. This is achieved by training a network G_θ that predicts a sequence of 2D displacement fields $\mathbf{D} = \{D^{(t)}\}_{t=1}^T$. The network G_θ encompasses all trainable components of the pipeline: the feature tokenizer, the temporal model, and the displacement field decoder. The entire process is formulated as a single, end-to-end optimization problem:

$$\hat{\theta} = \arg\min_\theta \mathcal{L}\left(\mathcal{R}_{\text{FDK}}\left(\mathcal{W}(\mathbf{P}_m, G_\theta(\mathbf{P}_m))\right), \mathbf{V}_c\right), \quad (1)$$

where T is the total number of projections, $\mathcal{W}$ is a differentiable warping function, $\mathcal{R}_{\text{FDK}}(\cdot)$ is a differentiable FDK reconstruction operator, $\mathcal{L}$ is a loss function (e.g., MSE), and θ are the learnable parameters of our LSTT network G_θ. The pipeline is illustrated in Fig. 1.

2.1 Projection Tokenization via VQ-VAE

To create a robust and compact representation, we initialize the feature extraction pipeline with the encoder from a pre-trained VQ-VAE. While pre-training provides a strong starting point, the encoder's weights are fine-tuned end-to-end along with the rest of the network. The encoder, $\mathcal{E}_{vq}$, maps each 2D projection $P_m^{(t)} \in \mathbb{R}^{H \times W}$ (with height H and width W) into a lower-dimensional latent representation $\mathbf{z}_e^{(t)} = \mathcal{E}_{vq}(P_m^{(t)})$. This continuous latent map is then discretized by finding the nearest neighbor for each latent vector in a learned codebook $\mathcal{C} = \{\mathbf{e}_k\}_{k=1}^K$, where each codebook entry $\mathbf{e}_k \in \mathbb{R}^d$, with K being the codebook size and d the dimension of each entry. This quantization step $\mathcal{Q}$ is a

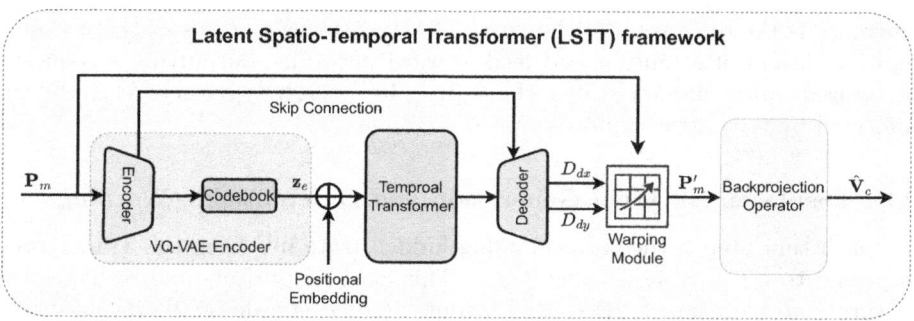

Fig. 1. The Latent Spatio-Temporal Transformer (LSTT) framework. Motion-corrupted projections $\mathbf{P}_m$ are fed into the network. Each projection is tokenized into a latent representation by a pre-trained VQ-VAE encoder. These sequences of tokens are then processed by a temporal Transformer, which models motion dynamics throughout the sequence. For each projection, a decoder subsequently generates a spatial displacement field $D_{dx,dy}$. The original projections are warped using these fields to produce a corrected sinogram $\mathbf{P}'_m$. A differentiable FDK layer reconstructs the volume $\hat{\mathbf{V}}_c$, and the objective function is computed against the ground-truth volume $\mathbf{V}_c$ for end-to-end training

non-differentiable operation. We enable gradient flow using the straight-through estimator (STE) [1]. The quantized latent map $\mathbf{z}_q^{(t)}$ is given by:

$$\mathbf{z}_q^{(t)}(i,j) = \mathcal{Q}(\mathbf{z}_e^{(t)}(i,j)) = \mathbf{e}_k \quad \text{where} \quad k = \arg\min_{k'} \|\mathbf{z}_e^{(t)}(i,j) - \mathbf{e}_{k'}\|_2^2. \quad (2)$$

This tokenization process abstracts away fine-grained texture, forcing the subsequent temporal model to focus on higher-level structural changes and geometric shifts that are indicative of motion. The resulting quantized latent map for each time step is flattened into a single feature vector $\mathbf{f}^{(t)}$.

2.2 Temporal Transformer for Motion Trajectory Modeling

The sequence of feature vectors $\mathbf{F} = [\mathbf{f}^{(1)}, \mathbf{f}^{(2)}, \ldots, \mathbf{f}^{(T)}]$ represents the time-series of anatomical states. To model the underlying motion dynamics, we feed this sequence into a standard Transformer encoder, which is a core trainable component of our network. The Transformer's self-attention mechanism is uniquely suited to capture global, long-range dependencies, allowing it to model complex, quasi-periodic motion patterns across the entire scan duration more effectively than recurrent or convolutional architectures. We augment the input sequence with a learnable positional embedding $\mathbf{P}_{pos}$ to provide the model with temporal context. The core operation is multi-head self-attention, where the input sequence $\mathbf{X} = \mathbf{F} + \mathbf{P}_{pos}$ is linearly projected into queries ($\mathbf{Q}$), keys ($\mathbf{K}$), and values ($\mathbf{V}$):

$$\text{Attention}(\mathbf{Q}, \mathbf{K}, \mathbf{V}) = \text{softmax}\left(\frac{\mathbf{Q}\mathbf{K}^T}{\sqrt{d_k}}\right)\mathbf{V}, \quad (3)$$

where d_k is the dimension of the keys. The Transformer processes this through multiple layers of attention and feed-forward networks, outputting a sequence of context-aware hidden states $\mathbf{H} = [\mathbf{h}^{(1)}, \mathbf{h}^{(2)}, \ldots, \mathbf{h}^{(T)}]$, where each $\mathbf{h}^{(t)}$ is informed by the entire motion sequence.

2.3 Displacement Field Generation and Differentiable Warping

For each time step t, the corresponding hidden state $\mathbf{h}^{(t)}$ from the Transformer is passed to a trainable decoder $\mathcal{D}_{disp}$. This decoder projects and reshapes the feature map back into a 2D spatial layout, predicting a dense 2D displacement vector field (DVF) $D^{(t)} \in \mathbb{R}^{H \times W \times 2}$ for each projection. This DVF provides an explicit and interpretable motion vector (dx, dy) for each pixel in the detector plane, offering valuable insight into the compensation process.

The predicted DVF is used to warp the original motion-corrupted projection $P_m^{(t)}$ to generate a motion-compensated projection $P_m'^{(t)}$. This is achieved via a differentiable spatial transformation layer $\mathcal{W}$ that performs backward warping for each pixel coordinate $\mathbf{u}$:

$$P_m'^{(t)}(\mathbf{u}) = P_m^{(t)}\left(\mathbf{u} + D^{(t)}(\mathbf{u})\right). \tag{4}$$

In practice, this operation is implemented using differentiable bilinear interpolation, allowing gradients to flow from the warped projection back to the DVF and thus to the decoder's parameters.

2.4 End-to-End Volumetric Supervision via Differentiable FDK

The sequence of corrected projections $\mathbf{P}_m'$ forms a motion-compensated sinogram. The crucial step is to connect this corrected data to the final volume quality. We achieve this by integrating a differentiable but parameter-free CBCT reconstruction operator, the Feldkamp-Davis-Kress (FDK) algorithm $\mathcal{R}_{\text{FDK}}$. This operator, implemented using the PYRO-NN toolkit [24], acts as a fixed, non-trainable layer in our network. It explicitly performs the inverse Radon transform on the corrected projections $\mathbf{P}_m'$ to produce the final volume $\hat{\mathbf{V}}_c$:

$$\hat{\mathbf{V}}_c = \mathcal{R}_{\text{FDK}}(\mathbf{P}_m') = \mathcal{B}\left(\mathcal{F}^{-1}\left(\mathcal{F}\left(\mathcal{C}_w(\mathbf{P}_m')\right) \cdot H_{\text{ramp}}\right)\right), \tag{5}$$

where $\mathcal{C}_w$ is the geometry-dependent cosine weighting, $\mathcal{F}$ and $\mathcal{F}^{-1}$ are the Fast Fourier Transform and its inverse, H_{ramp} is the Ram-Lak filter kernel, and $\mathcal{B}$ is the back-projection. Since every component is differentiable, we can backpropagate the error from the final volume through the fixed FDK operator all the way back to the trainable network parameters θ.

3 Experiments and Results

3.1 Data Generation

A major challenge in developing supervised motion compensation algorithms is the scarcity of paired clinical data. To address this issue, we constructed a high-fidelity digital phantom dataset with known ground truth by employing a physically principled simulation pipeline based on a publicly available 4D-CT scan.

Our simulation utilizes the thoracic 4D-CT dataset from the University of Texas MD Anderson Cancer Center [3], which consists of ten static 3D CT volumes of a patient's thorax, each representing a discrete phase of the respiratory cycle. To create a diverse dataset that captures realistic respiratory motion, we generated 500 unique dynamic volume sequences. For each sequence, we synthesized a continuous motion trajectory by performing nonlinear temporal interpolation among the static phases. To ensure broad variability, we randomized three key parameters for each of the 500 samples: motion amplitude (by scaling the deformations between phases), breathing rate (by varying the period of a sinusoidal temporal mapping), and the starting respiratory phase (by applying a random phase shift to the mapping).

For each of the 500 distinct motion trajectories, we created paired data using a differentiable forward projection operator (PYRO-NN [24]):

Motion-Corrupted Sinogram (P_m) : For every projection angle ϕ_t, the corresponding interpolated volume $V_{interp}^{(t)}$ was forward projected, yielding a sinogram that reflects the inconsistencies of a continuously deforming object.

Ground-Truth Volume (V_c) : The motion-free ground truth for all samples is the static end-exhalation phase from the original dataset, serving as the fixed reference for our supervised learning task.

In total, this process generated 500 unique (P_m, V_c) data pairs. The dataset was split into three subsets: 450 samples for training, 25 for validation, and 25 for testing. All simulations were conducted with a 360-degree circular trajectory comprising 360 projections, a 512×512 detector, and realistic geometric parameters.

3.2 Implementation Details and Baseline

The LSTT model was implemented in PyTorch 2.1.1. The VQ-VAE, pre-trained on a large set of individual projections, has an embedding dimension of 64 and a 512-entry codebook. The Transformer encoder consists of 4 layers and utilizes 4 attention heads. The entire model was trained for 500 epochs on our dataset of 450 training samples using a single NVIDIA A100 80GB GPU. We employed the AdamW optimizer with an initial learning rate of 1×10^{-3}, a cosine annealing scheduler [15], and MSE loss. We compare our model with the standard clinical baseline, which is an FDK reconstruction of the raw, motion-corrupted sinogram. We report three quantitative metrics calculated against the known ground-truth volume V_c: Peak Signal-to-Noise Ratio (PSNR), Structural Similarity Index (SSIM), and Root Mean Squared Error (RMSE).

3.3 Results and Analysis

The quantitative results, summarized in Table 1, demonstrate the clear benefit of our LSTT framework. The standard FDK reconstruction of the motion-corrupted data (No Compensation) is severely degraded, yielding a PSNR of only 22.94 dB and an RMSE of 0.071. Our LSTT provides a substantial improvement, increasing the PSNR by over 2.5 dB to 25.52 and reducing the RMSE to 0.053. This strong quantitative improvement, along with the increase in SSIM, indicates a clear restoration of both pixel-wise fidelity and structural integrity. The qualitative results presented in Fig. 2 visually corroborate these findings; the uncorrected image is plagued by significant blurring and unreal artifacts, particularly at high-contrast interfaces like the lung-diaphragm boundary. In contrast, our LSTT successfully restores sharp anatomical boundaries and eliminates these artifacts, yielding an image with significantly improved diagnostic quality.

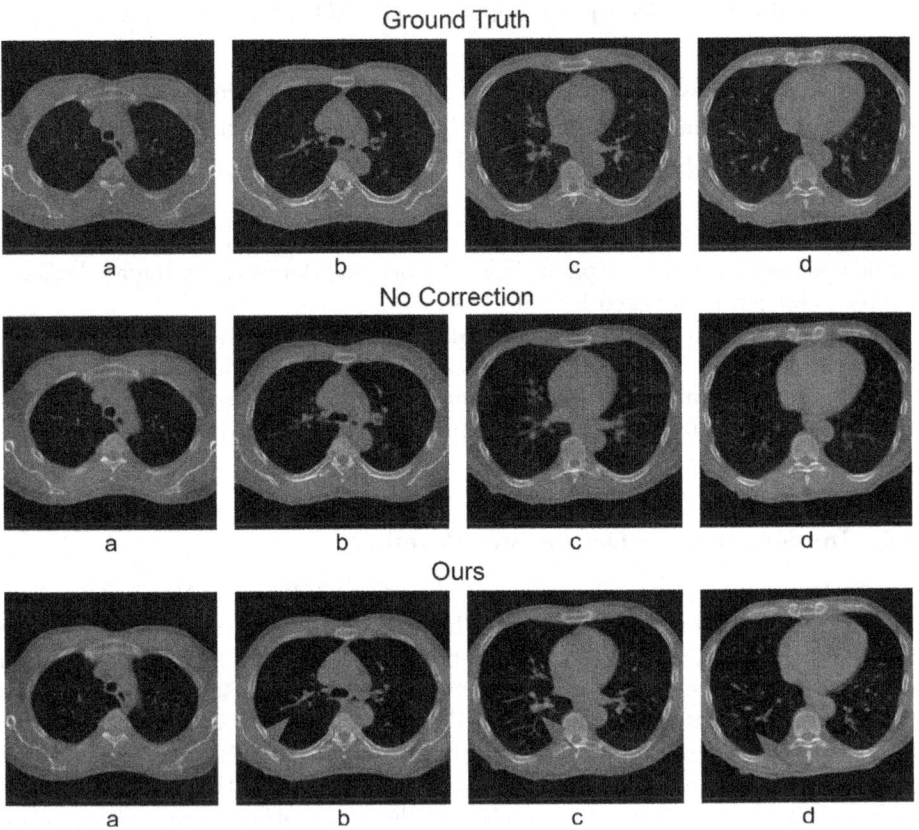

Fig. 2. Qualitative comparison of reconstructed axial slices. Our LSTT effectively removes the blurring and ghosting seen in the uncorrected FDK, closely matching the ground truth

Table 1. Quantitative comparison of our LSTT against the uncorrected baseline and an architectural ablation. Our full model significantly outperforms both. For PSNR/SSIM, higher is better (↑). For RMSE, lower is better (↓).

Method	PSNR (dB) ↑	SSIM ↑	RMSE ↓
No Compensation (FDK)	22.94	0.612	0.071
CNN-based (Ablation)	24.11	0.685	0.062
LSTT (Ours)	**25.52**	**0.741**	**0.053**

It is important to analyze these results in the context of the problem's inherent difficulty, which stems from two main sources. Firstly, the motion-corrupted sinogram contains anatomical information from a full respiratory cycle, while the supervised task forces the network to converge to a single, fixed respiratory phase (end-exhalation). This creates an information mismatch; for example, details of the anatomy at peak-inspiration exist in the input data but not in the ground-truth target. Secondly, the reconstruction itself is performed using the FDK algorithm with a 360-projection circular orbit. This trajectory does not satisfy Tuy's condition [9] for exact reconstruction, meaning that even a "perfect" FDK reconstruction of the static ground-truth volume will contain some inherent artifacts and deviations from the original data. These two factors-the information mismatch from motion and the physical limitations of the FDK algorithm-create a performance ceiling and explain why a perfect reconstruction is not achievable. Despite this challenge, the significant quantitative and qualitative improvements demonstrate that our LSTT effectively learns to correct the dominant motion components and produces a diagnostically superior image.

3.4 Ablation Study

To validate our architectural choice of using a Transformer for temporal modeling, we conducted a critical ablation study. We trained a baseline model where the Transformer encoder was replaced with a simple convolutional architecture that processes each projection's features independently, thus lacking any explicit temporal modeling. As shown in Table 1, the full LSTT significantly outperforms this architectural baseline (CNN-based) across all metrics. This result provides compelling evidence that explicitly modeling the global, long-range temporal dynamics of motion is crucial for achieving accurate, high-fidelity compensation.

4 Conclusion

In this work, we introduced the Latent Spatio-Temporal Transformer (LSTT), a novel end-to-end framework for CBCT motion compensation that operates directly on projection data without requiring external hardware. By combining latent space feature tokenization, a temporal Transformer for global motion modeling, and a differentiable FDK layer for end-to-end volumetric supervision, our method learns to predict highly accurate and interpretable displacement fields that effectively suppress non-rigid motion artifacts. Our results on a realistic phantom demonstrate a significant improvement over the standard uncorrected baseline, producing reconstructions with substantially higher quantitative scores and visual quality. While the current approach validates the architecture's potential, future work can extend it by addressing more complex phenomena. For instance, our model currently corrects for geometric shifts but does not explicitly account for motion-induced intensity changes from tissue compression or expansion, which in a clinical context provides information about lung ventilation. Future models could learn to predict a complementary intensity correction map alongside the displacement field. Furthermore, adapting this powerful architecture to unsupervised paradigms remains a key goal to eliminate the need for simulated ground-truth data and pave the way for clinical application. This work highlights the potential of integrating known CT operators into deep learning pipelines, a paradigm that enables networks to learn physically grounded motion compensation policies optimized for high-fidelity volumetric output.

Acknowledgements. This research was financed by the "Verbundprojekt 05D2022 - KI4D4E: Ein KI-basiertes Framework für die Visualisierung und Auswertung der massiven Datenmengen der 4D-Tomographie für Endanwender von Beamlines. Teilprojekt 5." (Grant number: 05D23WE1).

References

1. Bengio, Y., Léonard, N., Courville, A.: Estimating or propagating gradients through stochastic neurons for conditional computation. arXiv preprint arXiv:1308.3432 (2013)
2. Capostagno, S., Sisniega, A., Stayman, J., Ehtiati, T., Weiss, C., Siewerdsen, J.: Deformable motion compensation for interventional cone-beam CT. Phys. Med. Biol. **66**(5), 055010 (2021)
3. Castillo, R., et al.: A framework for evaluation of deformable image registration spatial accuracy using large landmark point sets. Phys. Med. Biol. **54**(7), 1849 (2009)
4. Chee, G., O'Connell, D., Yang, Y., Singhrao, K., Low, D., Lewis, J.: McSART: an iterative model-based, motion-compensated SART algorithm for CBCT reconstruction. Phys. Med. Biol. **64**(9), 095013 (2019)
5. Chen, G.P., Tai, A., Keiper, T.D., Lim, S., Li, X.A.: comprehensive performance tests of the first clinical real-time motion tracking and compensation system using MLC and jaws. Med. Phys. **47**(7), 2814–2825 (2020)
6. Schryver, T., et al.: Motion compensated micro-CT reconstruction for in-situ analysis of dynamic processes. Sci. Rep. **8**(1), 7655 (2018)
7. Goodfellow, I., et al.: Generative adversarial networks. Commun. ACM **63**(11), 139–144 (2020)
8. Hahn, B.N.: Motion compensation strategies in tomography. In: Time-Dependent Problems in Imaging and Parameter Identification, pp. 51–83. Springer (2021)

9. Herl, G., Maier, A., Zabler, S.: X-ray CT data completeness condition for sets of arbitrary projections. In: 7th International Conference on Image Formation in X-Ray Computed Tomography, vol. 12304, pp. 67–73. SPIE (2022)
10. Ko, Y., Moon, S., Baek, J., Shim, H.: Rigid and non-rigid motion artifact reduction in x-ray CT using attention module. Med. Image Anal. **67**, 101883 (2021)
11. Kyme, A.Z., Fulton, R.R.: Motion estimation and correction in Spect, pet and CT. Phys. Med. Biol. **66**(18), 18TR02 (2021)
12. Kyme, A.Z., Se, S., Meikle, S.R., Fulton, R.R.: Markerless motion estimation for motion-compensated clinical brain imaging. Phys. Med. Biol. **63**(10), 105018 (2018)
13. Lassen, M.L., et al.: Data-driven, projection-based respiratory motion compensation of pet data for cardiac pet/CT and pet/MR imaging. J. Nucl. Cardiol. **27**, 2216–2230 (2020)
14. Li, S., Nunes, J., Toumoulin, C., Luo, L.: 3D coronary artery reconstruction by 2D motion compensation based on mutual information. Irbm **39**(1), 69–82 (2018)
15. Loshchilov, I., Hutter, F.: SGDR: stochastic gradient descent with warm restarts. arXiv preprint arXiv:1608.03983 (2016)
16. Lossau, T., et al.: Motion estimation and correction in cardiac CT angiography images using convolutional neural networks. Comput. Med. Imaging Graph. **76**, 101640 (2019)
17. Lu, Y., et al.: Respiratory motion compensation for pet/CT with motion information derived from matched attenuation-corrected gated pet data. J. Nucl. Med. **59**(9), 1480–1486 (2018)
18. Maier, J., et al.: Deep learning-based coronary artery motion estimation and compensation for short-scan cardiac CT. Med. Phys. **48**(7), 3559–3571 (2021)
19. Polycarpou, I., Soultanidis, G., Tsoumpas, C.: Synergistic motion compensation strategies for positron emission tomography when acquired simultaneously with magnetic resonance imaging. Phil. Trans. R. Soc. A **379**(2204), 20200207 (2021)
20. Pretorius, P.H., Johnson, K.L., Dahlberg, S.T., King, M.A.: Investigation of the physical effects of respiratory motion compensation in a large population of patients undergoing TC-99m cardiac perfusion Spect/CT stress imaging. J. Nucl. Cardiol. **27**(1), 80–95 (2020)
21. Riblett, M.J., Christensen, G.E., Weiss, E., Hugo, G.D.: Data-driven respiratory motion compensation for four-dimensional cone-beam computed tomography (4D-CBCT) using groupwise deformable registration. Med. Phys. **45**(10), 4471–4482 (2018)
22. Ronneberger, O., Fischer, P., Brox, T.: U-net: Convolutional networks for biomedical image segmentation. In: Medical image computing and computer-assisted intervention–MICCAI 2015: 18th international conference, Munich, Germany, October 5-9, 2015, proceedings, part III 18, pp. 234–241. Springer (2015)
23. Sisniega, A., Thawait, G.K., Shakoor, D., Siewerdsen, J.H., Demehri, S., Zbijewski, W.: Motion compensation in extremity cone-beam computed tomography. Skeletal Radiol. **48**, 1999–2007 (2019)
24. Syben, C., Michen, M., Stimpel, B., Seitz, S., Ploner, S., Maier, A.K.: Technical Note: PYRO-NN: python reconstruction operators in neural networks. Med. Phys. **46**(11), 5110-5115 (2019)
25. Van Den Oord, A., Vinyals, O., et al.: Neural discrete representation learning. Adv. Neural Info. Process. Syst. **30** (2017)
26. Vaswani, A., et al.: Attention is all you need. Adv. Neural Info. Process. Syst. **30** (2017)

27. Zhang, Y., Huang, X., Wang, J.: Advanced 4-dimensional cone-beam computed tomography reconstruction by combining motion estimation, motion-compensated reconstruction, biomechanical modeling and deep learning. Vis. Comput. Ind. Biomed. Art **2**(1), 23 (2019)
28. Zhang, Z., Liu, J., Yang, D., Kamilov, U.S., Hugo, G.D.: Deep learning-based motion compensation for four-dimensional cone-beam computed tomography (4D-CBCT) reconstruction. Med. Phys. **50**(2), 808–820 (2023)

cIDIR: Conditioned Implicit Neural Representation for Regularized Deformable Image Registration

Sidaty El Hadramy[✉], Oumeymah Cherkaoui, and Philippe C. Cattin

Department of Biomedical Engineering, University of Basel, Basel, Switzerland
sidaty.elhadramy@unibas.ch

Abstract. Regularization is essential in deformable image registration (DIR) to ensure that the estimated Deformation Vector Field (DVF) remains smooth, physically plausible, and anatomically consistent. However, fine-tuning regularization parameters in learning-based DIR frameworks is computationally expensive, often requiring multiple training iterations. To address this, we propose cIDIR, a novel DIR framework based on Implicit Neural Representations (INRs) that conditions the registration process on regularization hyperparameters. Unlike conventional methods that require retraining for each regularization hyperparameter setting, cIDIR is trained over a prior distribution of these hyperparameters, then optimized over the regularization hyperparameters by using the segmentations masks as an observation. Additionally, cIDIR models a continuous and differentiable DVF, enabling seamless integration of advanced regularization techniques via automatic differentiation. Evaluated on the DIR-LAB [5,6] dataset, cIDIR achieves high accuracy and robustness across the dataset.

Keywords: Deformable Image Registration · Implicit Neural Representation · Regularization · Hyperparameter optimization

1 Introduction

Deformable image registration (DIR) is essential in medical imaging for aligning images across different views, modalities, time points, or patients. It enables image fusion and analysis, supporting applications in diagnosis, treatment planning, and intervention guidance [7,17]. To enhance the effectiveness of DIR, enforcing a diffeomorphic transformation is critical [24], as it ensures physical plausibility while minimizing artifacts and distortions that could affect medical image interpretation. Theoretically, a diffeomorphic transformation ϕ is both smooth and invertible [11]. To enforce these properties, various regularization strategies have been proposed. Rohlfing et al. [21] introduced a Jacobian regularization, where a negative determinant of the Jacobian matrix indicates a loss of invertibility. Burger et al. [4] proposed a hyperelastic regularization term to control variations in length, surface area, and volume. Alvarez et al. [1] introduced a regularization term enforcing the conservation of linear momentum. Additionally, Rueckert et al. [22] developed a Bending Energy regularization to further ensure the smoothness of the deformation vector field (DVF).

Recent advances in deep neural networks have led to the introduction of numerous DIR approaches [10]. These methods, supervised [16] or unsupervised [2], learn to predict DVF on a grid for unseen image pairs. However, these grid-based methods provide a discontinuous representation of the DVF, making it challenging to incorporate advanced regularization techniques that require accurate gradient computations. In particular, regularization methods that rely on second-order gradient calculations face challenges in this context. Therefore, the need to incorporate these regularization techniques has driven the development of methods aimed at learning continuous and differentiable representations of the DVF. One class of method that has gained attention relies on implicit neural representations (INRs) [19]. INRs use Multi-Layer Perceptrons (MLP) to encode information as a continuous generator function, mapping input coordinates to corresponding values within the defined space.

Building on INRs, Wolterink et al. [25] introduced IDIR, an implicit deformable image registration model that seamlessly integrates regularization techniques, facilitated by automatic differentiation techniques. While the method, validated on the DIR-LAB 4DCT dataset [5,6], demonstrated high accuracy when using the Bending Energy regularization [22], it requires hyperparameter tuning to balance the weights between data and regularization losses. Standard hyperparameter optimization methods, such as random search, grid search, and sequential search [3], are commonly used for this aim. More advanced techniques, including gradient-based tuning and Bayesian optimization, use probabilistic models to efficiently identify optimal values [18,23]. These methods are computationally intensive, as they require retraining the model multiple times to assess each hyperparameter choice. To avoid this, Hoopes et al. [14] introduced HyperMorph, which uses a hypernetwork to condition the main registration network (VoxelMorph [2]) on the loss hyperparameters. Thus, the hyperparameter tuning can be done at inference-time. However, training hypernetworks is often challenging due to the high difference between the network's input and output dimensions, which leads to slow convergence and high memory consumption [20].

In this work, we build upon IDIR by introducing cIDIR, a simple, yet effective approach that conditions an INR of the DVF on the hyperparameters of the loss functions. Similar to IDIR, cIDIR offers a continuous and differentiable representation of the DVF, making it easier to integrate advanced regularization techniques. The key difference, however, lies in how hyperparameter tuning is handled. While IDIR requires multiple training sessions to optimize the hyperparameters, cIDIR only needs to be trained once. Hyperparameter tuning is performed after training, allowing for real-time adjustments. This makes cIDIR more efficient and practical for applications where quick, on-the-fly registration is necessary. The paper is organized as follows: Sect. 2 introduces cIDIR and outlines its novelty. In Sect. 3, we evaluate cIDIR on the DIR-LAB 4DCT dataset

[5,6], demonstrating its accuracy and computational efficiency. Finally, Sect. 4 concludes the paper and discusses future research directions.

2 Methods

This section presents the proposed approach, starting with the formulation of the deformable image registration problem and key notations. We then describe the **cIDIR** architecture and its components, followed by a discussion of the regularization techniques used in this work and its hyperparameters optimization.

2.1 Deformable Image Registration

Let $\mathcal{M}$ and $\mathcal{F}$ be the moving and fixed images, respectively, defined over the domains $\Omega_0 \subset \mathbb{R}^d$ and $\Omega \subset \mathbb{R}^d$. The goal of deformable image registration is to estimate a DVF ϕ that aligns $\mathcal{M}$ with $\mathcal{F}$ by minimizing a given criterion. As in Eq. 1, this is formulated as an optimization problem. Where ϕ is sought to map features from the moving image to their corresponding homologous structures in the fixed image while preserving anatomical consistency.

$$\hat{\phi} = \arg\min_{\phi} (1 - \alpha)\mathcal{L}_{sim}(\mathcal{M} \circ \phi, \mathcal{F}) + \alpha\mathcal{L}_{reg}(\phi) \qquad (1)$$

In Eq. 1, ϕ is a DVF, $\mathcal{L}_{sim}$ denotes the similarity measure, $\mathcal{L}_{reg}$ is the regularization penalty, and α is the weighting factor for regularization. In this work, **we treat α as the hyperparameter to be tuned**, assuming a predefined similarity measure and a fixed regularization technique.

2.2 cIDIR

To address the optimization problem outlined in the previous section, we introduce a learning-based framework called cIDIR. Figure 1 shows an overview of the framework. cIDIR consists of two key components: the main network and the harmonizer network. The **main network** learns an implicit representation of the DVF ϕ, while the **harmonizer network** conditions it on the regularization weighting factor $\boldsymbol{\alpha}$. Both networks are trained in an end-to-end manner, with $\boldsymbol{\alpha}$ uniformly sampled from $[0, 1]$ during training.

The **main network** in cIDIR is a MLP with an input and output dimension of 3, corresponding to the spatial coordinates of the moving and fixed images. It consists of three hidden layers, each with 256 neurons, followed by an activation function. Since the network is designed as an INR, the choice of activation function is critical. Standard activations like ReLU are unsuitable, as they tend to bias the network toward low-frequency signals [8]. To address this, prior works have explored the use of periodic activation functions [25], which enable the network to capture high-frequency variations effectively. In this work, we employ a parameterized activation function σ, inspired by [15], which is formulated as:

$$\sigma(x) = \boldsymbol{a} \cdot \sin(\boldsymbol{b} \cdot \boldsymbol{x} + \boldsymbol{c}) + \boldsymbol{d} \qquad (2)$$

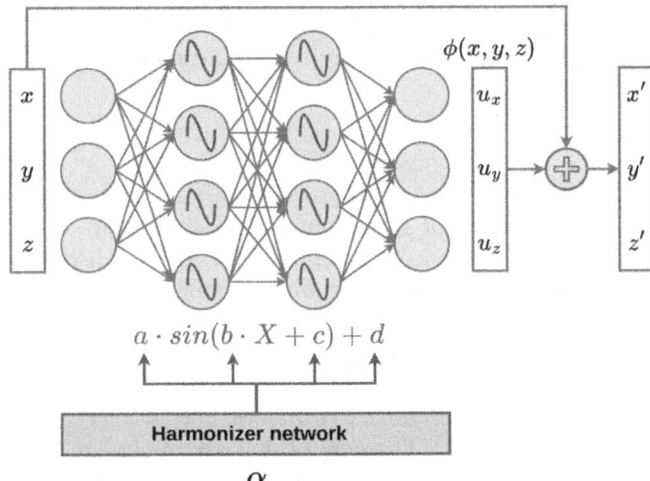

Fig. 1. Overview of cIDIR. The main network (in blue) learns an implicit representation of the Deformation Vector Field (DVF) ϕ, mapping coordinates (x, y, z) from the moving image to (x', y', z') in the fixed image. The network is conditioned on the regularization weighting factor α through the harmonization network, which predicts the parameters a, b, c, and d of the activation function used in the main network. (Color figure online)

In Eq. 2, the parameters of this activation function play distinct roles in shaping the output response of the network. a represents the amplitude scaling, determining the vertical scaling of the sinusoidal function. b represents the frequency scaling, influencing how rapidly the function oscillates. c represents the phase shift, it adjusts the horizontal displacement of the function along the x-axis. d is the vertical shift of the wave, acting as a baseline adjustment. By learning these parameters, the activation function becomes highly adaptive, enabling the network to model complex deformations.

The **harmonizer network** is an MLP designed to condition the main network on the regularization weighting factor α. It takes α as input and outputs four values corresponding to the parameters a, b, c, d of the main network's activation function. The network consists of three hidden layers with sizes 128, 64, and 32, respectively. Each hidden layer is followed by **layer normalization** to stabilize training and improve generalization, as well as a SiLU (Sigmoid Linear Unit) activation function, which enhances smooth gradient propagation and avoids vanishing gradients. The choice of the hidden layer dimensions was determined experimentally to balance model accuracy and computational efficiency. By dynamically predicting the activation function parameters, the harmonizer network enables adaptive control over the deformation field, allowing for a conditioning of the prediction over the weighting factor α.

2.3 Regularizations

This section briefly presents three common regularization techniques in DIR to ensure a smooth, physically plausible DVF. cIDIR's implicit representation enables seamless integration of these techniques, efficiently computing gradients of different orders for advanced regularization.

Jacobian Regularization, formulated in Eq. 3, ensures a diffeomorphic DVF, the Jacobian determinant must remain non-negative at each point x, as negative values indicate non-invertibility. Therefore, as proposed by [21], deviations from 1 are minimized to maintain a stable and realistic transformation.

$$\mathcal{L}_{jac}(\phi) = \int_\Omega |\det \nabla \phi - 1| \, dx \quad (3)$$

Hyperelastic Regularization [4] fromulated in Eq. 4. It regulates the length, surface area, and volume of the DVF. The Jacobian matrix governs length, while the cofactor matrix and determinant control area and volume.

$$\mathcal{L}^{\text{hyper}}[\phi] = \int_\Omega \left[\frac{1}{2} \alpha_l |\nabla u|^2 + \alpha_a \phi_c(\text{cof}\, \nabla \phi) + \alpha_v \psi(\det \nabla \phi) \right] dx \quad (4)$$

Bending Energy Penalty [22] enforces the smoothness of the DVF by penalizing large second derivatives, ensuring that the DVF remains smooth across the entire domain. This is formulated as:

$$\mathcal{L}^{\text{bending}}[\phi] = \frac{1}{8} \int_{-1}^{1} \int_{-1}^{1} \int_{-1}^{1} \left[\left(\frac{\partial^2 \phi}{\partial x^2}\right)^2 + \left(\frac{\partial^2 \phi}{\partial y^2}\right)^2 + \left(\frac{\partial^2 \Phi}{\partial z^2}\right)^2 \right.$$
$$\left. + 2\left(\frac{\partial^2 \phi}{\partial xy}\right)^2 + 2\left(\frac{\partial^2 \phi}{\partial xz}\right)^2 + 2\left(\frac{\partial^2 \phi}{\partial yz}\right)^2 \right] dx\, dy\, dz. \quad (5)$$

2.4 Optimization of the Regularization Weights

Upon cIDIR's training, a grid search over values of α in the range $[0, 1]$ is performed. For each α, a displacement field is generated and applied to the moving image to produce a **moved image**. Both the **moved and fixed images** are then segmented using the method from [13], and the resulting segmentations are used to compute a Dice score (DS). The α value that yields the highest DS is selected as the optimal. In this setup, **the segmentations serve as observations to guide the optimization of** α.

3 Results

3.1 Dataset and Implementation Details

We evaluate our method on the DIR-LAB 4DCT dataset [5,6], a widely used benchmark in DIR [9,12,25]. This dataset consists of 4D CT scans from 10 patients, where the registration task involves aligning inspiration-phase images with expiration-phase images. The challenge arises due to the combined effects of cardiac and respiratory motion, which cause significant deformations. For each patient, the dataset provides landmark coordinates in the fixed image along with their corresponding positions in the moving image. All networks are implemented in PyTorch, and experiments are conducted on an NVIDIA A100 GPU with 40GB of memory. cIDIR is trained using the Adam optimizer with a learning rate of 10^{-4}. The Bayesian Optimization (BO) is implemented with Scikit-Optimize and the Grid Search (GS) with Scikit-Learn.

3.2 Experiments

For each of the 10 patients in the DIR-LAB 4DCT dataset, cIDIR was trained for 50K epochs, with the regularization weight α varying between 0 and 1 using Bending Energy regularization. In each epoch, a batch of 10K points was sampled from the lung region of the moving image, identified using a segmentation mask computed with the method from Hofmanninger et al. [13]. These point coordinates, along with a sampled α value, were input into cIDIR to predict their corresponding positions in the fixed image. cIDIR is trained by minimizing the voxel intensity differences between the predicted and actual points in the fixed image using the Normalized Cross-Correlation (NCC) as $\mathcal{L}_{sim}$ loss in Eq. 1. After training, the optimal value of α is obtained using the approach described in 2.4. Table 1 presents the results, comparing cIDIR using Bending Energy regularization with state-of-the-art learning-based methods, including IDIR [25], and CNN [9]. Additionally, we report the initial **Displacement** error. For IDIR, we performed experiments using a Bending Energy regularization with $\alpha = 10$, as suggested in the original paper [25]. When reproducing IDIR's results, we observed improved performance for patients 01, 02, 03, 04, and 06 with $\alpha = 10$ compared to the results reported in their paper. However, for patients 05, 07, 08, 09, and 10, we were unable to match the accuracy reported in their work. Table 1 demonstrates that cIDIR, by selecting the optimal α in real-time during inference, outperforms IDIR and CNN [9] in terms of average Target Registration Error (TRE) over the 10 patients. This is because cIDIR allows for real-time grid search to find the optimal α for each patient after training. In contrast, the fixed $\alpha = 10$ used in IDIR might not be optimal for all patients, and optimizing it is time-consuming, requiring multiple training.

To further evaluate cIDIR's adaptability across different regularization techniques, we conducted experiments using both hyperelasticity and Jacobian regularizations. For IDIR, we trained the model with a fixed value of $\alpha = 0.5$ across all 10 patients. cIDIR was trained on a uniform distribution of α over $[0, 1]$, and at inference, we selected $\alpha = 0.5$. As shown in Table 2, both methods achieved comparable results for the Jacobian regularization, while cIDIR outperformed IDIR with hyperelasticity. This difference can be attributed to the fact that the optimal α value is not necessarily the same for both methods, even when using the same regularization technique. Since cIDIR incorporates a harmonizer network that conditions the registration process on α, its architecture is more

Table 1. TRE in mm of cIDIR compared to state-of-the-art learning-based methods: IDIR [25], CNN [9], and Displacement (TRE before registration). Both IDIR and cIDIR use Bending Energy regularization, with $\alpha = 10$ for IDIR as proposed in their paper. For cIDIR, the value of α is optimized per patient.

Scan	cIDIR (ours)	IDIR [25]	CNN [9]	Displacement
4DCT 01	0.66 (1.25)	0.52 (1.11)	1.21 (0.88)	4.01 (2.91)
4DCT 02	0.76 (1.33)	0.55 (1.15)	1.13 (0.65)	4.65 (4.09)
4DCT 03	0.68 (1.23)	0.76 (1.32)	1.32 (0.82)	6.73 (4.21)
4DCT 04	1.18 (1.3)	0.82 (1.47)	1.84 (1.76)	9.42 (4.81)
4DCT 05	1.17 (1.86)	1.29 (1.78)	1.80 (1.60)	7.10 (5.14)
4DCT 06	0.82 (1.84)	0.86 (1.40)	2.30 (3.78)	11.10 (6.98)
4DCT 07	1.35 (1.65)	1.76 (2.29)	1.91 (1.65)	11.59 (7.87)
4DCT 08	1.44 (3.05)	2.54 (4.30)	3.47 (5.00)	15.16 (9.11)
4DCT 09	3.72 (2.59)	3.54 (2.65)	1.47 (0.85)	7.82 (3.99)
4DCT 10	1.61 (2.14)	1.50 (1.94)	1.79 (2.24)	7.63 (6.54)
Average	**1.33**	1.47	1.83	8.52

complex than IDIR. This added complexity not only allows cIDIR to generalize over a range of α values but also affects how the regularization influences the learned deformation field, potentially leading to different optimal α for both methods.

3.3 Computation Time

cIDIR offers significant benefits in terms of computation time. Table 3 presents training and fine-tuning time per regularization technique for both cIDIR and IDIR over a patient. While a single IDIR training is faster than cIDIR due to its shorter training duration, cIDIR is training longer (50K epochs) with α varying over a uniform distribution $[0, 1]$. However, after cIDIR's training, α is optimized in near real-time, requiring only a single prediction per evaluation, which is performed in about 1ms. This allows a **Grid Search** (GS) over multiple values of α in less than two seconds (see Table 3). In contrast, IDIR requires a full training run for each α evaluation. To optimize IDIR over α, we employed **Bayesian Optimization** (BO) with a Gaussian Process as a surrogate model to efficiently search for the best α within $[0, 1]$. Instead of exhaustively testing all values, the Gaussian Process models the performance landscape, guiding the search towards promising regions and reducing the number of required training runs. However, as shown in Table 3, at least 20 training iterations of IDIR were needed for convergence to an optimal α, making its fine-tuning process longer than the total training and fine-tuning time of cIDIR.

Table 2. Comparison of TRE in mm for cIDIR and IDIR on the DIR-LAB [5,6] dataset using Hyperelastic [4] and Jacobian [21] regularizations. IDIR is trained with a fixed $\alpha = 0.5$, while cIDIR is trained over $\alpha \in [0, 1]$ and evaluated with $\alpha = 0.5$.

Scan	Hyperelastic ($\alpha = 0.5$)		Jacobian ($\alpha = 0.5$)	
	cIDIR	IDIR	cIDIR	IDIR
4DCT 01	0.73 (1.16)	7.10 (5.48)	1.46 (1.77)	1.61 (2.18)
4DCT 02	0.63 (1.14)	3.34 (3.09)	2.78 (2.31)	2.55 (3.06)
4DCT 03	1.32 (1.80)	3.99 (2.81)	4.11 (2.78)	3.17 (3.59)
4DCT 04	1.11 (1.44)	7.83 (4.90)	5.65 (4.65)	3.95 (5.72)
4DCT 05	1.42 (1.79)	6.35 (5.36)	2.86 (3.13)	2.29 (2.99)
4DCT 06	1.08 (1.35)	8.70 (6.64)	3.68 (3.32)	5.95 (6.99)
4DCT 07	2.48 (2.48)	6.97 (7.37)	6.40 (5.32)	6.20 (5.18)
4DCT 08	6.00 (5.71)	7.89 (7.98)	11.4 (10.9)	11.01 (11.51)
4DCT 09	3.59 (3.10)	7.23 (6.02)	3.97 (2.41)	3.15 (2.49)
4DCT 10	2.17 (2.09)	5.17 (4.52)	6.91 (4.4)	6.26 (5.34)
Average	2.05	6.45	4.92	4.61

Table 3. Training and fine-tuning time per regularization technique for cIDIR and IDIR on 4DCT 04. We use a **Bayesian Optimization** (BO) to fine-tune α for IDIR, which requires multiple full training runs. In contrast, cIDIR selects the optimal α using a simple **Grid Search** (GS) (See Sect. 2.4)

Scan	Hyperelastic [4]		Jacobian [21]		Bending [22]	
	Training	Fine-tuning	Training	Fine-tuning	Training	Fine-tuning
IDIR [25] + BO	3 min	60 min	1 min	20 min	10 min	200 min
cIDIR + GS	43 min	2 sec	20 min	2 sec	90 min	2 sec

4 Conclusion

In this work, we introduced cIDIR, a conditioned implicit neural representation (INR) for regularized deformable image registration. cIDIR is patient-specific and leverages an INR to model a continuous deformation vector field, enabling the integration of advanced regularization techniques that require higher-order gradients. By conditioning the activation functions of the INR on the regularization weighting factor, cIDIR allows for real-time hyperparameter optimization after training, eliminating the need for expensive retraining. Our experiments highlight cIDIR's accuracy, computational efficiency, and robustness across different regularization techniques and patient data. Despite these advantages, cIDIR has limitations. Its patient-specific nature requires a dedicated training phase for each new subject, leading to long training times that may hinder its practical deployment, particularly in time-sensitive clinical settings. Another limitation is the assumption of a well-defined prior distribution for regularization parameters, which may not always align with the optimal settings for every case. Future work will focus on reducing training time to enhance practicality, exploring strategies to improve generalization across patients and datasets, and extending cIDIR to broader applications, such as multi-modal image registration. Finally, expanding cIDIR to handle multiple hyperparameters could allow for the efficient integration of diverse regularization techniques within a single training process, further strengthening diffeomorphism enforcement.

Acknowledgment. This work was financially supported by the Werner Siemens Foundation through the MIRACLE II project.

References

1. Alvarez, P., Cotin, S.: Deformable image registration with stochastically regularized biomechanical equilibrium. In: 2024 IEEE International Symposium on Biomedical Imaging (ISBI), vol. 33, pp. 1–5. IEEE (2024)
2. Balakrishnan, G., Zhao, A., Sabuncu, M.R., Guttag, J., Dalca, A.V.: VoxelMorph: a learning framework for deformable medical image registration. IEEE Trans. Med. Imaging **38**(8), 1788–1800 (2019)
3. Bergstra, J., Bengio, Y.: Random search for hyper-parameter optimization. J. Mach. Learn. Res. **13**(null), 281–305 (2012)
4. Burger, M., Modersitzki, J., Ruthotto, L.: A hyperelastic regularization energy for image registration. SIAM J. Sci. Comput. **35**(1), B132–B148 (2013)
5. Castillo, E., Castillo, R., Martinez, J., Shenoy, M., Guerrero, T.: Four-dimensional deformable image registration using trajectory modeling. Phys. Med. Biol. **55**(1), 305–327 (2010)
6. Castillo, R., et al.: A framework for evaluation of deformable image registration spatial accuracy using large landmark point sets. Phys. Med. Biol. **54**(7), 1849–1870 (2009)
7. El hadramy, S., Verde, J., Padoy, N., Cotin, S.: Intraoperative CT augmentation for needle-based liver interventions. In: Lecture Notes in Computer Science, pp. 291–301. Lecture notes in computer science, Springer Nature Switzerland, Cham (2023)
8. Essakine, A., et al.: Where do we stand with implicit neural representations? A technical and performance survey. Submitted to Transactions on Machine Learning Research (2024). under review
9. Fechter, T., Baltas, D.: One-shot learning for deformable medical image registration and periodic motion tracking. IEEE Trans. Med. Imaging **39**(7), 2506–2517 (2020)
10. Fu, Y., Lei, Y., Wang, T., Curran, W.J., Liu, T., Yang, X.: Deep learning in medical image registration: a review. Phys. Med. Biol. **65**(20), 20TR01 (2020)
11. Glaunès, J., Qiu, A., Miller, M.I., Younes, L.: Large deformation diffeomorphic metric curve mapping. Int. J. Comput. Vis. **80**(3), 317–336 (2008)
12. Hering, A., Häger, S., Moltz, J., Lessmann, N., Heldmann, S., van Ginneken, B.: CNN-based lung CT registration with multiple anatomical constraints. Med. Image Anal. **72**(102139), 102139 (2021)
13. Hofmanninger, J., Prayer, F., Pan, J., Röhrich, S., Prosch, H., Langs, G.: Automatic lung segmentation in routine imaging is primarily a data diversity problem, not a methodology problem. Eur. Radiol. Exp. **4**(1), 50 (2020)

14. Hoopes, A., Hoffmann, M., Fischl, B., Guttag, J., Dalca, A.V.: HyperMorph: amortized hyperparameter learning for image registration. In: Lecture Notes in Computer Science, pp. 3–17. Lecture notes in computer science, Springer International Publishing, Cham (2021)
15. Kazerouni, A., Azad, R., Hosseini, A., Merhof, D., Bagci, U.: Incode: implicit neural conditioning with prior knowledge embeddings. In: Proceedings of the IEEE/CVF Winter Conference on Applications of Computer Vision, pp. 1298–1307 (2024)
16. Lafarge, M.W., Moeskops, P., Veta, M., Pluim, J.P.W., Eppenhof, K.A.J.: Deformable image registration using convolutional neural networks. In: Angelini, E.D., Landman, B.A. (eds.) Medical Imaging 2018: Image Processing. SPIE (2018)
17. Maintz, J.B., Viergever, M.A.: A survey of medical image registration. Med. Image Anal. **2**(1), 1–36 (1998)
18. Mockus, J.: On Bayesian methods for seeking the extremum. In: Proceedings of the IFIP Technical Conference, pp. 400–404. Springer-Verlag, Berlin, Heidelberg (1974)
19. Molaei, A., et al.: Implicit neural representation in medical imaging: a comparative survey. In: 2023 IEEE/CVF International Conference on Computer Vision Workshops (ICCVW), pp. 2373–2383. IEEE (2023)
20. Ortiz, J.J.G., Guttag, J.V., Dalca, A.V.: Magnitude invariant parametrizations improve hypernetwork learning. In: ICLR (2024)
21. Rohlfing, T., Maurer, C.R., Jr., Bluemke, D.A., Jacobs, M.A.: Volume-preserving nonrigid registration of MR breast images using free-form deformation with an incompressibility constraint. IEEE Trans. Med. Imaging **22**(6), 730–741 (2003)
22. Rueckert, D., Sonoda, L.I., Hayes, C., Hill, D.L., Leach, M.O., Hawkes, D.J.: Nonrigid registration using free-form deformations: application to breast MR images. IEEE Trans. Med. Imaging **18**(8), 712–721 (1999)
23. Snoek, J., Larochelle, H., Adams, R.P.: Practical Bayesian optimization of machine learning algorithms. In: Proceedings of the 26th International Conference on Neural Information Processing Systems - Volume 2. NIPS'12, vol. 2, pp. 2951–2959. Curran Associates Inc., Red Hook, NY, USA (2012)
24. Vercauteren, T., Pennec, X., Perchant, A., Ayache, N.: Non-parametric diffeomorphic image registration with the demons algorithm. In: Medical Image Computing and Computer-Assisted Intervention – MICCAI 2007, pp. 319–326. Lecture notes in computer science, Springer Berlin Heidelberg, Berlin, Heidelberg (2007)
25. Wolterink, J., Zwienenberg, J.C., Brune, C.: Implicit neural representations for deformable image registration. In: International Conference on Medical Imaging with Deep Learning (2022)

Self-supervised Motion-Compensated Reconstruction for Cardiac Cine MRI

Siying Xu[1](✉), Aya Ghoul[1], Kerstin Hammernik[2], Jens Kuebler[3], Patrick Krumm[3], Andreas Lingg[3], Daniel Rueckert[2,4,5], Sergios Gatidis[1,6], and Thomas Küstner[1]

[1] Medical Image and Data Analysis (MIDAS.lab), Department of Diagnostic and Interventional Radiology, University Hospital of Tuebingen, Tuebingen, Germany
siying.xu@med.uni-tuebingen.de
[2] School of Computation, Information and Technology, Technical University of Munich, Munich, Germany
[3] Department of Diagnostic and Interventional Radiology, University Hospital of Tuebingen, Tuebingen, Germany
[4] TUM University Hospital, Technical University of Munich, Munich, Germany
[5] Department of Computing, Imperial College London, London, UK
[6] Department of Radiology, Stanford University, Stanford, CA, USA

Abstract. Cardiac Cine MRI is limited by prolonged acquisition times and motion-related artifacts. Existing deep learning-based reconstruction methods typically depend on fully-sampled ground truth, which is often difficult to acquire in practice. In this work, we propose SSL-MoCo, a fully self-supervised motion-compensated reconstruction framework for cardiac Cine MRI that eliminates the need for fully-sampled references. SSL-MoCo adopts a two-stage design: a transformer-based registration network estimates non-rigid inter-frame motion in a self-supervised manner, followed by a physics-based unrolled reconstruction network that integrates the estimated motion fields in the data consistency steps. Evaluations on an in-house dataset of 120 subjects (including 82 patients) demonstrate that SSL-MoCo significantly outperforms other self-supervised methods, particularly during challenging systolic phases. The integrated motion compensation enhances temporal coherence, resulting in more accurate myocardial morphology, which is crucial for clinical functional assessment. Our results suggest that SSL-MoCo provides an effective solution for dynamic MRI reconstruction in data-constrained settings.

Keywords: MRI Reconstruction · Self-supervised learning · Motion-compensated reconstruction

1 Introduction

Magnetic Resonance Imaging (MRI) is a widely used non-invasive imaging modality, renowned for its excellent soft tissue contrast in the absence of ionizing radiation. However, MRI acquisition is inherently time-consuming, particularly in dynamic applications such as cardiac Cine MRI, which serves as the clinical gold standard for assessing cardiac function. Conventional protocols typically require multiple breath-holds, leading to increased susceptibility to motion artifacts. Over the past decades, significant efforts have been made to accelerate acquisition while preserving diagnostic quality.

Traditional techniques, such as Parallel Imaging (PI) [1–3] and Compressed Sensing (CS) [4–6], accelerate the imaging process by exploiting multi-coil redundancies and image sparsity, respectively. Nonetheless, their performance degrades under high acceleration factors and typically requires carefully tuned regularization parameters along with computationally intensive iterative optimization.

Recently, deep learning has emerged as a powerful alternative, substantially improving both reconstruction quality and computational efficiency. Representative methods include k-space learning [7,8], image enhancement [9,10], physics-based unrolled networks [11–13], and hybrid learning [14–16]. Despite their success, two fundamental limitations remain: (i) Most existing methods are supervised, relying on fully-sampled reference data, which are often difficult, or even infeasible, to obtain in practice, particularly in dynamic imaging where prolonged acquisition times and motion-related artifacts (e.g., from respiration or involuntary movements) pose significant challenges. (ii) Dynamic MRI presents unique challenges in motion handling. Many methods attempt to learn motion implicitly through spatiotemporal representations. Although some works have demonstrated promising reconstruction quality, performance often degrades during phases with larger motion, e.g., systolic phases.

Self-supervised learning (SSL) has emerged as a promising paradigm to alleviate the dependency on fully-sampled data. One representative method is Self-Supervision via Data Undersampling (SSDU) [17], which splits the acquired undersampled k-space data into two disjoint sets: one for enforcing data consistency (DC) and the other for defining the training loss. Multi-mask SSDU [18] further improves performance by cycling through different data partitions during training. More recently, advanced SSL strategies, such as contrastive learning, have been applied to extract semantically meaningful features that facilitate reconstruction [19,20]. However, most of these methods have been developed and validated primarily on static imaging.

To better resolve dynamic information, motion-compensated reconstruction has been proposed to explicitly incorporate motion estimates into the reconstruction process. Several prior works [21,22] adopt a joint optimization strategy, learning motion and reconstruction in an end-to-end framework. However, this tightly coupled design often comes with a resource-intensive training, especially for high-resolution or long dynamic sequences. Moreover, joint optimization is prone to error accumulation, particularly during early training stages or under high acceleration rates, and thus requires careful tuning and regularization. Insufficient convergence of the reconstruction network may lead to suboptimal motion estimation, which in turn can impair reconstruction performance and vice versa.

To address these limitations, we propose SSL-MoCo, a self-supervised motion-compensated reconstruction framework for dynamic MRI. The framework adopts a two-stage design: (1) a self-supervised image registration network learns non-rigid inter-frame motion, (2) the estimated motion is incorporated into a self-supervised reconstruction network. This decoupled strategy enables dynamic learning in two complementary ways: explicit motion learning and implicit spatiotemporal regularization. Importantly, SSL-MoCo is fully self-supervised and does not require any fully-sampled reference data. Experiments on cardiac Cine MRI demonstrate that SSL-MoCo achieves high-quality reconstructions and significantly improves morphological fidelity during challenging motion phases.

2 Method

The proposed SSL-MoCo framework consists of two stages: (1) self-supervised image registration and (2) self-supervised motion-compensated reconstruction. The overall pipeline is illustrated in Fig. 1.

2.1 Self-supervised Image Registration

The registration network follows our previous work, GMA-RAFT [23], which adopts a dual-branch architecture to estimate dense motion between fixed and moving images. The first branch uses a residual feature encoder to extract spatial representations from both images, which are then combined via pairwise dot products to form a 4D correlation volume. This volume is further encoded into motion features. In parallel, the second branch processes the fixed image along with its two adjacent frames (together forming I_{context}) through a ResNet-based denoiser to suppress artifacts, followed by context feature encoding. A global motion aggregation (GMA) module applies self-attention over the context features to refine the motion features. Temporal coherence is enforced through an iterative refinement process using gated recurrent units (GRUs), which fuse local and global representations to produce a sequence of N progressively refined and dense motion fields $\{u_1, ..., u_N\}$.

Different from the previous work [23], a self-supervised training strategy that does not require fully-sampled data is adopted here, which incorporates three loss components. The photometric loss $\mathcal{L}_{\text{photo},i}$ measures the l_1 difference between the motion-corrected moving image $T(I_{mov}, u_i)$ and the reconstructed fixed image I_{fix}, which is reconstructed from undersampled data using iterative SENSE [24]. Here, T denotes bilinear warping, and i indexes the GRU refinement iteration. To encourage spatial coherence, a smoothness term $\mathcal{L}_{\text{smooth},i}$ penalizes the l_1 norm of the spatial gradients of the estimated motion, i.e., ∇u_i. Additionally, a denoising loss $\mathcal{L}_{\text{denoiser}}$ is applied as the mean squared error (MSE) between the denoised image and the reconstructed context image I_{context}.

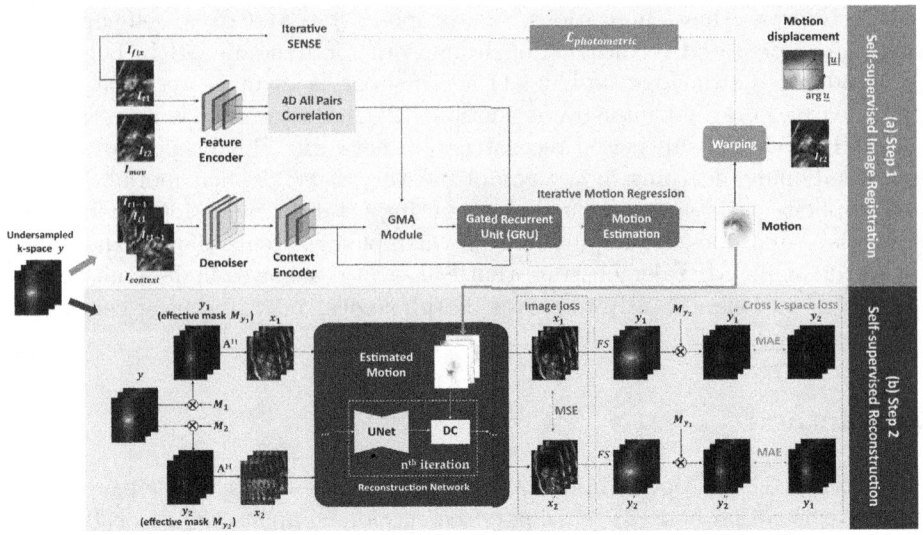

Fig. 1. Proposed SSL-MoCo framework. (a) Step 1: Self-supervised image registration network [23] with the photometric loss to minimize the differences between the reconstructed fixed image and the motion-corrected moving image. (b) Step 2: Self-supervised reconstruction network. Estimated motion fields from the first step are incorporated into the data consistency of the physics-based unrolled network using the forward and adjoint warp operations. The loss is composed of an image consistency loss and a cross k-space loss

The total loss is defined in Eq. 1, with $N = 12$, $\delta = 0.5$ and $\gamma = 0.1$:

$$\mathcal{L}_{\text{registration}} = \delta \mathcal{L}_{\text{denoiser}} + \sum_{i=1}^{N} \left(\mathcal{L}_{\text{photo},i} + \gamma \mathcal{L}_{\text{smooth},i} \right). \tag{1}$$

2.2 Self-supervised Motion-Compensated Reconstruction

In the second stage, we train a physics-based unrolled reconstruction network that integrates the estimated motion fields into the DC operation to reconstruct images $\mathbf{x}$ from the undersampled k-space measurements $\mathbf{y}$ by minimizing:

$$\arg \min_{\mathbf{x}} \|\mathbf{y} - \mathbf{MFSUx}\|_2 + \lambda \mathcal{R}(\mathbf{x}) \tag{2}$$

for undersampling mask $\mathbf{M}$, coil sensitivity maps $\mathbf{S}$, Fourier transformation $\mathbf{F}$, and estimated motion $\mathbf{U}$. The term $\mathcal{R}(\mathbf{x})$ describes the learnable regularizer represented as the UNet in Fig. 1. To enable self-supervision, the initial undersampled k-space measurement $\mathbf{y}$ is further re-undersample using two independent re-undersampling masks $\mathbf{M}_1$ and $\mathbf{M}_2$:

$$\mathbf{y}_1 = \mathbf{M}_1 \mathbf{y}, \quad \mathbf{y}_2 = \mathbf{M}_2 \mathbf{y}, \tag{3}$$

with the effective sampling masks denoted as $\mathbf{M}_{\mathbf{y}_1}$ and $\mathbf{M}_{\mathbf{y}_2}$. Re-undersampled k-spaces $\mathbf{y}_k$ are transformed to the input images $\mathbf{x}_k = \mathbf{A}_k^H \mathbf{y}_k$ using the multi-coil encoding operator $\mathbf{A}_k = \mathbf{M}_{\mathbf{y}_k} \mathbf{FS}$ with undersampling masks $\mathbf{M}_{\mathbf{y}_k}$, $k \in \{1, 2\}$.

The reconstruction network consists of m unrolled iterations, each containing a UNet-based regularizer $\mathcal{R}$ and a motion-compensated DC layer employing a generalized matrix decomposition [25]. Specifically, each DC layer utilizes the estimated motion fields from cardiac phase t_i to its $\pm k$ neighboring frames to warp the current frame t_i with bilinear interpolation, resulting in a total of $2k + 1$ warped frames. Under the assumption of ideal motion estimates, these warped frames are all aligned. The applied temporally incoherent sampling, i.e., each frame exhibits a distinct sampling pattern, enables the network to exploit complementary k-space measurements from neighboring frames.

The network independently reconstructs two output images $\mathbf{x}_1'$ and $\mathbf{x}_2'$. To train the model without fully-sampled ground truth, we employ a self-supervised loss composed of two terms: an image consistency loss and a cross k-space loss. The image consistency loss encourages similarity between the two reconstructions, as both originate from the same acquisition, defined as the MSE between $\mathbf{x}_1'$ and $\mathbf{x}_2'$. To further ensure data fidelity in the frequency domain, we introduce a cross k-space loss. Specifically, each reconstructed image is transformed into k-space via the forward operator, in which the effective mask of the other input is applied. The cross k-space loss is then defined as the mean absolute error (MAE) between the masked reconstructed k-space and the corresponding acquired measurements. The overall reconstruction loss is formulated as:

$$\mathcal{L}_{recon} = \mathcal{L}_{img} + \mathcal{L}_{ksp} = \left\| \mathbf{x}_1' - \mathbf{x}_2' \right\|_2^2 + \left\| \mathbf{A}_2 \mathbf{x}_1' - \mathbf{y}_1 \right\|_1 + \left\| \mathbf{A}_1 \mathbf{x}_2' - \mathbf{y}_2 \right\|_1. \quad (4)$$

3 Dataset and Experiments

We evaluated the proposed method on an in-house 2D cardiac Cine dataset comprising 82 patients with cardiovascular disease and 38 healthy volunteers. The dataset was split into 106 subjects for training (72 patients and 34 volunteers) and 14 for testing (10 patients and 4 volunteers). All scans were acquired on a 1.5T Siemens Aera MRI scanner using a balanced steady-state free precession (bSSFP) sequence with the following parameters: TE/TR = 1.06/2.12 ms, flip angle = 52°, resolution = $1.9 \times 1.9 \, \text{mm}^2$, and slice thickness = 8 mm. Undersampling masks were generated using the variable density incoherent spatiotemporal acquisition (VISTA) strategy [26]. For both the registration and reconstruction stages, the initial undersampled k-space had an $R = 2$ undersampled VISTA sampling. During the reconstruction network training, additional re-undersampling masks were applied with randomly selected acceleration rates ranging from $R = 2$ to $R = 24$ at each step.

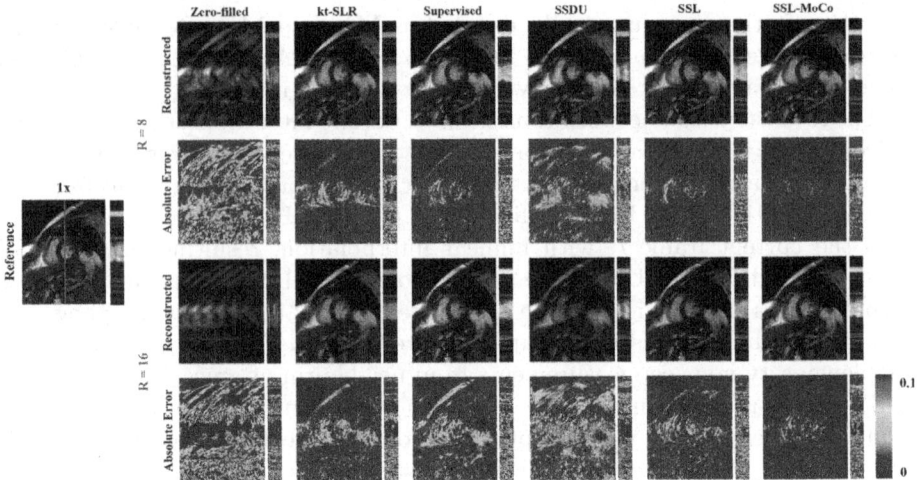

Fig. 2. Comparison of the proposed SSL-MoCo framework in the spatial (x-y) and spatial-temporal (y-t) plane to zero-filled, kt-SLR [30], supervised reconstruction (without motion compensation), SSDU [17], and ablated SSL reconstruction for a healthy subject. The dynamic performance in the y-t plane corresponds to the blue line in the reference x-y plane image. The inputs were retrospectively undersampled using VISTA [26]. Both R = 8 (top) and R = 16 (bottom) reconstructions are shown alongside the corresponding absolute error maps

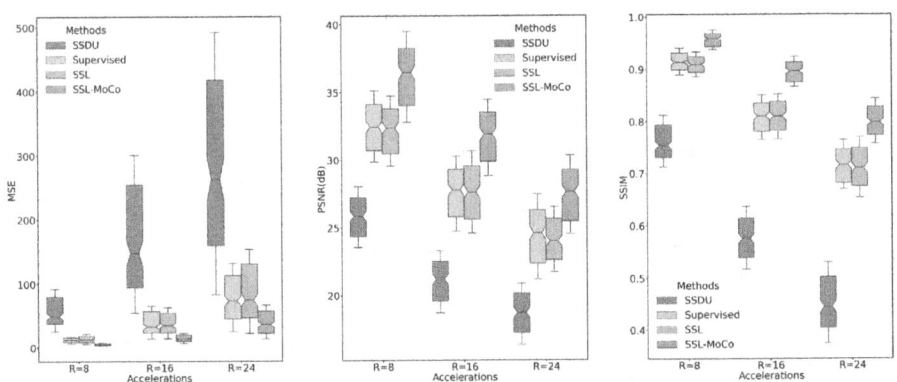

Fig. 3. Quantitative analysis in terms of MSE, PSNR, and SSIM between SSDU [17], supervised reconstruction, ablated SSL, and the proposed SSL-MoCo. Metrics are calculated for all subjects in the test dataset under $R = 8, 16, 24$ retrospective VISTA [26] undersampling. Results are depicted as box plots (horizontal line: median, box: 25% and 75% percentile, whiskers: 0.3 ∗ interquartile range)

The registration network adopts the original architecture described in [23]. For the unrolled reconstruction network, each iteration comprises a UNet followed by a motion-compensated DC layer. The UNet starts with 12 convolu-

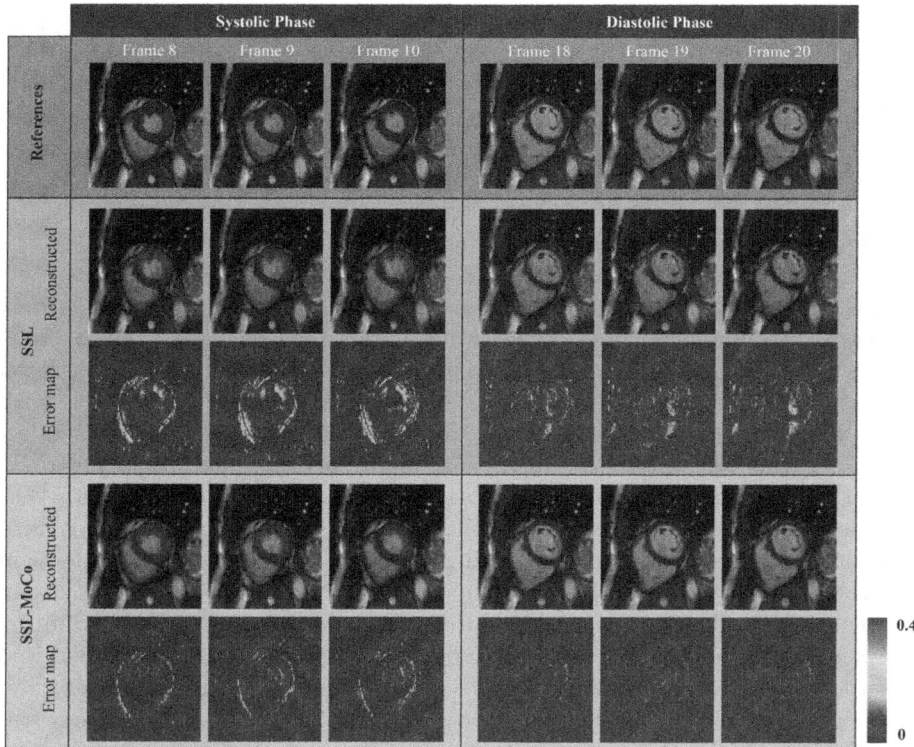

Fig. 4. Ablation study: Reconstructions of the proposed SSL-MoCo framework compared to the ablated self-supervised reconstruction (SSL) without motion compensation. The healthy subject was retrospectively undersampled with R = 12 VISTA [26] sampling. Three cardiac systolic phases and three diastolic phases are depicted in the left and right columns, respectively, along with the absolute error maps. Fully-sampled reference images are shown for comparison in the first row

tional filters, which are doubled after each downsampling layer in the encoder and halved during upsampling in the decoder. To facilitate spatiotemporal feature aggregation, we employ 2D+t convolutions with a kernel size of 5 in the spatial dimensions and 3 along the temporal axis. The entire reconstruction network is implemented in the complex domain, where complex-valued convolutions [27] and ModRelu activations [28] are realized using the MERLIN v0.3 [29].

We compared the proposed SSL-MoCo framework against three baselines: the traditional low-rank method kt-SLR [30], the self-supervised reconstruction method SSDU [17], and a supervised variant of our reconstruction network (without motion compensation) trained using fully-sampled data. Additionally, we conducted an ablation study using the same self-supervised reconstruction strategy but without motion compensation, denoted as SSL in the results. To ensure a fair comparison, all methods were trained and evaluated using the same VISTA sampling technique. For deep learning-based methods, we used consistent hyperparameters: batch size of 1, learning rate of 4×10^{-4} with Adam optimizer. Each adopted the same network architecture with $m = 2$ unrolls and trained until convergence. For motion compensation in SSL-MoCo, we utilized the motion fields estimated from ± 1 neighboring frames for each target cardiac phase. The source code is publicly available: https://github.com/midas-tum/SSL-MoCo.

4 Results and Discussion

Figure 2 illustrates representative reconstruction results for a healthy subject during a systolic phase under acceleration factors $R = 8$ and $R = 16$. Systolic frames pose a greater challenge for dynamic reconstruction due to pronounced myocardial motion and reduced data redundancy, especially in datasets with a predominance of patients exhibiting limited contractility. At both accelerations, kt-SLR [30] tends to oversmooth anatomical details, indicating over-regularization. SSDU [17] exhibits noticeable residual aliasing artifacts that become more pronounced at higher undersamplings. The ablation study SSL performs on par with supervised learning and exhibits slightly fewer artifacts in this representative case. Importantly, both models are trained without motion compensation and share the same architecture, differing only in the use of fully-sampled references for supervision. These results suggest that our self-supervised reconstruction strategy can effectively reconstruct images in the absence of fully-sampled data. Building on this, the proposed SSL-MoCo further improves reconstruction quality by incorporating motion fields, particularly in preserving fine structures and suppressing noise and artifacts.

This advantage is further highlighted in the reconstructed y-t plane, where temporal aliasing manifests as vertical striping artifacts. Unlike other methods, SSL-MoCo exhibits improved temporal consistency, effectively suppressing these artifacts and yielding smoother cardiac dynamics across frames. This improvement can be attributed to the explicit modeling of inter-frame motion, which enables the reconstruction network to access temporally aligned information rather than relying solely on implicit spatiotemporal regularization.

To provide a comprehensive evaluation, we quantitatively compared SSL-MoCo, SSL, supervised learning, and SSDU [17] across three acceleration rates: $R = 8, 16, 24$. Evaluation metrics include MSE, peak signal-to-noise ratio (PSNR), and structural similarity index measure (SSIM), computed over all slices from all test subjects. The results are presented as box plots in Fig. 3. Across all accelerations, SSL-MoCo consistently yields the lowest MSE and the highest PSNR and SSIM. Taking SSIM as the most perceptually relevant metric, SSL-MoCo improves over SSDU, supervised reconstruction, and the ablated SSL by approximately 70.5%, 10.7%, and 10.0% at $R = 16$, respectively. Moreover, the relatively narrow box of metric distributions of SSL-MoCo further indicates stable behavior across subjects, which is essential for clinical reliability. These results are in line with the qualitative findings from Fig. 2 and further confirm the benefit of integrating explicit motion modeling into self-supervised reconstruction.

To assess the impact of motion compensation, we compare the proposed SSL-MoCo with its ablated variant (SSL) that omits motion estimation. Figure 4 presents representative frames from both the systolic (frames 810) and diastolic (frames 1820) phases. Notably, SSL-MoCo provides more anatomically accurate myocardial morphology, especially during the systolic phase, which is critical for clinical assessment, such as ejection fraction (EF) calculation. The proposed method reconstructs sharper endocardial boundaries and more realistic myocardial contours, which can directly impact downstream segmentation tasks. Moreover, in both cardiac phases, SSL-MoCo exhibits reduced aliasing artifacts compared to SSL, indicating that motion-guided regularization not only improves reconstruction accuracy but also enhances robustness throughout the cardiac cycle. These observations support the broader applicability of SSL-MoCo in dynamic imaging scenarios where motion is a dominant confounder.

5 Conclusion

In this work, we proposed SSL-MoCo, a self-supervised motion-compensated reconstruction framework for dynamic MRI. The proposed method explicitly decouples motion estimation and image reconstruction into two self-supervised stages, allowing accurate motion learning and effective motion-compensated reconstruction without requiring fully-sampled reference data. Through extensive experiments, SSL-MoCo demonstrates superior reconstruction performance compared to both supervised and self-supervised reconstructions, particularly under high acceleration rates. The integration of motion compensation notably improves morphology accuracy during systolic phases, which is essential for clinical assessment. These results highlight the potential of SSL-MoCo as a scalable and clinically relevant solution for dynamic MRI in data-constrained settings.

Acknowledgements. This work was supported by the Deutsche Forschungsgemeinschaft (DFG, German Research Foundation) under Germany's Excellence Strategy EXC 2064/1 Project number 390727645.

Disclosure of Interests. The authors have no competing interests to declare that are relevant to the content of this article.

References

1. Pruessmann, K.P., Weiger, M., Scheidegger, M.B., Boesiger, P.: SENSE: sensitivity encoding for fast MRI. Magn. Reson. Med. **42**(5), 952–962 (1999)
2. Griswold, M.A., et al.: Generalized autocalibrating partially parallel acquisitions (GRAPPA). Magn. Reson. Med. **47**(6), 1202–1210 (2002)
3. Uecker, M., Lai, P., Murphy, M.J., Virtue, P., Elad, M.: ESPIRiT-an eigenvalue approach to autocalibrating parallel MRI: where SENSE meets GRAPPA. Magn. Reson. Med. **71**(3), 990–1001 (2014)

4. Lustig, M., Donoho, D., Pauly, J.M.: Sparse MRI: the application of compressed sensing for rapid MR imaging. Magn. Reson. Med. **58**(6), 1182–1195 (2007)
5. Block, K.T., Uecker, M., Frahm, J.: Undersampled radial MRI with multiple coils. Iterative image reconstruction using a total variation constraint. Magn. Reson. Med. **57**(6), 1086–1098 (2007)
6. Lustig, M., Donoho, D.L., Santos, J.M., Pauly, J.M.: Compressed sensing MRI. IEEE Signal Process. Mag. **25**(2), 72–82 (2008)
7. Lee, D., Jin, K.H., Kim, E.Y., Park, S.H., Ye, J.C.: Acceleration of MR parameter mapping using annihilating filter-based low rank Hankel matrix (ALOHA). Magn. Reson. Med. **76**(6), 1848–1864 (2016)
8. Akçakaya, M., Moeller, S., Weingärtner, S., Uğurbil, K.: Scan-specific robust artificial-neural-networks for k-space interpolation (RAKI) reconstruction: database-free deep learning for fast imaging. Magn. Reson. Med. **81**(1), 439–453 (2019)
9. Hauptmann, A., Arridge, S., Lucka, F., Muthurangu, V., Steeden, J.A.: Real-time cardiovascular MR with spatio-temporal artifact suppression using deep learning–proof of concept in congenital heart disease. Magn. Reson. Med. **81**(2), 1143–1156 (2019)
10. Kofler, A., Dewey, M., Schaeffter, T., Wald, C., Kolbitsch, C.: Spatio-temporal deep learning-based undersampling artefact reduction for 2D radial cine MRI with limited training data. IEEE Trans. Med. Imaging **39**(3), 703–717 (2019)
11. Schlemper, J., Caballero, J., Hajnal, J.V., Price, A.N., Rueckert, D.: A deep cascade of convolutional neural networks for dynamic MR image reconstruction. IEEE Trans. Med. Imaging **37**(2), 491–503 (2017)
12. Hammernik, K., et al.: Learning a variational network for reconstruction of accelerated MRI data. Magn. Reson. Med. **79**(6), 491–503 (2018)
13. Küstner, T., et al.: CINENet: deep learning-based 3D cardiac CINE MRI reconstruction with multi-coil complex-valued 4D spatio-temporal convolutions. Sci. Rep. **10**(1), 13710 (2020)
14. Eo, T., Jun, Y., Kim, T., Jang, J., Lee, H.J., Hwang, D.: KIKI-net: cross-domain convolutional neural networks for reconstructing undersampled magnetic resonance images. Magn. Reson. Med. **80**(5), 2188–2201 (2018)
15. El-Rewaidy, H., et al.: Multi-domain convolutional neural network (MD-CNN) for radial reconstruction of dynamic cardiac MRI. Magn. Reson. Med. **85**(3), 1195–1208 (2021)
16. Xu, S., et al.: Attention incorporated network for sharing low-rank, image and k-space information during MR image reconstruction to achieve single breath-hold cardiac Cine imaging. Comput. Med. Imaging Graph. **120**, 102475 (2025)
17. Yaman, B., Hosseini, S.A.H., Moeller, S., Ellermann, J., Uğurbil, K., Akakaya, M.: Self-supervised learning of physics-guided reconstruction neural networks without fully sampled reference data. Magn. Reson. Med. **84**(6), 3172–3191 (2020)
18. Yaman, B., et al.: Multi-mask self-supervised learning for physics-guided neural networks in highly accelerated magnetic resonance imaging. NMR Biomed. **35**(12), e4798 (2022)
19. Wang, S., et al.: PARCEL: physics-based unsupervised contrastive representation learning for multi-coil MR imaging. IEEE/ACM Trans. Comput. Biol. Bioinf. **20**(5), 2659–2670 (2022)
20. Xu, S., et al.: Self-supervised feature learning for cardiac Cine MR image reconstruction. IEEE Transactions on Medical Imaging (2025)

21. Pan, J., Rueckert, D., Küstner, T., Hammernik, K.: Learning-based and unrolled motion-compensated reconstruction for cardiac MR CINE imaging. Int. Conf. Med. Image Comput. Comput. Assist. Interv. 686–696 (2022)
22. Pan, J., Hamdi, M., Huang, W., Hammernik, K., Kuestner, T., Rueckert, D.: Unrolled and rapid motion-compensated reconstruction for cardiac CINE MRI. Med. Image Anal. **91**, 103017 (2024)
23. Ghoul, A., et al.: Attention-aware non-rigid image registration for accelerated MR imaging. IEEE Transactions on Medical Imaging (2024)
24. Uecker, M., et al.: Berkeley advanced reconstruction toolbox. Proc. Intl. Soc. Mag. Reson. Med **23**(2486), 9 (2015)
25. Batchelor, P.G., Atkinson, D., Irarrazaval, P., Hill, D.L.G., Hajnal, J., Larkman, D.: Matrix description of general motion correction applied to multishot images. Magn. Reson. Med. **54**(5), 1273–1280 (2005)
26. Ahmad, R., Xue, H., Giri, S., Ding, Y., Craft, J., Simonetti, O.P.: Variable density incoherent spatiotemporal acquisition (VISTA) for highly accelerated cardiac MRI. Magn. Reson. Med. **74**(5), 1266–1278 (2015)
27. Trabelsi, C., et al.: Deep complex networks. arXiv preprint arXiv:1705.09792 (2017)
28. Arjovsky, M., Shah, A., Bengio, Y.: Unitary evolution recurrent neural networks. Int. Conf. Mach. Learn. 1120–1128 (2016)
29. Hammernik, K., Küstner, T.: Machine enhanced reconstruction learning and interpretation networks (MERLIN). In: Proceedings of the International Society for Magnetic Resonance in Medicine (ISMRM 2022)
30. Lingala, S.G., Hu, Y., DiBella, E., Jacob, M.: Accelerated dynamic MRI exploiting sparsity and low-rank structure: k-t SLR. IEEE Trans. Med. Imaging **30**(5), 1042–1054 (2011)

Generating Realistic Synthetic Motion Curves for MRI Retrospective Motion Correction

Hristo Georgiev[1,3](✉)[iD], Jakob Sheye[1], Madeleine Wyburd[1,2], Jens Petersen[1,3], Vincent Beliveau[4], and Melanie Ganz[1,2]

[1] Department of Computer Science, University of Copenhagen, Copenhagen, Denmark
hrge@di.ku.dk
[2] Neurobiology Research Unit, Copenhagen University Hospital, Copenhagen, Denmark
[3] Department of Clinical Oncology, Center for Cancer and Organ Diseases, Copenhagen University Hospital, Copenhagen, Denmark
[4] Department of Neurology, Medical University of Innsbruck, Innsbruck, Austria

Abstract. Patient motion during MRI acquisition causes image artifacts that distort the scans. Retrospective motion correction (MoCo) is a post-processing method that reduces such artifacts. Deep learning models are a promising approach towards retrospective MoCo. However, these models require large datasets of motion-free and motion-corrupted scans, which are often unavailable. Current methods simulate motion artifacts using random transformations in the spatial or frequency domain of the scan. Such random transformations can result in artifacts that are not physically realistic. An alternative approach would use synthetic motion curves that mimic patient movement. This project evaluates two generative models, `TimeVAE` and `Fourier Flow`, trained on motion curves from the ABCD study. We assess their ability to generate realistic motion curves using statistical metrics and a novel benchmarking protocol (`TSGBench`). Both models capture key distribution characteristics but struggle with abrupt motions such as coughing. Nonetheless, the results are promising for future research.

Keywords: MRI · Generative models · Motion artifacts · Motion correction · MRI reconstruction · Time-VAE · Fourier Flow · TSGBench

1 Introduction

Magnetic resonance imaging (MRI) is a non-invasive modality with excellent soft-tissue contrast, making it the standard choice for brain imaging tasks [1–3]. Distinct features of MRI protocols are their lengthy duration (30–60 minutes)

H. Georgiev and J. Sheye—Equal contribution.

© The Author(s), under exclusive license to Springer Nature Switzerland AG 2026
L. Felsner et al. (Eds.): RIME 2025/GRAIL 2025, LNCS 16703, pp. 108–117, 2026.
https://doi.org/10.1007/978-3-032-06103-4_11

and their sensitivity to motion. Voluntary and involuntary patient movements at acquisition time cause motion artifacts that hinder image analysis [4,5].

Scan quality can be improved by applying retrospective motion correction (MoCo) [6]. Retrospective MoCo can be achieved by training a convolutional neural network (CNN) on a dataset of paired motion-free and motion-corrupted images. The goal during inference is for the network to map the motion-corrupted images to motion-free ones. As large datasets of real motion curves are often unavailable, many image-based denoising algorithms rely on random transformations for motion simulation [6].

We aim to address the issue with unrealistic motion simulation and patient data privacy by training generative models to produce realistic synthetic motion curves. For this purpose, we use data provided by the Adolescent Brain Cognitive Development (ABCD) study [7]. Thus, we hypothesize that the generated motion curves will be based on real motion patterns, rather than random transformations.

2 Background

2.1 Motion During an MRI Brain Scan

Head motion can be estimated using six degrees of freedom (DOFs) - translations along and rotations around the x, y, and z axes. Since a functional MRI (fMRI) scan is a time series of n images, motion can be inferred by registering a floating image n_f to a target image n_t. Therefore, motion can be described as the total displacement at image n_f WRT n_t, where n_t is often the first image of the time series. Previous studies have found that translations along y and z, and rotations around x are the dominant sources of motion for children [8,9]. Those movements also have a natural explanation, as translations along y and z correspond to the body drifting out of the scanner bore and the head sinking into the pillow, respectively, while rotations around the x axis correspond to a nodding motion, often caused by swallowing saliva.

2.2 Generative Models

The early attempts at synthetic time series generation have predominantly focused on generative adversarial networks (GANs) [10–13]. Although capable of generating realistic data, GANs often suffer from slow and unstable training due to their network architecture, optimization function, and algorithm [14,15]. This work evaluates two models that aim to resolve the limitations of GANs - TimeVAE [16] and Fourier Flow [17].

TimeVAE (T-VAE) [16]: T-VAE is a variational autoencoder (VAE) composed of an encoder and decoder. The encoder takes an input of shape $(N \times T \times D)$, where N is the batch size, T is the number of time steps, and D is the number of feature dimensions. The encoder passes the input through a series of $1D$ convolution layers with a ReLU activation, before flattening it, and passing it through a dense layer. The dense layer has $2m$ neurons, where m is the number of latent dimensions.

`T-VAE` comes with two types of decoders: a base and an interpretable one. The base decoder samples a vector z using the reparameterization trick [18], passes it through a dense layer, and reshapes it into a $3D$ vector. The reshaped vector will go through a few transposed $1D$ convolution layers, before passing through a dense layer. We utilize an interpretable decoder, with a single parallel trend and seasonality block, and the base decoder. In essence, the trend block is a monotonic function that models the trend polynomials, while the seasonality blocks are defined by the number of seasons s and their duration d. We use one block with $s = 6$ and $d = 380$ (6 curves, each 380 frames long).

`Fourier Flow (F-Flow)` [17]: `F-Flow` is a flow-based model, and unlike GANs and VAEs, it does not train a pair of networks. Instead, it utilizes the discrete Fourier transform (DFT). In essence, `F-Flow` is composed of two layers: 1) a frequency transform and 2) a spectral filter. The frequency layer will perform temporal zero-padding to ensure that the time series and their frequency spectrum representations will both have a fixed length of N. The DFT will then convert each time series into a real and imaginary matrix. Finally, the symmetric property of the DFT is used to discard redundant components from the real and imaginary matrices. `F-Flow`'s second layer is an affine coupling layer that performs spectral filtering of the data. The layer will apply the real component to a filter H with a transfer function, where the transfer function depends on the imaginary component. Finally, the mapping from the imaginary component to the transfer functions is implemented by a recurrent neural network.

3 Methods

Figure 1 shows an overview of our training and evaluation protocol. We designed a two-stage pipeline: in stage one, we perform a grid search to find an optimal hyperparameter setting for both models; in stage two, the models are retrained with the optimal hyperparameter settings and evaluated on the complete set of metrics. Code is available at https://github.com/melanieganz/MoCoProject_motionGen.

3.1 Dataset and Preprocessing

The ABCD study is a long-term study of the biological and behavioral development of adolescents in the USA [7]. The cohort consists of 11,878 children aged 8 to 11, including male, female, and inter-gender children. It is available through the ABCD study (see https://abcdstudy.org/scientists/data-sharing/ and https://nda.nih.gov/abcd/abcd-citing). 8 and 11-year-old (-yo.), and intergender children, are underrepresented (a total of 208 subjects) compared to other subgroups at baseline. Therefore, we excluded these groups from the dataset. See Table. 1 for the final demographics and data split.

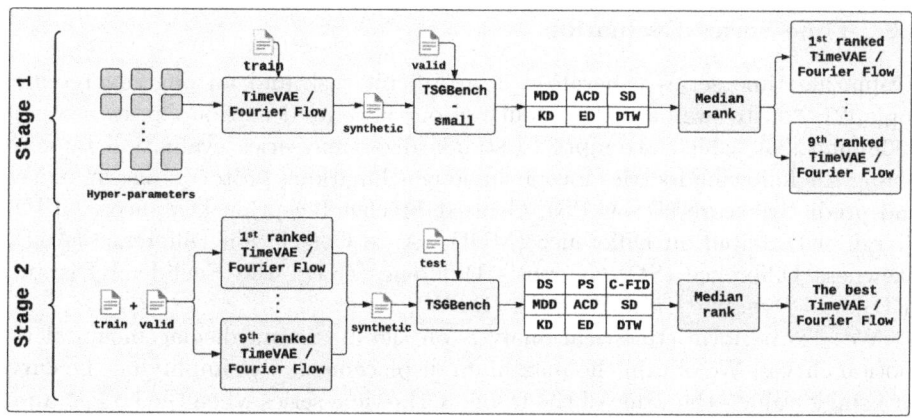

Fig. 1. A diagram of our training procedure. In Stage 1, we perform a grid-search over a fixed set of hyper-parameters for both models, and we select the nine best-performing configurations based on a subset of TSGBench's metrics. In Stage 2, we retrain the nine best-performing configurations and evaluate them on the full set of TSGBench metrics to select the best-performing T-VAE and F-Flow

We focus on the first run (baseline) of resting-state fMRI (rs-fMRI) in ABCD's acquisition protocol [19]. We hypothesize that this run is most representative of the children's natural behavior inside the scanner, given their unfamiliarity with the procedure at this point. Furthermore, rs-fMRI allows us to extract motion curves by registering images to a target. We use those motion curves as input to T-VAE and F-Flow during training. The preprocessing performed by the ABCD data analysis group using the ABCD-BIDS pipeline [20] was used to extract motion curves from the rs-fMRI scan. The preprocessing utilizes pairwise rigid registration between the different rs-fMRI images, with a repetition time of $800ms$.

We resolve varying durations of the rs-fMRI scans by limiting them to the first 380 images, corresponding to a temporal duration of 304 seconds. We also remove points below the 2.5^{th} and 97.5^{th} percentiles of the $3D$ displacement distributions, as they correspond to unrealistic movements of up to $100mm$, which might be the result of miss-registration.

Table 1. A table of the demographics per data split (m - male, f - female)

Data split	9-yo. m.	9-yo. f.	10-yo. m.	10-yo. f.	Total per split
Train	1,455	1,418	1,461	1,354	5,688
Validation	169	174	163	126	632
Test	378	372	444	387	1,581
Total per group	2,002	1,964	2,068	1,867	7,901

3.2 Time-Series Evaluation

Evaluating time series generation is a difficult task and an ongoing research topic [21,22]. To evaluate the quality of our generated motion curves, we used TSGBench [23], which attempts to standardize time-series evaluation by combining the following metrics into a single benchmarking protocol: discriminative and predictive score (DS & PS), Context Fréchet Inception Distance (C-FID), Marginal Distribution Difference (MDD), Auto Correlation Difference (ACD), Skewness Difference (SD), Kurtosis Difference (KD), the Euclidean Distance (ED), and Dynamic Time Warping (DTW).

We also perform statistical analysis on the maximum displacements of the motion curves. We obtain the maximum displacements by simplifying the curve as a single value - the value of the frame in the time-series when the largest absolute displacement WRT to the patient's starting position occurred. The maximum displacement is of interest, as excessive movements can lead the patient out of the scanner's field of view, hence, completely ruining the scan. We compare the maximum displacements of the motion curves generated by T-VAE and F-Flow with the maximum displacements of the curves in the testing dataset with violin plots. Moreover, we perform Spearman correlation analysis on the maximum displacement to investigate any existing relations between the different DOFs. Finally, we report the results in 6×6 heatmaps, where blank squares denote statistically insignificant correlations ($p > 0.05$).

3.3 Training

In stage one, we do a grid search over a fixed number of hyperparameters (see Table 2). We keep the models' default batch sizes - 32 for T-VAE and 128 for F-Flow [16,17]. We trained for a fixed number of epochs (5,000), but T-VAE, unlike F-Flow, implements early stopping (patience of 10 epochs) [16,17]. Both models are trained and evaluated on the training and validation sets, respectively (see Table 1). During grid search, every model generates a distribution of synthetic curves that matches the validation set in size. We then compare the synthetic distribution WRT the validation set on the MDD, ACD, SD, KD, and ED metrics of TSGBench [23]. Note that we omit DS, PS, and C-FID as they require training a model, which is time-consuming. We used the median rank to select the nine best-performing sets of hyperparameters for both models.

In stage two, we retrain the nine models using the hyperparameter setting that we found to be the best in stage one (batch size and epochs remain the same). This time, the models are trained on the combined training and validation set and evaluated on the testing set, respectively (see Table 1). We prompt the models to generate a synthetic distribution that matches the testing set in size. This time, however, we include DS, PS, and C-FID as evaluation metrics. We again use the median ranking to determine the best-performing hyperparameter set for both models.

Table 2. Hyperparameter search space. Optimal values are in **bold** font

Model	Hyper-parameters	Values
T-VAE	Layer size	$(50, 100, 200), (75, 150, 300), \mathbf{(100, 200, 400)}$
	Latent dimensions	$4, \mathbf{8}, 12$
	Trend polynomials	$2, \mathbf{4}, 6$
	Learning rate	$1e-3, \mathbf{1e\text{-}4}, 1e-5$
F-Flow	Layer size	$150, \mathbf{200}, 250$
	Number of flows	$5, \mathbf{10}, 15$
	Learning rate	$\mathbf{1e\text{-}3}, 1e-4, 1e-5$

4 Results

Table 3 shows that T-VAE's and F-Flow's performance is comparable across most metrics, but F-Flow has a clear advantage on C-FID. The lower C-FID value indicates that F-Flow's synthetic curves are more diverse and have higher fidelity. This observation is confirmed by the violin plots in Fig. 2, which show that although both models capture the dominant sources of motion, F-Flow does a better job of modeling the real data's maximum displacements than T-VAE.

The heatmaps in Fig. 3 also show that the correlations in F-Flow's synthetic data are closer in magnitude to the real distribution compared to T-VAE's synthetic data. As T-VAE and F-Flow have been exposed to the training set but not the testing set, their synthetic distributions' correlations resemble the training set correlations' magnitudes and directions. For example, neither of the models captures the existing positive correlation between translations along and rotations around z that, albeit weak, is present in the testing set. Another interesting observation is that T-VAE tends to over-correlate the data, hence, introducing correlations between DOFs that do not exist in either the training or the testing set (see rotations around x and y, for example). Nonetheless, both models struggle with recreating abrupt movements (see Fig. 4).

Table 3. The TSGBench metric values of T-VAE's and F-Flow's synthetic motion curves compared to the testing distribution. Better scores are underlined. The standard deviation (±) for DS, PS, and C-FID is based on five runs [23]

Model	DS	PS	C-FID	MDD	ACD	SD	KD	ED	DTW
T-VAE	<u>0.185</u>	0.088	1.368	0.072	1.028	0.616	10.262	<u>6.255</u>	<u>18.702</u>
	(±0.033)	(±0.001)	(±0.101)	N/A	N/A	N/A	N/A	N/A	N/A
F-Flow	0.208	0.088	<u>0.336</u>	<u>0.046</u>	<u>0.911</u>	<u>0.515</u>	<u>6.778</u>	6.970	19.659
	(±0.079)	(±0.001)	(±0.012)	N/A	N/A	N/A	N/A	N/A	N/A

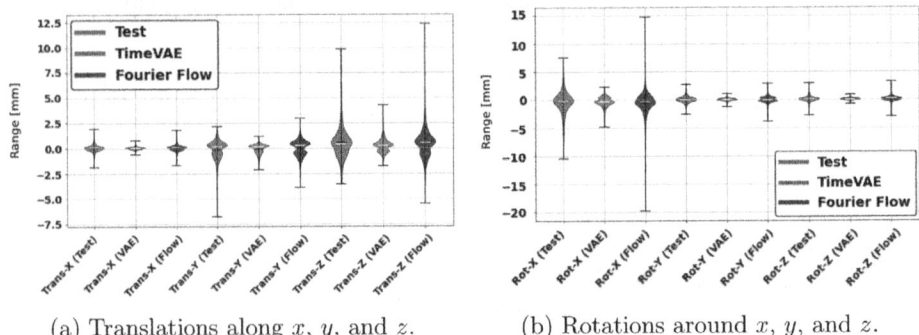

(a) Translations along x, y, and z. (b) Rotations around x, y, and z.

Fig. 2. Violin plots of the maximum displacements of the test set and the distributions generated by T-VAE and F-Flow

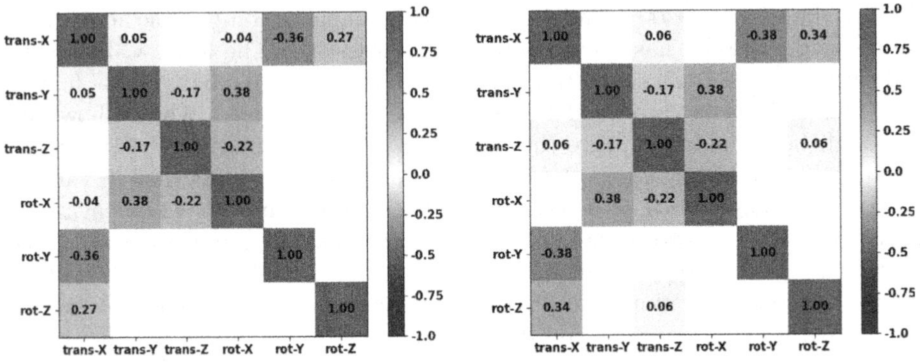

(a) Training set. (b) Testing set

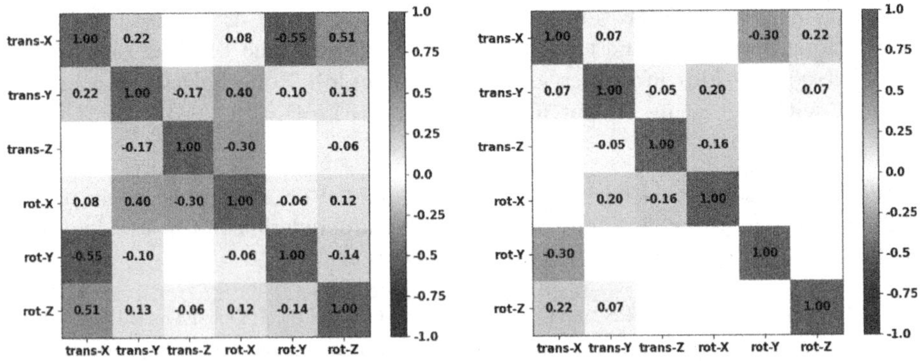

(c) T-VAE's distribution (d) F-Flow's distribution.

Fig. 3. The maximum displacement Spearmann correlations of the (a) training & (b) testing set, and (c) T-VAE's & (d) F-Flow's distribution. Statistically significant ($p < 0.05$) results are shown in colored squares. The value inside the squares is the correlation's magnitude, while the color indicates its direction

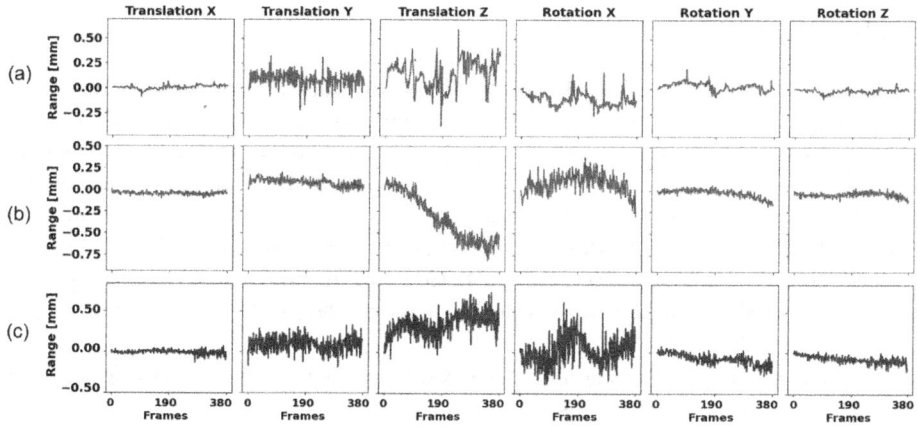

Fig. 4. The first sample from the (a) testing set, (b) T-VAE's distribution, and (c) F-Fow's distribution

5 Discussion

Our results demonstrate that generative models can be successfully used to generate motion curves. Both T-VAE and F-Flow demonstrate the ability to capture the correlations between the dominant sources of motion (translation along y and z, and rotations around x). Upon visual inspection (Fig. 4), we notice that neither model recreates abrupt displacements that might be caused by involuntary movements, such as coughing or sneezing. What separates F-Flow from T-VAE is that it did not overfit the synthetic data, and that the generated curves' maximum displacements are closer to the real data.

A limitation of our work is the lack of further evaluation of the generated curves. It is worth investigating whether using generated curves for retrospective MoCo is better than applying random transformations by applying them to motion-affected images and evaluating the quality of the transformed image. However, we hypothesize that using synthetic motion curves is a step in the right direction, as they at least mimic naturally occurring movements, such as drifting out of the scanner bore and sinking into the pillow.

An application study like ours would also benefit from a more thorough investigation of model architectures. For example, it would be beneficial to include GAN models, multiple variational auto-encoders, etc. A limitation of T-VAE and F-Flow for this task is their inability to produce gender or age-specific motion curves, which poses an issue, as it is established that there is a difference in motion between different genders and age groups [8]. This is not a criticism of the models, as they were not designed to tackle the issue we apply them to [16,17]. However, it raises the question of whether retrospective MoCo could benefit from a generative model that aims to produce gender and age-specific motion curves.

6 Conclusion

To the best of our knowledge, this is the first study that uses generative models to produce realistic motion curves for retrospective MoCo. The main contributions of this work include 1) the use of a large and accessible dataset provided by the ABCD study [7] and 2) a robust and reproducible framework for model selection based on a two-stage training procedure and thorough evaluation using TSGBench [23], correlation analysis, and visual inspection. Further investigation is necessary to conclude whether synthetic motion curves are a better way to simulate motion artifacts than random transformations.

Disclosure of Interests. The authors have no competing interests to declare that are relevant to the content of this article.

References

1. Roberts, T.P., Rowley, H.A.: Diffusion weighted magnetic resonance imaging in stroke. Eur. J. Radiol. **45**(3), 185–194 (2003)
2. Raichle, M.E., Mintun, M.A.: Brain work and brain imaging. Annu. Rev. Neurosci. **29**, 449–476 (2006)
3. Işın, A., Direkoğlu, C., Şah, M.: Review of MRI-based brain tumor image segmentation using deep learning methods. Procedia Comput. Sci. **102**, 317–324 (2016)
4. Andre, J.B., et al.: Toward quantifying the prevalence, severity, and cost associated with patient motion during clinical MR examinations. J. Am. Coll. Radiol. **12**(7), 689–695 (2015)
5. Gilmore, A.D., Buser, N.J., Hanson, J.L.: Variations in structural MRI quality significantly impact commonly used measures of brain anatomy. Brain Inf. **8**, 1–15 (2021)
6. Spieker, V., et al.: Deep learning for retrospective motion correction in MRI: a comprehensive review. IEEE Transactions on Medical Imaging (2023)
7. Casey, B.J., et al.: The adolescent brain cognitive development (ABCD) study: imaging acquisition across 21 sites. Dev. Cogn. Neurosci. **32**, 43–54 (2018)
8. Afacan, O., et al.: Evaluation of motion and its effect on brain magnetic resonance image quality in children. Pediatr. Radiol. **46**(12), 1728–1735 (2016). https://doi.org/10.1007/s00247-016-3677-9
9. Eichhorn, H., et al.: Characterisation of children's head motion for magnetic resonance imaging with and without general Anaesthesia. Front. Radiol. **1**, 789632 (2021)
10. Esteban, C., Hyland, S.L., Rätsch, G.: Real-valued (medical) time series generation with recurrent conditional GANs. arXiv preprint arXiv:1706.02633 (2017)
11. Yoon, J., Jarrett, D., Van der Schaar, M.: Time-series generative adversarial networks. Adv. Neural Inf. Process. Syst. **32** (2019)
12. Goodfellow, I., et al.: Generative adversarial nets. Adv. Neural Inf. Process. Syst. **27** (2014)
13. Mogren, O.: C-RNN-GAN: Continuous recurrent neural networks with adversarial training. arXiv preprint arXiv:1611.09904 (2016)
14. Saxena, D., Cao, J.: Generative adversarial networks (GANs) challenges, solutions, and future directions. ACM Comput. Surv. (CSUR) **54**(3), 1–42 (2021)

15. Pavan Kumar, M., Jayagopal, P.: Generative adversarial networks: a survey on applications and challenges. Int. J. Multimedia Inf. Retrieval **10**(1), 1–24 (2021)
16. Desai, A., Freeman, C., Wang, Z., Beaver, I.: TimeVAE: A variational auto-encoder for multivariate time series generation. arXiv preprint arXiv:2111.08095 (2021)
17. Alaa, A., Chan, A.J., van der Schaar, M.: Generative time-series modeling with fourier flows. In: International Conference on Learning Representations (2020)
18. Kingma, D.P., Welling, M., et al.: Auto-encoding variational bayes (2013)
19. ABCD Study: ABCD study: Brochure protocol baseline (2019). https://abcdstudy.org/wp-content/uploads/2019/12/Brochure_Protocol-Baseline-eg.pdf. Accessed 24 Jan 2025
20. Feczko, E., et al.: Adolescent brain cognitive development (ABCD) community MRI collection and utilities. BioRxiv, pp. 2021–07 (2021)
21. Creswell, A., White, T., Dumoulin, V., Arulkumaran, K., Sengupta, B., Bharath, A.A.: Generative adversarial networks: an overview. IEEE Signal Process. Mag. **35**(1), 53–65 (2018)
22. Xu, Q., et al.: An empirical study on evaluation metrics of generative adversarial networks. arXiv preprint arXiv:1806.07755 (2018)
23. Ang, Y., Huang, Q., Bao, Y., Tung, A.K., Huang, Z.: TSGBench: Time series generation benchmark. arXiv preprint arXiv:2309.03755 (2023)

Neural Space-Time Modeling for Motion-Corrected MR Reconstruction

Aizada Nurdinova[1(✉)], Wenqi Huang[2], Daniel Raz Abraham[4], Jaehyeok Bae[4], Yimeng Lin[4], Kawin Setsompop[1,4], and Brian Andrew Hargreaves[1,3,4]

[1] Department of Radiology, Stanford University, Stanford, USA
nurdaiza@stanford.edu
[2] Institute of AI in Healthcare and Medicine, Technical University of Munich, Munich, Germany
[3] Department of Bioengineering, Stanford University, Stanford, USA
[4] Department of Electrical Engineering, Stanford University, Stanford, USA

Abstract. Motion remains a significant challenge in MRI, particularly for body imaging and prolonged multi-contrast acquisitions. Although numerous classical and deep learning approaches have been developed for nonrigid motion correction, many require additional calibration, pre-training, or may have limited generalizability across protocols and subjects. Inspired by implicit neural representations (INR), we explore the application of a Neural Space-Time Model (NSTM) for motion-corrected MRI reconstruction without the need for priors or external training data. In this approach, the image and motion fields are modeled as continuous functions of spatial and temporal coordinates, with separate learnable time encodings used to disentangle contrast dynamics from motion evolution. The two INRs are trained jointly by minimizing a loss formulated directly on the acquired k-space measurements. Preliminary results in motion simulations and cardiac in vivo data show that NSTM recovers plausible nonrigid motion fields and multi-frame images. In *in vivo* dynamic contrast-enhanced MRI (DCE-MRI), it effectively captures contrast dynamics at a temporal resolution of 1.5 s, highlighting its potential as a tool for robust dynamic MRI reconstruction. The code repository with the full model implementation is available at: https://github.com/nurdinova/nstm_mri_moco

Keywords: Motion correction · Implicit neural representations · Neural space-time modeling

1 Introduction

Magnetic Resonance Imaging (MRI) is susceptible to motion artifacts due to long acquisition times. This may be particularly problematic in high-resolution multi-contrast head imaging, where even small head or neck movements disrupt quantitative analysis. In body imaging, involuntary motions such as respiratory and cardiac motion introduce complex nonrigid deformations that limit achievable resolution and image quality. Although rigid-body models are commonly applied in head MRI, they often break down near the neck, while motion in the body is inherently nonrigid.

Many MRI motion correction techniques rely on navigator signals or external sensors to estimate displacements. Navigators can take the form of low-resolution images [22], dedicated k-space signals [4,11,24], or signals from external devices such as optical tracking systems [25] and pilot tones [13]. However, these approaches often require careful calibration, reduce scan efficiency, and may become intractable when extended to nonrigid motion correction.

To address nonrigid motion without navigation, approaches such as XD-GRASP [8] sort data into respiratory and cardiac bins and jointly reconstruct images, leveraging the periodic nature of motion and binning based on the signal in k-space center over time. Recent deep learning pipelines such as RANGR and Movienet [15,16] automate binning and achieve higher accelerations but need supervised pretraining. Other data-driven methods, including LAP(A)Net [9,12], MedGAN [2], BladeNet [20] have shown strong performance on highly-undersampled data, however, they require pre-training on large high-quality simulated or in vivo datasets.

Recent implicit neural representation (INR) approaches have shown strong performance in dynamic MRI reconstructions [3,10,19,21]. Specifically, the work [6,7] models DCE images as continuous functions of spatiotemporal coordinates, demonstrating promising results for navigator-free dynamic MRI reconstruction without relying on priors or pre-training. Building upon these concepts, we propose an unsupervised neural space-time framework that jointly represents motion fields and image content using INRs. In context of existing INR-based methods, we explicitly separate motion via a dedicated motion network and appropriate regularization, resulting in an interpretable model that disentangles nonrigid motion from other dynamic effects, such as contrast variations. This flexible approach enables modeling complex, aperiodic motion and is broadly applicable across different anatomies, contrasts, and sampling trajectories.

2 Methods

2.1 Neural Space-Time Model

We formulated motion-corrected MRI reconstruction as the joint learning of a static *scene* and dynamic *motion* within a unified neural space-time framework, illustrated in Fig. 1. The motion network, represented by a coordinate-based neural field D_ϕ (a multi-layer perceptron, MLP), predicted displacement vectors at each spatiotemporal coordinate:

$$M(x,y,t) = \begin{bmatrix} \delta x_t & \delta y_t \end{bmatrix}^T = D_\phi(\text{hash}(x,y), \Psi_\omega(t)), \tag{1}$$

where ϕ and ω are learnable parameters, Ψ is a learnable temporal encoding mapping t onto a higher-dimensional feature space, and hash denotes multi-resolution hash encoding of spatial inputs. This encoding enhances capacity and efficiently captures spatial high-frequency details [14]. Spatial and temporal coordinates were normalized to $[0,1]^3$ before encoding.

Inspired by [17], we interpreted the motion mapping as transforming observation-space coordinates into a canonical space. The contrast-modulated 2D MR image at frame index t was modeled by a second neural field F_θ, which predicted complex image intensities from warped spatial coordinates:

$$I(x, y, t) = F_\theta(\text{hash}(x + \delta x_t, y + \delta y_t), \Phi_\eta(t)), \qquad (2)$$

where θ and η are learnable parameters, Φ is a temporal encoding for contrast, and hash encoding was applied to the warped coordinates.

We employed separate temporal embeddings implemented through dedicated temporal MLPs instead of encoding time as an additional spatial dimension. Because the motion and scene networks capture fundamentally distinct temporal dynamics displacement fields and contrast variations, respectively we trained distinct temporal embeddings for each, enabling more effective disentanglement of motion and contrast evolution.

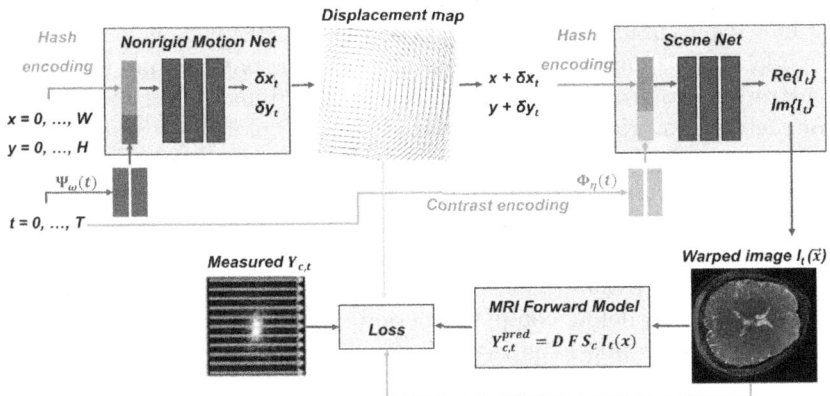

Fig. 1. The Neural Space-Time Model (NSTM) performs motion-corrected MRI reconstruction by jointly learning motion and scene implicit functions, each represented by an MLP. Both INRs take hash-encoded spatial coordinates and temporal features derived from a dedicated MLP embedding, with separate temporal features used for the motion and image networks to capture displacement fields and contrast variations, respectively. The motion network outputs pixel-wise displacement vectors that warp the input coordinates fed to the scene network, which then predicts complex image intensities at these warped locations. The resulting images pass through a multi-coil undersampled MRI forward model, and both networks are trained end-to-end by minimizing the Mean Squared Error (MSE) between the predicted and measured k-space signals.

2.2 MRI Forward Model and Gridding

We trained NSTM by minimizing the mean squared error (MSE) between the measured multi-coil k-space data and the predictions of a differentiable MRI forward model. This forward model operates on temporally resolved images $\hat{\mathbf{x}}$ generated by the scene network, containing both motion and contrast dynamics.

In discrete form by stacking multi-coil and temporal signals and applying operators for coil sensitivities S, the discrete Fourier transform F, and a binary k-space sampling mask P, the predicted k-space signals can be written as:

$$\mathbf{y_{pred}} = PFS\hat{\mathbf{x}}. \tag{3}$$

To accelerate training, we converted the non-Cartesian radial acquisition into Cartesian k-space using implicit GROG [1]. Calibration was performed on a multi-channel 32×32 central k-space region extracted from the first 300 spokes, which was also used to estimate coil sensitivity maps via ESPIRiT [23]. Gridding the full dataset allowed treating k-space data and sampling masks as Cartesian, significantly reducing computational load from approximately 10 s per training epoch to under one second.

2.3 Loss Function and Regularization

We jointly trained the scene and motion networks using a data fidelity term enforcing consistency with the acquired multi-coil k-space data:

$$\mathcal{L}_{\text{data}} = \|\mathbf{y}_{\text{pred}} - \mathbf{y}_{\text{meas}}\|_2^2, \tag{4}$$

where $\mathbf{y}_{\text{pred}}$ and $\mathbf{y}_{\text{meas}}$ denote predicted and measured k-space data with dimensions $N_c \times N_t \times N_x \times N_y$, corresponding to the number of coils, temporal frames, and k-space dimensions.

To stabilize nonrigid motion map estimation, we introduced regularization on the displacement field to encourage spatial smoothness and sparsity:

$$\mathcal{L}_{\text{nonrigid}} = \lambda_m \left(0.1 \cdot \|\nabla_{xy}\mathbf{M}\|_2 + \|\mathbf{M}\|_1 \right), \tag{5}$$

where ∇_{xy} denotes finite differences along spatial dimensions x and y, and displacement field $\mathbf{M}$ was evaluated at all coordinates, hence, had dimensions of $N_t \times N_x \times N_y \times 2$.

The total loss function combined these terms as

$$\mathcal{L}_{\text{total}} = \mathcal{L}_{\text{data}} + \mathcal{L}_{\text{nonrigid}}. \tag{6}$$

2.4 Implementation Details

The nonrigid motion network was implemented as a three-layer MLP with 128 units per layer, while the image network used five layers of the same size. Temporal encodings were represented by two-layer MLPs, of sizes 4 and $3 \times N_t$ for the motion and image networks, respectively. These architectural choices were empirically tuned.

Model parameters were optimized using Adam with a learning rate of 10^{-4}, without scheduling. Training was conducted per slice and typically converged within 3000 iterations.

To improve convergence and stability, we employed a multi-resolution hash encoding strategy for both networks. At each training iteration i, the granularity parameter was set as $\alpha = i$, and the hash features were weighted by

$$w_i = \frac{1}{2} - \frac{1}{2}\cos(\pi \operatorname{trunc}(\alpha N - i)), \tag{7}$$

where trunc truncated its argument to the range $[0, 1]$. This annealing scheme encourages the models to first learn coarse, global structures, followed by higher-resolution features and finer local displacements.

For datasets with heavier undersampling, stronger regularization was applied, initialized at $\lambda_m \approx 10^{-4}$ and linearly reduced to $\lambda_m \approx 10^{-6}$ during training.

2.5 Experiments

To validate NSTM, we used a free-breathing liver DCE dataset from https://cai2r.net/resources. The dataset consisted of 12 slices of a 3D golden-angle radial stack-of-stars acquisition, with data from 8 coils, matrix size 256×256, and 1144 spokes per slice, resulting in a total scan time of approximately 180 s.

To evaluate the nonrigid displacement fields estimated by our method, we simulated multi-coil k-space data using images from four motion states and eleven contrast phases obtained from XD-GRASP reconstructions. We used the corresponding coil sensitivity maps and undersampling masks with 25 radial spokes per frame. This controlled setup enabled us to quantitatively compare the motion fields estimated by our model against displacement maps derived from registering the XD-GRASP reference images.

To evaluate NSTM displacement fields in vivo, we used a publicly available fully sampled cardiac cine dataset [5]. Raw k space measurements of 200 spokes per frame were acquired at 25 cardiac phases and 16 coil channels. The non Cartesian kspace data were gridded, coil sensitivity maps were estimated from the central 32×32 region, and a root sum of squares coil combination was performed for the fully sampled reconstruction. For NSTM, the fully sampled radial spokes were uniformly undersampled to 50 spokes per frame, using different sets of spokes at consecutive frames.

For in vivo evaluation, we applied the proposed method at a temporal resolution of 8 spokes per frame (approximately 1.4 s) and compared it to standard Conjugate Gradient SENSE (CG-SENSE) [18] and XD-GRASP reconstructions. XD-GRASP was performed at a temporal resolution of 25 spokes per frame (approximately 4 s) using publicly available code (https://cai2r.net/resources/xd-grasp-matlab-code/). For the three methods, images were reconstructed across multiple contrast frames, and signal intensity-time curves were extracted from manually segmented regions-of-interest (ROIs) in the aorta (AO) and portal vein (PV) to facilitate quantitative evaluations.

3 Results

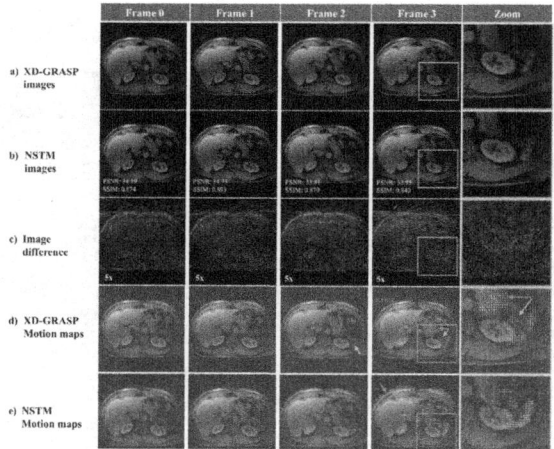

Fig. 2. Validation of NSTM on multi-coil undersampled k-space simulations using multi-contrast, motion-resolved XD-GRASP images. Reconstructions shown for contrast index 4 (of 11) across four motion states: a) XD-GRASP, b) NSTM, c) difference maps (x5). PSNR and SSIM confirm NSTM closely matches XD-GRASP images. Motion fields from d) XD-GRASP registration and e) NSTM show similar trends, with yellow arrows marking missed motion and orange arrows marking overestimated displacement.

3.1 Simulations

Figure 2 shows reconstructed images at contrast index 4 (out of 11) across four motion states for (a) XD-GRASP and (b) NSTM, along with their difference maps in (c). The difference maps, scaled by a x5 factor, reveal mostly noise-like variations between the two reconstruction methods, with slightly greater discrepancies in the contrast-enhanced kidney region, and with more differences at motion frame 4. Structural similarity (SSIM) and peak signal-to-noise ratio (PSNR) metrics calculated between (a) and (b), ranging from 0.84–0.90 and 33–35 respectively, indicate that the two methods produce highly similar reconstructions across motion frames.

We further evaluated the displacement fields by registering all XD-GRASP motion images to the first motion frame and comparing these fields to those directly obtained from NSTM, which were similarly normalized to the first time frame. Overlaid displacement maps in (d) and (e) of Fig. 2 show overall agreement between the two methods. However, noisy estimates are evident in the anterior subcutaneous fat with NSTM (orange arrows) and above the left kidney with the registration-based approach (also highlighted with orange arrows). Additionally, some displacements visible in the reconstructed images were not fully captured by either method, as indicated by yellow arrows.

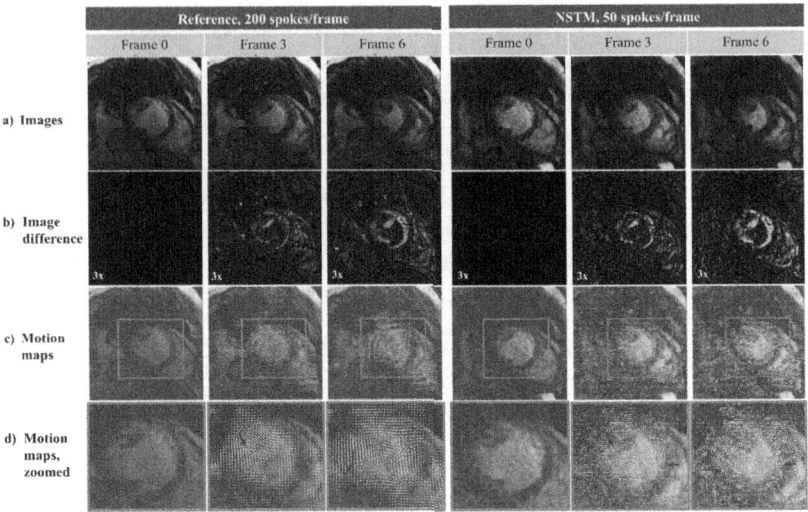

Fig. 3. Validation of NSTM on an in vivo cardiac cine dataset from [5]: Reconstructions from a fully-sampled radial acquisition are compared to undersampled (R=4) NSTM reconstructions. Panels a) and b) show cine frames and their differences relative to the reference frame = 0. Panels c) and d) present motion maps from registered fully-sampled frame images and motion maps derived from NSTM. Both reconstructed images and motion maps from NSTM are matching the fully-sampled results (green arrows), although noise is evident in both the motion and image estimates.

3.2 In Vivo

For in vivo evaluation, NSTM was applied to an undersampled cardiac cine dataset [5] with 50 radial spokes/frame (R=4) and compared with the fully-sampled reference (200 spokes/frame). As shown in Fig. 3, NSTM reconstructions exhibit image quality close to that of the fully-sampled data, although residual undersampling noise is visible. Image differences relative to frame = 0 are structurally similar to the reference, but show slightly higher and more blurred residuals for NSTM. Motion fields estimated by NSTM remain noisy and less spatially smooth than those from registration, yet still capture displacement trends consistent with the reference, as indicated by the green arrows.

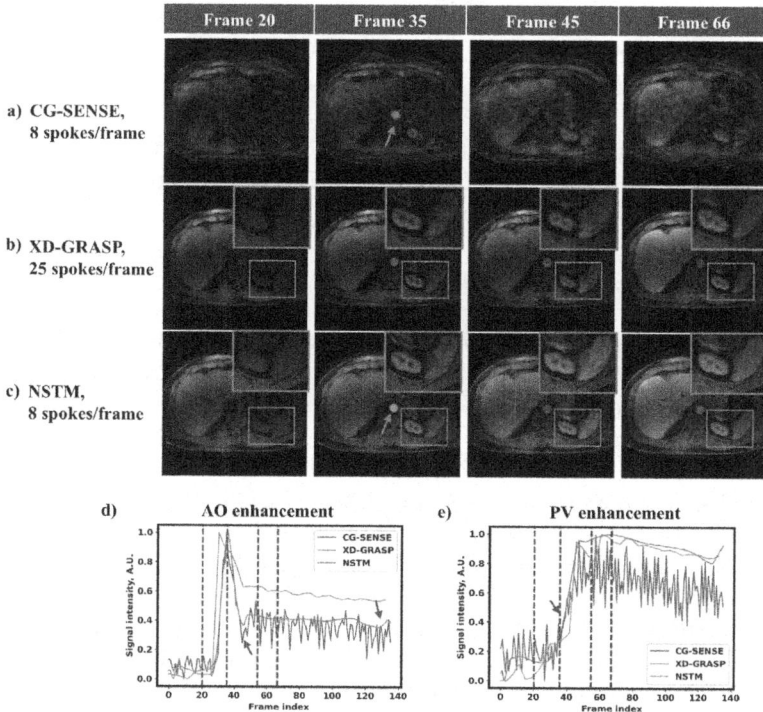

Fig. 4. Validation of NSTM on an in vivo free-breathing DCE liver dataset. Panels a) and c) show CG-SENSE and NSTM reconstructions at 8 spokes/frame, compared to the reference b) XD-GRASP at 25 spokes/frame. While NSTM images are generally noisier than XD-GRASP, they maintain good fidelity. Contrast differences at frames 35 and 45 are due to different temporal resolutions. Signal intensity-time curves for the segmented aorta (AO) and portal vein (PV) in d) and e) show NSTM closely follows CG-SENSE but remains smooth. Green arrows mark areas where NSTM captures CG-SENSE dynamics well, while red arrows indicate edge effects where NSTM may overfit. Motion fields estimated by NSTM were zeros, not resolved at the achieved temporal rate of 1.5 s. (Color figure online)

We present preliminary validation of NSTM on an in vivo liver DCE dataset in Fig. 4. Due to failure to reconstruct images with XD-GRASP at higher temporal rates, we used images reconstructed with CG-SENSE (a) as a reference for signal intensitytime curves at a higher temporal resolution of 8 spokes per frame. Images from (b) XD-GRASP and (c) NSTM show that NSTM reconstructions at 8 spokes per frame are noisier than XD-GRASP at 25 spokes per frame, yet still demonstrate reasonable overall fidelity. The image contrast at frames 35 and 45 does not perfectly align between the two methods because of their differing temporal resolutions. However, the NSTM contrast in the aorta region (green arrow) closely matches the CG-SENSE reconstruction across all frames. This is further highlighted by the signal intensitytime curves in (d) for the aorta and

(e) for the portal vein. Across most frame indices, the NSTM contrast in both segmented areas agrees well with CG-SENSE estimates (green arrows), except at the first and last few frames where edge effects are observed with NSTM (red arrows).

In this case study, the motion fields estimated by NSTM were effectively zero, suggesting that the achieved temporal resolution of 8 spokes (approximately 1.4 s) was insufficient to detect motion dynamics.

4 Discussion and Conclusion

We presented preliminary validation of a neural space-time model (NSTM) designed for motion-corrected MRI reconstruction. Our framework addresses the reconstruction problem by jointly learning displacement fields and image content through coordinate-based neural networks. Simulation studies demonstrated the model's ability to distinguish between nonrigid displacement and other temporal dynamics, such as contrast evolution. On in vivo cardiac cine data, NSTM produced reasonable motion maps and image reconstructions, though further improvements are needed for smoother displacement fields and reduced noise. When applied to free-breathing liver dynamic contrast-enhanced MRI (DCE-MRI), the NSTM successfully captured contrast dynamics, highlighting its potential as a robust yet straightforward approach for fully unsupervised dynamic MRI reconstruction, maintaining inherent spatiotemporal continuity.

In the current study, the achieved temporal resolution was insufficient for reliably estimating motion fields from free-breathing data. This indicated the need for higher temporal resolution reconstructions (e.g., at 13 spokes per frame). Future work will therefore prioritize implementing these enhanced temporal resolutions, optimizing GPU utilization, and achieving practical training times. It will also be important to test the method on other body regions and imaging types, such as brain-neck MR Fingerprinting (MRF) and breast DCE imaging.

In context of previous motion-informed INR methods in MRI, which encoded temporal changes implicitly or through latent variables [7,8], our proposed approach explicitly models displacement fields. This explicit representation potentially facilitates flexible regularization, enables more interpretable motion analysis, and reduces the complexity of learning motion-related variability within the image network. Nevertheless, further research is required to enhance robustness of this framework, identify optimal time encodings for accurately modeling motion dynamics, and refine training strategies. Additionally, appropriate image regularization methods are critical for improving the optimization process and quality of the resulting motion fields and reconstructed images.

Disclosure of Interests. The authors have no competing interests in the paper.

References

1. Abraham, D., Nishimura, M., Cao, X., Liao, C., Setsompop, K.: Implicit representation of grappa kernels for fast MRI reconstruction (2023)

2. Armanious, K., et al.: MedGAN: medical image translation using GANs. Comput. Med. Imaging Graph. **79**, 101684 (2018). the official journal of the Computerized Medical Imaging Society
3. Baik, D., Yoo, J.: Dynamic-aware spatio-temporal representation learning for dynamic MRI reconstruction. arXiv preprint arXiv:2501.09049 (2025). https://arxiv.org/abs/2501.09049
4. Cheng, J., et al.: Free-breathing pediatric MRI with nonrigid motion correction and acceleration. J. Magn. Reson. Imaging **42** (2015)
5. El-Rewaidy, H.: Replication data for: Multi-domain convolutional neural network (MD-CNN) for radial reconstruction of dynamic cardiac MRI (2020). https://doi.org/10.7910/DVN/XJXQRX, version 1
6. Fend, J., et al.: Calibration-free DCE-MRI with sub-second temporal resolution using interpretable implicit neural representation. In: Proceedings of the 33nd Annual Meeting of ISMRM. Hawaii (2025). abstract #0123
7. Feng, J., et al.: Spatiotemporal implicit neural representation for unsupervised dynamic MRI reconstruction. IEEE Trans. Med. Imaging **44**, 2143–2156 (2022)
8. Feng, L., Axel, L., Chandarana, H., Block, K.T., Sodickson, D.K., Otazo, R.: XD-GRASP: golden-angle radial MRI with reconstruction of extra motion-state dimensions using compressed sensing. Magn. Reson. Med. **75** (2016)
9. Ghoul, A., et al.: Highly efficient non-rigid registration in k-space with application to cardiac magnetic resonance imaging. ArXiv abs/2410.18834 (2024)
10. Huang, W., et al.: Subspace implicit neural representations for real-time cardiac cine MR imaging. arXiv:2412.12742 (2024). https://doi.org/10.48550/arXiv.2412.12742
11. van der Kouwe, A., Benner, T., Dale, A.M.: Real-time rigid body motion correction and shimming using cloverleaf navigators. Magn. Reson. Med. **56** (2006)
12. Kustner, T., et al.: LAPNet: non-rigid registration derived in k-space for magnetic resonance imaging. IEEE Trans. Med. Imaging **40**, 3686–3697 (2021)
13. Ludwig, J., Speier, P., Seifert, F., Schaeffter, T., Kolbitsch, C.: Pilot tone–based motion correction for prospective respiratory compensated cardiac cine MRI. Magn. Reson. Med. **85**, 2403–2416 (2020)
14. Müller, T., Evans, A., Schied, C., Keller, A.: Instant neural graphics primitives with a multiresolution hash encoding. ACM Trans. Graph. (TOG) **41**, 1–15 (2022)
15. Murray, V., et al.: Movienet: deep space–time-coil reconstruction network without k-space data consistency for fast motion-resolved 4D MRI. Magn. Reson. Med. **91**, 600–614 (2023)
16. Nario, J.J.Q., Murray, V., Mekhanik, A., Otazo, R.: RANGR: deep learning auto navigation of free-breathing golden-angle radial abdominal MRI. In: Proceedings of the 32nd Annual Meeting of ISMRM. Singapore (2024). abstract #2794
17. Park, K., et al.: Nerfies: deformable neural radiance fields. In: 2021 IEEE/CVF International Conference on Computer Vision (ICCV), pp. 5845–5854 (2020)
18. Pruessmann, K.P., Weiger, M., Scheidegger, M.B., Boesiger, P.: SENSE: sensitivity encoding for fast MRI. Magn. Reson. Med. **42** (1999)
19. Shao, H.C., Mengke, T., Deng, J., Zhang, Y.: 3D cine-magnetic resonance imaging using spatial and temporal implicit neural representation learning. Phys. Med. Biol. **69**(9), 095007 (2024). https://doi.org/10.1088/1361-6560/ad33b7
20. Shimron, E., et al.: BladeNet: rapid propeller acquisition and reconstruction for high spatio-temporal resolution abdominal MRI. ISMRM Annual Meeting
21. Spieker, V., et al.: PISCO: Self-supervised k-space regularization for improved neural implicit k-space representations of dynamic MRI. arXiv:2501.09403 (2025). https://doi.org/10.48550/arXiv.2501.09403

22. Tisdall, M.D., Hess, A.T., Reuter, M., Meintjes, E.M., Fischl, B.R., van der Kouwe, A.: Volumetric navigators for prospective motion correction and selective reacquisition in neuroanatomical MRI. Magn. Reson. Med. **68** (2012)
23. Uecker, M., et al.: ESPIRiT—an eigenvalue approach to autocalibrating parallel MRI: where sense meets grappa. Magn. Reson. Med. **71** (2014)
24. Ulrich, T., Riedel, M., Pruessmann, K.P.: Servo navigators: linear regression and feedback control for rigid-body motion correction. Magn. Reson. Med. **91**, 1876–1892 (2024)
25. Zaitsev, M., Maclaren, J.R., Herbst, M.: Motion artifacts in MRI: a complex problem with many partial solutions. J. Magn. Reson. Imaging **42** (2015)

Proceedings of Graphs in BiomedicAl Image anaLysis (GRAIL), Topology- and Graph-Informed Imaging Informatics (TGI), and Hypergraph Computation for Medical Image Analysis (HGMIA)

Skip Priors and Add Graph-Based Anatomical Information, for Point-Based Couinaud Segmentation

Xiaotong Zhang, Alexander Broersen, Gonnie C.M. van Erp, Silvia L. Pintea, and Jouke Dijkstra(✉)

Radiology Department, Leiden University Medical Center, Albinusdreef 2, 2333 Leiden, ZA, The Netherlands
j.dijkstra@lumc.nl

Abstract. The preoperative planning of liver surgery relies on Couinaud segmentation from computed tomography (CT) images, to reduce the risk of bleeding and guide the resection procedure. Using 3D point-based representations, rather than voxelizing the CT volume, has the benefit of preserving the physical resolution of the CT. However, point-based representations need prior knowledge of the liver vessel structure, which is time consuming to acquire. Here, we propose a point-based method for Couinaud segmentation, without explicitly providing the prior liver vessel structure. To allow the model to learn this anatomical liver vessel structure, we add a graph reasoning module on top of the point features. This adds implicit anatomical information to the model, by learning affinities across point neighborhoods. Our method is competitive on the *MSD* and *LiTS* public datasets in Dice coefficient and average surface distance scores compared to four pioneering point-based methods. Our code is available at https://github.com/ZhangXiaotong015/GrPn.

Keywords: Couinaud segmentation · 3D graph reasoning · Point net

1 Introduction

Effective treatment for primary liver cancer relies on two main procedures: liver resection and radiofrequency ablation [12]. Both approaches depend on accurate Couinaud segmentation, to reduce the risk of main vessel puncture, and to guide the placement of ablation needles [1,13]. Couinaud segmentation divides the liver into eight functionally independent segments. The right-, middle- and left-hepatic veins divide the liver into four sections. These sections are then further split by the horizontal plane defined by the portal vein, as shown in Fig. 1.

Prior work for automatically Couinaud segmentation, voxelizes the liver CT to be used in 3D convolutional neural networks (CNNs) [3,9,21,22]. More recently [28], computes point embeddings from sampled 3D points within the liver area. Using sample 3D points has the added value that they preserve the

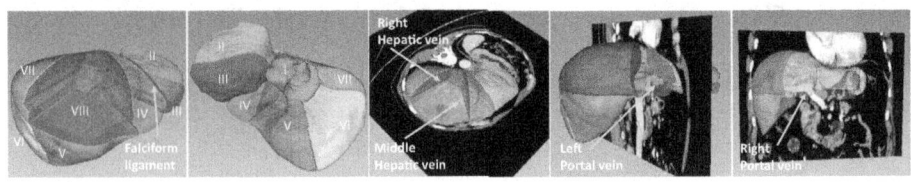

Fig. 1. Couinaud segmentation is challenging because it requires prior knowledge of the liver vessels. I–VIII indicate different Couinaud segments.

physical CT resolution, without the need to resize or crop along the axial direction. Therefore, here we restrict our focus to point-based methods. While relying on 3D point-based representations, our proposed method does not need prior liver vessel information, unlike Zhang et al. [28]. Yet, without prior liver vessel knowledge, we lose anatomical information. To incorporate this anatomical information, we add a graph reasoning module, learning affinities between the 3D point embeddings.

To summarize: (i) We propose a 3D point-based method for Couinaud segmentation that *removes the need for prior knowledge* of the liver vessel structure; (ii) We *extend 2D graph reasoning to a 3D version*, and use it to learn affinities between points along the liver, thus, adding implicitly anatomical liver structure; (iii) We evaluate on two public datasets: *MSD* and *LiTS* and *show competitive accuracy* when compared to four popular point-based segmentation methods.

2 Related Work

Couinaud Segmentation. Prior work on automatic Couinaud segmentation creates liver atlases [5,16], divides the liver into voxels to be used in 3D convolutional neural networks (CNNs) [3,9,21,22], or builds deep models on top of sampled 3D points [28]. Atlas-based and partial CNN-based methods [3,5,16,22] require manual landmarks along the hepatic veins, whereas the other prior-free CNN-based methods [9,21] on CT images need to resize the CT volume to a fixed grid size which changes the physical resolution of the CT images. Point-based models [28] address the limitation of voxelized methods, while still requiring prior liver vessel information. Here, we build on 3D point-based methods, while discarding the need for prior anatomical information, and learning this implicitly via dynamic graph reasoning.

Dynamic Graph Reasoning. Dynamic graph reasoning is widely used in both image-based [7,15,23,27] as well as point-based semantic segmentation methods [14,24,29], to capture long-range dependencies. Most image-based methods [7,15,23] consider all position pairs when calculating affinities, resulting in high complexity. Unlike these methods, DGMN [27] proposed an adaptive sampling method that considers only limited positions. Similarly, point-based methods also suffer from high complexity of affinity calculation. K-nearest neighbor (k-NN) is typically used for the complexity reduction, as in [24,29]. Alternatively, Ma

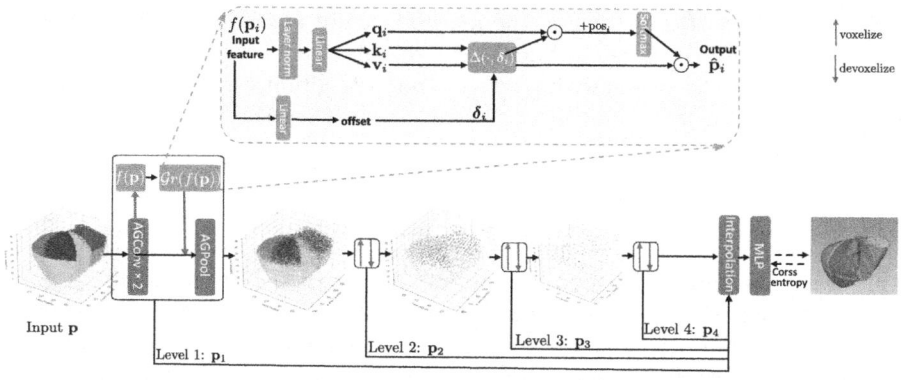

Fig. 2. Network architecture. We build on the design of [25], and extend this with the green blocks: grid feature embeddings, $f(\mathbf{p})$ adding anatomical information; and graph reasoning model, $\mathcal{G}r(f(\mathbf{p}))$, learning dynamic affinities in neighborhood areas.

et al. [14] propose to learn channel dependencies instead of dependencies between nodes to capture global contextual information while reducing the computations. Here, we take advantage of both k-NN in the point domain, and the adaptive sampling method in [27] to reduce computations.

3 Couinaud Liver Segmentation

Our model starts from a set of 3D points, $\mathbf{p}$, sampled from the liver region over the complete CT volume, and their associated intensities. We follow the design of Adaptive Graph CNN (AGCNN) [25], processing the points at four levels, as in Fig. 2. We extend AGCNN with the green blocks: the grid feature embeddings, $f(\mathbf{p})$, enhancing the anatomical information; and graph reasoning model, $\mathcal{G}r(f(\mathbf{p}))$, which dynamically learns affinities across points. At the last level, we interpolate the point embeddings, $\mathbf{p}_4$, and feed the result to an MLP with two layers ($\{64, 8\}$). We predict the eight Couinaud segments, and use the cross-entropy loss to train the model.

3.1 Relation to Adaptive Graph CNN

At the first level, AGCNN starts from a set of 3D points, $\mathbf{p}_1$, and their associated CT intensities. Before training, AGCNN precomputes a set of $K \ll N_i$ neighbors for each point at different levels $i \in \{1, 2, 3, 4\}$, where N_i is the number of points at level i. We compute the K neighbors using a ball-query sampling scheme, with radius r, as in Fig. 3. At the first level the neighbors of point j, $\mathbf{p}_1^{ne(j)}$, are taken from the initial set of points $\mathbf{p}_1$, while at lower resolution levels, the neighbors are taken from the previous level points, $\mathbf{p}_{i-1}, i \in \{2, 3, 4\}$.

To add implicit anatomical vessel structure to AGCNN, we use a graph reasoning module $\mathcal{G}$. Specifically, at each level i we voxelize all N_i points, $\mathbf{p}_i$ into a grid of size $[M_i \times M_i \times M_i]$, using the method of Liu et al. [11]. In this voxelized space, the graph reasoning module $\mathcal{G}r(\cdot)$ dynamically learns the informative voxels. And it selects 3^3 voxels with which to compute affinities. Intuitively, AGCNN computes affinities of points in the current level with the previous level, while our graph reasoning module computes affinities among the current level points.

3.2 Graph-Based Implicit Anatomical Information

Grid Feature Embeddings. The inputs to the model are a set of 3D points in $\mathbb{R}^3$, together with their corresponding CT intensity values. We revoxelize the point features and obtain the voxelized point embeddings, $f(\mathbf{p}_i)$ at each level, i, by performing two 3D residual convolutions.

Dynamic Graph Reasoning. The grid-space $[M_i \times M_i \times M_i]$ at each level i contains many voxels $\left(\geq \left(\frac{32}{2^{i-1}}\right)^3\right)$ at different levels. This causes memory bottlenecks when calculating the affinity matrix. To address this, we draw inspiration from DGMN [27] and extend their method from 2D to 3D. Specifically, we consider only a subset of $3^3 (= 27)$ voxels out of the $[M_i \times M_i \times M_i]$ grid, when computing affinities. Following DGMN [27], we use self-attention [10] to define affinities between voxelized points. Given the voxelized point feature $f(\mathbf{p}_i)$ at level i, we first project this to *query* $\mathbf{q}_i$, *key* $\mathbf{k}_i$, and *value* $\mathbf{v}_i$, via a shared linear layer. Additionally, we extend the offsets to learnable 3D offsets $\boldsymbol{\delta}_i$ pointing to a set of 3^3 voxels. These 3^3 voxels, to which the offsets are pointing, should contain all the useful anatomical information encoded by neighboring voxels. Similar to DGMN [27], we use a deformable unfold layer [30], $\Delta(\cdot, \boldsymbol{\delta}_i)$, to adapt the keys and values—$\mathbf{k}_i$, $\mathbf{v}_i$. The output of the graph reasoning is simply a 3D self-attention block over deformed keys and values, as shown in Fig. 2:

$$\hat{\mathbf{p}}_i = \text{softmax}(\mathbf{q}_i \cdot \Delta(\mathbf{k}_i, \boldsymbol{\delta}_i) + \text{pos}_i) \cdot \Delta(\mathbf{v}_i, \boldsymbol{\delta}_i), \quad (1)$$

where $\boldsymbol{\delta}_i = \text{Linear}(f(\mathbf{p}_i))$, $\boldsymbol{\delta}_i \in \mathbb{R}^{3 \times 3^3}$, $\text{pos}_i \in \mathbb{R}^{3^3}$ are the positional embeddings for the query $\mathbf{q}_i$ of input features [20]. Finally, we devoxelize $\hat{\mathbf{p}}_i$ back to point representations using the coordinate-based interpolation [11].

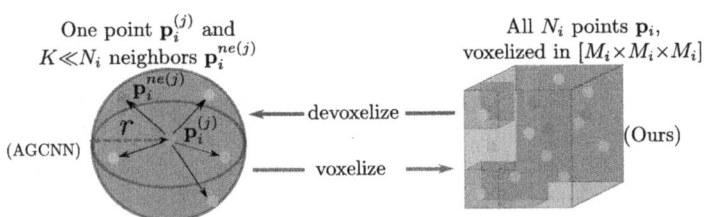

Fig. 3. Left: AGCNN computes offline a set of $K \ll N_i$ neighbors $\mathbf{p}_i^{ne(j)}$ for each point indexed by j, $\mathbf{p}_i^{(j)}$ in level i. **Right:** Our approach voxelizes all N_i points, $\mathbf{p}_i$ in level i, into a grid $[M_i \times M_i \times M_i]$, and dynamically focuses on 3^3 voxels when learning affinities.

4 Experiments on Couinaud Segmentation

Dataset Description. We evaluate our method on two public datasets: *MSD* [2] and *LiTS* [4]. Given that there are no Couinaud segment annotations in these two datasets, we use the annotations of Tian et al. [21] and Zhang et al. [28]. The *MSD* and *LiTS* datasets contain 192 and 131 annotated CT scans, respectively, with in-plane resolutions of 0.57–0.98 mm and 0.56–1.00 mm, and interplanar resolutions of 1.25–7.50 mm and 0.70–5.00 mm.

Implementation Details. We reoriented all CT scans to left-posterior-inferior (LPI) and maintain the original CT image origin and spacing for all experiments. We divide both datasets into training/validation/test sets following the ratio 10/3/7. The CT values are truncated to the range of $[-100, 300]$ Hounsfield units and then normalized to $[0, 1]$. We consider each voxel, $\mathbf{v}$, of a CT scan as a point, and compute the point coordinates as: $\mathbf{p}=sd\mathbf{v}+o$ where s, d and o are the physical spacing, direction and origin parameters recorded in the CT. We also normalize the physical coordinates of the points to $[0, 1]$. The r in the ball-query sampling (Fig. 3) is equal to $\frac{1}{2 \cdot 64}$ and $\frac{1}{2 \cdot 32}$, and the first-scale grid size is 64^3, and 32^3 for the *MSD* and *LiTS* datasets, respectively. The number of neighbors ($K \leq 100$) of a point depends on the r used in ball-query sampling. We use the same point down-sampling ratio as AGCNN [25]. We randomly sample 10% points ($\approx$50K) for each training iteration. We use 400 epochs, the SGD optimizer with a momentum of 0.98, and a learning rate of 0.01. For all the experiments, we use an NVIDIA A100 (40GB) GPU. We consider point-based baselines: PointNet [17], PointNet++ [18], AGCNN [25], and Zhang et al.'s [28] method, using their default settings and we follow their official implementation.

Evaluation Metrics. We evaluate all methods only on the liver region, by masking out other areas. The same liver masks were used for all experiments. We report *Dice* coefficient and average surface distance (*ASD*) in our evaluation. We use *Torchmetrics* [8] and *MONAI* [6] to calculate the metrics. These results are different from Zhang et al. [28], because they use their own implementation for the metrics. We also report inference times and GFLOPs [26].

4.1 Quantitative Evaluation

Tables 1 and 2 show the quantitative comparison in *Dice* coefficient and *ASD* scores. In Table 1, our method achieves the highest *Dice* score for each Couinaud segment. Both our method and *PointNet* have lower *ASD* scores compared to the other methods. *PointNet* has a low *ASD* score because it uses a voting scheme at inference to reduce false positives. However, both *PointNet++* and *PointNet* have large segmentation errors for segment (I). This may be due to the segment having the lowest volume fraction in the liver. In Table 2, our method has the highest *Dice* coefficient and the lowest *ASD* averaged over all segments. This is especially positive, given the high axial resolution (*i.e.* $\leq$1.0 mm) of the *LiTS* dataset, which increases the voxel-wise class imbalance.

PointNet++, AGCNN, and our method use multi-scale point sampling. This results in longer inference times compared to PointNet and Zhang et al., which

Table 1. Quantitative evaluation on *MSD*: We report results per segments (I – VIII), as well as the average. Our method achieves the highest average in *Dice* coefficient, and comparable average in *ASD* with *PointNet*, demonstrating the effectiveness of the added implicit knowledge. (We denote with * the use of extra vessel-priors.)

	MSD (Dice %) ↑					MSD (ASD mm) ↓				
	PointNet [17]	PointNet++ [18]	AGCNN [25]	Zhang et al. [28]*	Ours	PointNet [17]	PointNet++ [18]	AGCNN [25]	Zhang et al. [28]*	Ours
(I)	62.26	72.26	63.96	80.45	**83.31**	4.04	5.01	2.76	2.96	**1.97**
(II)	81.53	80.27	77.79	82.39	**86.71**	1.67	**1.02**	6.19	4.75	1.89
(III)	69.80	72.75	64.55	75.06	**79.35**	3.31	**1.96**	5.78	3.30	2.51
(IV)	61.74	68.04	63.26	69.83	**73.26**	5.76	5.90	6.52	5.46	**4.12**
(V)	70.55	71.78	68.78	71.49	**75.89**	**3.56**	5.61	11.75	5.91	6.36
(VI)	75.88	75.29	69.91	72.51	**79.51**	**2.68**	4.57	8.20	5.23	5.15
(VII)	82.38	82.24	80.35	80.56	**85.11**	**2.47**	3.02	5.62	5.57	3.27
(VIII)	75.86	76.03	73.72	75.74	**80.16**	**4.10**	4.19	6.79	4.82	4.84
Avg	72.50	74.83	70.29	76.00	**80.41**	3.45	3.91	6.70	4.75	3.76
	Time (s) per case					GFLOPs				
	2.81	21.55	6.13	1.73	10.03	302.38	64.65	638.54	608.85	771.15

do not employ multi-scale point sampling. The GFLOPs of AGCNN and our method depend on the radius used in the ball-query sampling.

4.2 Qualitative Evaluation

We visualize the Couinaud segmentations on the *MSD* and *LiTS* datasets, as shown in Figures (4-5). In Fig. 4, we show three different cases with varying axial spacing from 1.5 mm to 7.5 mm, in the *MSD* dataset. For the cases with lower axial resolution (5 mm and 7.5 mm), we show the plane in axial view, and use the red dotted bounding-box to highlight the boundaries between segments. We also show one case with relatively high axial resolution (1.5 mm) in a coronal view, in the last row of Fig. 4. All boxes are on the same location, in the corresponding CT image. On the last row, both our method and Zhang *et al.*'s [28] boundaries follow the box centerline. However, unlike Zhang *et al.*'s [28], our method correctly recognizes segment (III) in this case, as seen in the right-upper corner.

In Fig. 5, we show the boundary comparison for three cases in three views (axial, coronal and sagittal). Similar to the results in Fig. 4, the boundaries of our segments are located on the box centerline. In the second and third rows, ours method show the most precise predictions.

4.3 Model Ablation Experiments

To verify that all the components of our model contribute to the segmentation, we perform model ablations in Table 3. We consider four cases: (a) the baseline AGCNN [25]; (b) our method with grid feature embeddings, $f(\mathbf{p})$, but without graph reasoning, $\mathcal{G}r(f(\mathbf{p}))$; (c) our model with graph reasoning, $\mathcal{G}r(f(\mathbf{p}))$, but

Table 2. Quantitative evaluation on *LiTS*: Results per Couinaud segment (I – VIII) and the overall average. Our method, without prior vessel knowledge, achieves the highest average in *Dice* coefficient and the lowest average in *ASD*.

	LiTS (Dice %) ↑					LiTS (ASD mm) ↓				
	PointNet [17]	PointNet++ [18]	AGCNN [25]	Zhang et al. [28]*	Ours	PointNet [17]	PointNet++ [18]	AGCNN [25]	Zhang et al. [28]*	Ours
(I)	49.80	37.85	69.20	**73.64**	68.84	8.39	8.58	**4.36**	6.54	6.97
(II)	70.69	72.78	82.62	82.82	**86.17**	9.39	5.31	4.58	4.81	**3.18**
(III)	58.26	65.68	76.09	72.46	**80.82**	15.18	7.42	6.04	5.09	**3.18**
(IV)	53.87	70.19	73.88	**75.40**	75.24	8.46	9.69	10.00	**7.88**	8.38
(V)	80.46	80.51	78.97	81.92	**83.03**	**5.04**	5.85	7.25	6.49	6.20
(VI)	77.69	79.23	73.45	79.28	**79.29**	4.55	6.02	5.76	6.67	**3.78**
(VII)	79.92	81.71	80.42	**83.40**	82.79	**4.38**	4.93	6.75	4.63	4.60
(VIII)	77.20	79.29	77.53	80.02	**80.26**	6.45	6.67	7.71	**5.84**	7.71
Avg	68.49	70.90	76.52	78.62	**79.56**	7.73	6.81	6.56	6.00	**5.50**
	Time (s) per case					GFLOPs				
	10.62	106.97	19.81	6.52	13.83	302.38	64.65	256.10	608.85	141.56

Table 3. Model ablation study in *Dice* coefficient and average surface distance (*ASD*) on the *MSD* [2] and *LiTS* [4] datasets. (a) AGCNN [25] baseline; (b) without graph reasoning, $\mathcal{G}r(f(\mathbf{p}))$, in our model; (c) without grid feature embeddings, $f(\mathbf{p})$, in our proposed model; (d) our proposed model. All our model components contribute to the final model's scores.

	MSD		LiTS	
	Dice ↑ (%)	ASD ↓ (mm)	Dice ↑ (%)	ASD ↓ (mm)
(a)	70.29 (± 16.69)	6.70 (± 5.80)	76.52 (± 8.90)	6.56 (± 2.65)
(b)	78.02 (± 11.06)	4.00 (± 3.39)	77.46 (± 10.87)	6.58 (± 4.03)
(c)	68.98 (± 14.58)	7.73 (± 5.19)	72.67 (± 10.30)	9.81 (± 5.44)
(d)	**80.41 (± 10.74)**	**3.76 (± 4.04)**	**79.56 (± 6.95)**	**5.50 (± 2.42)**

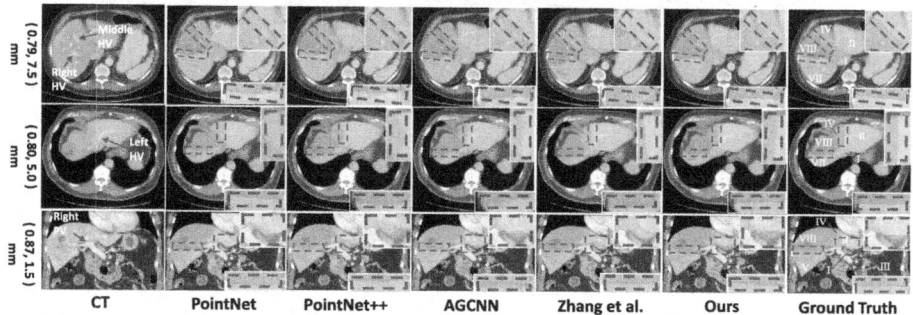

Fig. 4. Qualitative evaluation on *MSD*. The anatomical landmarks [19] are marked with red arrows in the CT image, as a reference. The red dotted bounding-box highlights the boundary between segments. Our method predicts segment boundaries that are closest to the ground truth. We show to the left the in-plane and interplane resolutions. (HV: heaptic vein, PV: portal vein)

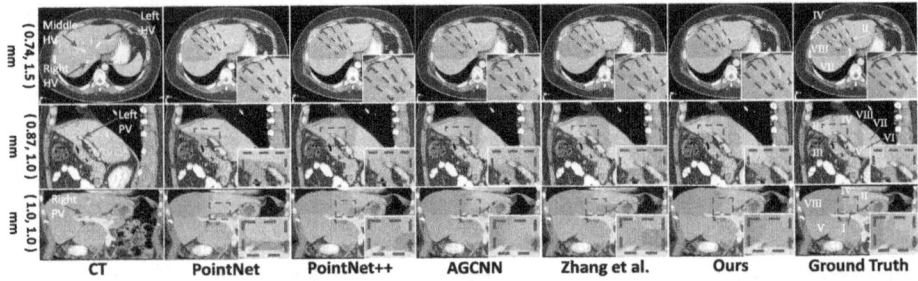

Fig. 5. Qualitative evaluation on *LiTS*. Three cases in axial, sagittal, and coronal views. The red arrows mark the anatomical landmarks [19], and the bounding-boxes highlight segment boundaries. Our method displays the most accurate boundaries.

without grid feature embeddings, $f(\mathbf{p})$; (d) our complete model. Grid feature embeddings $f(\mathbf{p})$ are an important bridge between features in 3D point-space, $\mathbf{p}$, and the graph reasoning module $\mathcal{G}r(f(\mathbf{p}))$. As seen for the setting (c) in Table 3, the graph reasoning module fails to work directly, on the voxelized point-features. In addition, the graph reasoning module in itself contributes to the model predictions, as seen in the setting (b) of Table 3. Removing $\mathcal{G}r(f(\mathbf{p}))$ decreases the dice scores by $\approx 2\%$ on both the *MSD* and *LiTS* datasets. On both the *LiTS* data and the *MSD* dataset, all model components prove useful.

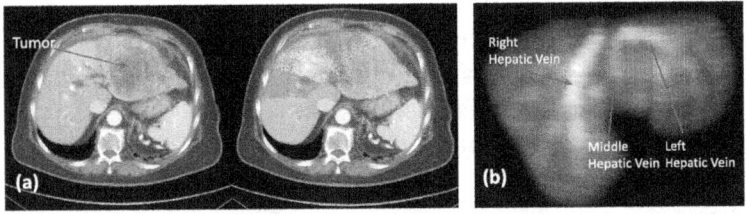

Fig. 6. Failure case analysis. (a) The anatomical landmarks disappear in the tumor region, marked by the red arrow. (b) Output of $\mathcal{G}r(f(\mathbf{p}))$ of a successful case. The model learns the graph affinities based on the anatomical landmarks in the CT images.

5 Discussion and Conclusion

The proposed method does not need explicit vessel priors, unlike [28]. Moreover, as shown in Fig. 6 (b), it can learn anatomical liver landmarks in the graph $\mathcal{G}r(f(\mathbf{p}))$. However, the proposed method cannot accurately discriminate segments in the liver when a large tumor appears, as shown in Fig. 6 (a). Here, the partial anatomical landmarks in the CT image are absent, such as the left hepatic vein and the left portal vein. To conclude, we propose a 3D point-based method for Couinaud segmentation for CT images. Our model incorporates implicitly anatomical information, by learning affinities between voxels in the CT volume in a dynamic graph reasoning module. This implicit anatomical information makes our model competitive compared to prior point-based methods, where we exceed *PointNet*, *PointNet++*, *AGCNN* and Zhang *et al.*'s [28] method on the *MSD* and *LiTS* public benchmarks. This approach avoids the time-consuming definition of the prior vessel structure in the liver, while still showing competitive performance.

Acknowledgments. This work was supported by China Scholarship Council under Grant 202108310010.

References

1. Alirr, O.I., Rahni, A.A.A.: Survey on liver tumour resection planning system: steps, techniques, and parameters. J. Digit. Imaging **33**(2), 304–323 (2020)
2. Antonelli, M., et al.: The medical segmentation decathlon. Nat. Commun. **13**(1), 4128 (2022)
3. Arya, Z., Ridgway, G., Jandor, A., Aljabar, P.: Deep learning-based landmark localisation in the liver for couinaud segmentation. In: Medical Image Understanding and Analysis: 25th Annual Conference, MIUA 2021, Oxford, United Kingdom, July 12–14, 2021, Proceedings 25, pp. 227–237. Springer (2021). https://doi.org/10.1007/978-3-030-80432-9_18
4. Bilic, P., et al.: The liver tumor segmentation benchmark (LiTS). Med. Image Anal. **84**, 102680 (2023)
5. Boltcheva, D., Passat, N., Agnus, V., Jacob-Da Col, M.A., Ronse, C., Soler, L.: Automatic anatomical segmentation of the liver by separation planes. In: Medical Imaging 2006: Visualization, Image-Guided Procedures, and Display, vol. 6141, pp. 383–394. SPIE (2006)
6. Cardoso, M.J., et al.: MONAI: an open-source framework for deep learning in healthcare. arXiv preprint arXiv:2211.02701 (2022)
7. Chen, Y., Rohrbach, M., Yan, Z., Shuicheng, Y., Feng, J., Kalantidis, Y.: Graph-based global reasoning networks. In: Proceedings of the IEEE/CVF Conference on Computer Vision and Pattern Recognition, pp. 433–442 (2019)
8. Detlefsen, N.S., et al.: TorchMetrics-measuring reproducibility in PyTorch. J. Open Sour. Softw. **7**(70), 4101 (2022)
9. Jia, X., et al.: Boundary-aware dual attention guided liver segment segmentation model. KSII Trans. Internet Inf. Syst. (TIIS) **16**(1), 16–37 (2022)
10. Liu, Z., et al.: Swin transformer: hierarchical vision transformer using shifted windows. In: Proceedings of the IEEE/CVF International Conference on Computer Vision, pp. 10012–10022 (2021)
11. Liu, Z., Tang, H., Lin, Y., Han, S.: Point-voxel CNN for efficient 3D deep learning. In: Advances in Neural Information Processing Systems, vol. 32 (2019)
12. Llovet, J.M., et al.: Hepatocellular carcinoma (primer). Nat. Rev. Dis. Primers. **7**(1), 6 (2021)
13. Luo, M., Jiang, H., Shi, T.: Multi-stage puncture path planning algorithm of ablation needles for percutaneous radiofrequency ablation of liver tumors. Comput. Biol. Med. **145**, 105506 (2022)

14. Ma, Y., Guo, Y., Liu, H., Lei, Y., Wen, G.: Global context reasoning for semantic segmentation of 3D point clouds. In: Proceedings of the IEEE/CVF Winter Conference on Applications of Computer Vision, pp. 2931–2940 (2020)
15. Manessi, F., Rozza, A., Manzo, M.: Dynamic graph convolutional networks. Pattern Recogn. **97**, 107000 (2020)
16. Pla-Alemany, S., Romero, J.A., Santabárbara, J.M., Aliaga, R., Maceira, A.M., Moratal, D.: Automatic multi-atlas liver segmentation and couinaud classification from CT volumes. In: 2021 43rd Annual International Conference of the IEEE Engineering in Medicine & Biology Society (EMBC), pp. 2826–2829. IEEE (2021)
17. Qi, C.R., Su, H., Mo, K., Guibas, L.J.: PointNet: deep learning on point sets for 3D classification and segmentation. In: Proceedings of the IEEE Conference on Computer Vision and Pattern Recognition, pp. 652–660 (2017)
18. Qi, C.R., Yi, L., Su, H., Guibas, L.J.: PointNet++: deep hierarchical feature learning on point sets in a metric space. In: Advances in Neural Information Processing Systems, vol. 30 (2017)
19. Rutkauskas, S., Gedrimas, V., Pundzius, J., Barauskas, G., Basevicius, A.: Clinical and anatomical basis for the classification of the structural parts of liver. Medicina (Kaunas) **42**(2), 98–106 (2006)
20. Shaw, P., Uszkoreit, J., Vaswani, A.: Self-attention with relative position representations. In: Walker, M., Ji, H., Stent, A. (eds.) Proceedings of the 2018 Conference of the North American Chapter of the Association for Computational Linguistics: Human Language Technologies, Volume 2 (Short Papers), pp. 464–468. Association for Computational Linguistics, New Orleans, Louisiana (2018)
21. Tian, J., Liu, L., Shi, Z., Xu, F.: Automatic couinaud segmentation from CT volumes on liver using GLC-UNet. In: International Workshop on Machine Learning in Medical Imaging, pp. 274–282. Springer (2019)
22. Wang, M., Jin, R., Lu, J., Song, E., Ma, G.: Automatic CT liver couinaud segmentation based on key bifurcation detection with attentive residual hourglass-based cascaded network. Comput. Biol. Med. **144**, 105363 (2022)
23. Wang, X., Girshick, R., Gupta, A., He, K.: Non-local neural networks. In: Proceedings of the IEEE Conference on Computer Vision and Pattern Recognition, pp. 7794–7803 (2018)
24. Wang, Y., Sun, Y., Liu, Z., Sarma, S.E., Bronstein, M.M., Solomon, J.M.: Dynamic graph CNN for learning on point clouds. ACM Trans. Graph. (tog) **38**(5), 1–12 (2019)
25. Wei, M., et al.: AGConv: adaptive graph convolution on 3D point clouds. IEEE Trans. Pattern Anal. Mach. Intell. **45**(8), 9374–9392 (2023)
26. Ye, X.: calflops: a FLOPs and params calculate tool for neural networks in PyTorch framework (2023). https://github.com/MrYxJ/calculate-flops.pytorch
27. Zhang, L., Chen, M., Arnab, A., Xue, X., Torr, P.H.: Dynamic graph message passing networks for visual recognition. IEEE Trans. Pattern Anal. Mach. Intell. (2022)
28. Zhang, X., et al.: Robust and smooth couinaud segmentation via anatomical structure-guided point-voxel network. Comput. Biol. Med. **182**, 109202 (2024)
29. Zhou, W., Wang, Q., Jin, W., Shi, X., He, Y.: Graph transformer for 3D point clouds classification and semantic segmentation. Comput. Graph. **124**, 104050 (2024)
30. Zhu, X., Hu, H., Lin, S., Dai, J.: Deformable convnets v2: more deformable, better results. In: Proceedings of the IEEE/CVF Conference on Computer Vision and Pattern Recognition, pp. 9308–9316 (2019)

Prompt-Driven Multi-view Representation Learning for Clinical Progression Prediction of Significant Memory Concern

Cui Wang[1,2], Yongheng Sun[2], Minhui Yu[2], Yuzhen Gao[1,2], and Mingxia Liu[2(✉)]

[1] School of Information Science and Engineering, Shandong Agriculture and Engineering University, Ji'nan 250100, China
[2] Department of Radiology and Biomedical Research Imaging Center, University of North Carolina at Chapel Hill, Chapel Hill, NC 27599, USA
mingxia_liu@med.unc.edu

Abstract. Significant memory concern (SMC), a preclinical phase of Alzheimer's disease (AD), is at increased risk of underlying AD pathology. Monitoring the progression of SMC is crucial for timely intervention of AD and related disorders. Learning-based neuroimage analysis provides a non-invasive and objective solution for SMC prognosis. However, existing studies usually focus on utilizing imaging data, without considering subjects' demographic information which is essential for individual-level analysis. In addition, due to the characteristics of SMC progression requiring longitudinal analysis (*e.g.*, 2 years), the data used for SMC analysis are usually very limited (*e.g.*, tens), which poses a huge challenge to model training. To address these limitations, we propose a prompt-driven multi-view learning (PML) framework for predicting the clinical progression of SMC by integrating T1-weighted MRI with demographic information. Specifically, PML comprises four key components: (1) *data-driven MRI feature extraction* using a residual neural network to learn representative features from 3D MRI scans; (2) *handcrafted MRI feature extraction* to incorporate domain knowledge on brain tissues (*e.g.*, cortical thickness); (3) *demographic feature encoding* using a prompt-based strategy through a contrastive language-image pretraining encoder; and (4) *feature fusion and classification* to jointly model multimodal information. To alleviate data scarcity challenges, we initialize our model with pretrained weights and employ transfer learning to enhance performance. Experimental results on a total of 469 subjects demonstrate the efficacy of PML in predicting SMC progression.

Keywords: Prompt-driven · Preclinical Progression Prediction · Multi-View Representation Learning · Alzheimer's Disease

1 Introduction

Significant memory concern (SMC), also called subjective cognitive decline (SCD), is a preclinical phase of Alzheimer's disease (AD), which occurs before

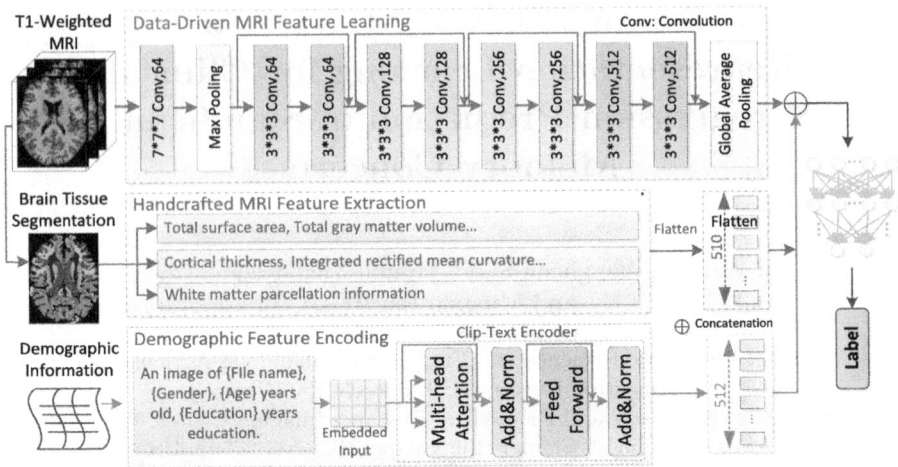

Fig. 1. Illustration of PML framework, consisting of (1) MRI feature extraction through data-driven feature learning, (2) handcrafted MRI feature extraction with brain tissue segmentation, and (3) demographic feature encoding with Contrastive Language-Image Pre-Training (CLIP) model, and (4) multi-view feature fusion and classification. The ResNet18 encoder used in MRI feature learning was fine-tuned during training, while the CLIP text encoder was used as a fixed feature extractor without fine-tuning.

deficits could be detected by cognitive tests. There is increasing evidence that subjects with SMC are at increased risk of underlying AD pathology [1,16,18,20]. It is essential to monitor the future progression of SMC to prodromal AD (i.e., mild cognitive impairment, MCI) and clinical AD, facilitating drug development and timely intervention for AD and AD-related disorders.

Many learning-based methods have been developed for automated prediction of SMC progression with neuroimaging data. Some methods rely solely on brain MRI scans [23], while others integrate both MRI and positron emission tomography (PET) data to improve prediction accuracy [12]. However, these approaches typically ignore individual-specific demographic information, such as gender, age, and education years, which is essential for subject-level analysis. Additionally, as SMC progression necessitates extended longitudinal analysis (e.g., over 2 years) compared to MCI, the data available for SMC analysis tend to be highly limited (e.g., tens or hundreds), which significantly hinders effective model training.

To this end, we propose a prompt-driven multi-view learning (PML) framework to predict the clinical progression of SMC. As shown in Fig. 1, the PML consists of (1) data-driven MRI feature extraction using a residual neural network to learn representative features from 3D MRI scans, (2) handcrafted MRI feature extraction to incorporate domain knowledge (e.g., cortical thickness) with brain tissue segmentation, (3) demographic feature encoding using a prompt-based strategy by leveraging rich semantic knowledge from the Contrastive Language-Image Pre-training (CLIP) encoder [25], and (4) multi-view feature fusion and classification. To handle the small-sample-size issue, in our feature encoder, we use the weights of Med3D [4] which was pre-trained on 1474 T1-weighted (T1w) MRIs and CT images through segmentation of multiple organs. Experimental results on a total of 469 subjects from two cohorts demonstrate the efficacy of PML in predicting SMC progression. To the best of our knowledge, this represents one of the initial efforts to utilize CLIP-encoded demographic data for brain MRI-based SMC progression prediction.

2 Methodology

We reasonably assume that the effective fusion of data-driven MRI representations, handcrafted features, and subjects' demographic information can promote accurate prediction of cognitive changes in SMC. As shown in Fig. 1, the proposed PML consists of four components: (1) data-driven MRI feature learning with a residual neural network (ResNet), (2) handcrafted MRI feature extraction to leverage domain knowledge on brain changes conveyed in MRI such as cortical thickness and gray matter volume, (3) demographic feature encoding by utilizing the CLIP model, and (4) feature fusion and classification.

(1) MRI Feature Learning. To explore imaging biomarkers from MRIs in a data-driven manner, we propose to employ a ResNet18 [9] as a backbone for MRI feature extraction. As shown in the top left panel of Fig. 1, ResNet18 takes 3D MRI scans as input for feature learning. It consists of four parts: a 3D convolutional layer (kernel size: $7 \times 7 \times 7$, stride: 2), a maxpooling layer (kernel size: $3 \times 3 \times 3$, stride: 2), four residual network blocks (RNBs), and a Global Average Pooling (GAP) layer. Each RNB consists of two convolutional layers, both with a kernel size of 3, a stride of 1, and zero padding. Batch normalization (BN) and rectified linear unit (ReLU) activation are applied to the first convolutional layer, whereas only BN is applied to the second convolutional layer. A skip connection is incorporated to sum the input and output of each residual block, followed by the application of ReLU activation to the summed output. The number of channels in the four RNBs is 64, 128, 256, and 512, respectively. Following the final RNB, a GAP operation is applied to the resulting feature maps, generating a compact feature vector with a dimension of 512 for each subject.

(2) Handcrafted MRI Feature Extraction. Given that handcrafted features encapsulate valuable domain expertise regarding anatomical structures of the brain, like gray matter volume and cortical thickness, combining them with data-driven features makes intuitive sense for accurately capturing the essence of each MR image. In this work, three types of handcrafted features are extracted from each MRI scan, including 1) structural statistics, such as surface area and gray matter volume; 2) attribute information, including cortical thickness and mean curvature of the cortical surface; and 3) white matter parcellation data, such as radial diffusivity. These features are computed through brain tissue segmentation using FreeSurfer [8]. For each MRI, we can obtain a 510-dimensional handcrafted feature vector for representing a subject.

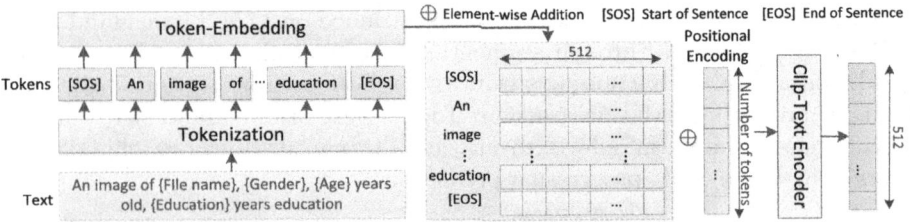

Fig. 2. Workflow of demographic feature encoding.

(3) Demographic Feature Encoding with CLIP. In this study, we used the CLIP text encoder to embed demographic attributes such as gender, age, and years of education. Compared to conventional encodings (e.g., one-hot or normalized values), CLIP offers semantically rich and unified representations that facilitate integration with other modalities. Although the prompts used in this work are relatively simple and the immediate performance gain may be limited, the approach is extensible to more complex or unstructured demographic data (e.g., clinical notes or electronic health records). Thus, this serves as an initial step toward a more generalizable multimodal fusion framework.

Prior research has demonstrated that individuals over the age of 65 have an increased risk of developing AD [3,7]. Additionally, a higher level of education has been shown to delay the progression of the disease [22]. Therefore, demographic information, such as age, gender, and education years, also has a critical impact on the clinical progression prediction of SMC. To incorporate these attributes into the proposed framework, we construct textual descriptions for demographic feature encoding using the following format:

An image of {*filename*}, {*gender*}, {*age*} years old, {*education*} years education.

These descriptions are then processed using CLIP's text encoder.

As shown in Fig. 2, the text input is first tokenized into discrete tokens, followed by padding or truncation to ensure alignment with a fixed sequence length. Tokens of the text sequence are bracketed with [SOS] and [EOS] tokens, all tokens (including [SOS] and [EOS]) will be mapped to a vector space of fixed dimension (512) through Token-Embedding. Then, a learnable positional encoding is added to ensure that the Transformer pays attention to the order of the tokens. The tokenized text is subsequently fed into a Transformer-based text encoder in CLIP [19], where it is transformed into vector representations through an embedding layer (see Clip-Text Encoder in the lower right of Fig. 1). The [EOS] token is extracted as the global representation of the text. Finally, a linear projection layer maps the extracted features into a 512-dimensional text embedding.

(4) Feature Fusion and Classification. After the above operations, we have image features extracted by ResNet18, handcrafted MRI features, and CLIP-encoded demographic information features. These three types of data need to be fused and then sent to the modified ResNet18's Linear layer to complete the classification. Specifically, the image features extracted by ResNet18 and the handcrafted features are flattened into a vector, which contains 512 and 510 elements respectively. And then we concatenate the demographic information data encoded by CLIP and the flattened features to obtain a feature vector with a dimension of 1,534. We remove the last fully connected layer of ResNet18 and modify the linear layer to make sure its input feature dimension is 1,534. At last, we feed the 1,534-dimension feature vector to this linear layer to get a probability score vector that indicates the probability of a specific category.

Considering the characteristics of data used in this experiment, we use focal loss [13] to partly address the problem of class imbalance. It modifies the standard cross-entropy loss by introducing a scaling factor that reduces the loss contribution from well-classified samples while emphasizing hard misclassified samples. The focal loss for binary classification is formulated as:

$$FL(p_t) = -\alpha(1 - p_t)^\gamma \log(p_t), \tag{1}$$

where p_t is the predicted probability for a target sample, and α is a balancing factor to handle class imbalance (typically $\alpha \in [0,1]$). γ is a focusing parameter, controlling how much to down-weight easy examples (commonly set to $\gamma = 2$). The term $(1 - p_t)^\gamma$ is the modulating factor, which reduces the loss for well-classified samples (p_t close to 1) while focusing more on misclassified ones.

3 Experiments

Studied Subjects and Image Pre-processing. This work involves T1-weighted MRI scans acquired from two studies. (1) Alzheimer's Disease Neuroimaging Initiative (ADNI) [10], with predominantly Caucasian brain magnetic resonance images of 291 MCI subjects and 102 age- and gender-match subjects with significant memory complaints (SMC). (2) Longitudinal Aging Study (LAS) with brain MRIs from 24 progressive SCD (PSCD) subjects that convert to MCI within 7 years and 52 SCD subjects who remain stable over 7 years (SSCD). All 3D MR images are preprocessed through standard procedures, containing skull stripping, intensity inhomogeneity correction, resampling to have the resolution of $1 \times 1 \times 1mm^3$, and spatial normalization to the Montreal Neurological Institute (MNI) space. To remove the uninformative background, we crop each MRI to $176 \times 208 \times 176$ and ensure that the entire brain remains within each volume. TorchIO [17] is used for histogram normalization and Z-normalization is performed to make the data have a mean of 0 and a standard deviation of 1, to avoid the influence of multi-site data heterogeneity.

Experimental Setup. Two tasks are performed in this work, including (1) SMC vs. MCI classification on ADNI, and (2) PSCD vs. SSCD classification on LAS. We use five metrics to evaluate the prediction performance, including the area under the ROC curve (AUC), accuracy (ACC), sensitivity (SEN), specificity (SPE), and F1 scores (F1s). A 5-fold cross-validation (CV) strategy is used in the experiments. Specifically, we first randomly split all subjects into 5 groups for each category. Then, one group is alternatively used as a test set and the remaining 4 groups are used as the training set. The experiments are repeated 5 times independently to avoid any bias introduced by random data split. The mean and standard deviation results were recorded. The parameter α in the focal loss is set as 0.3, and γ is set as 2. Adam optimizer is used and the learning rate is set as 10^{-4}.

Competing Methods. We compare the proposed PML with the most popular machine learning methods, including support vector machine (SVM) [21], random forest (RF) [2] and XGBoost (XGB) [24]. In SVM, the radial basis function kernel is used with the default regularization parameter $C=1$. The RF method uses 100 decision trees with a depth of 5. In XGB, a grid search strategy is used to find a good combination of hyperparameters (i.e., number of boosting rounds, maximum tree depth for base learners, learning rate). We use XGB with boosting rounds of 200, a maximum tree depth of 4, and a learning rate of 0.2. These three classifiers take 510-dimensional handcrafted MRI features as input.

We also compare PML with state-of-the-art deep learning methods, including ResNet [9] and ViT [5] with 3D MRI scans as input and their variants with multi-view features. We denote ResNet+HF and ViT+HF as ResNet and ViT variants that take both MR images and 510-dimensional handcrafted features as input. Similarly, we denote ViT+HF+CLIP as another ViT variant that uses MR images, handcrafted features, and CLIP-encoded demographic features as input. In ResNet, a batch size of 2 is used for model training. We download a pretrained Med3D model (https://github.com/Tencent/MedicalNet) for ResNet initialization so that ResNet can use the pretrained parameters without having to train from scratch. The ViT consists of a Transformer encoder and a multilayer perceptron (MLP) for classification. In ViT and its variants (ViT+HF and ViT+HF+CLIP), the iteration of the Transformer encoder is set to 6, the hidden size and number of attention heads are set to 128 and 16 respectively, while the patch size and the batch size are set to $8\times8\times8$ and 6, respectively. The dimension of the hidden layer in the MLP block is set as 512, and an average pooling strategy is used. The Adam optimizer with a learning rate of 10^{-4} is used. We employ a model pre-trained on BRATS2023, IXI and OASIS3 [11] for ViT initialization.

Classification Results. The results obtained from all methods in SMC progression prediction are presented in Table 1 (SMC vs. MCI classification), and Table 2 (PSCD vs. SSCD classification), paired t-test is made between our PML and several other methods, $p < 0.05$ via paired t-test denotes the difference between our PML and a competing method is statistically significant. It can be seen from Table 1 and Table 2 that our PML outperforms the competing methods in the two different tasks. For instance, the AUC of PML (i.e., 97.40%) is 1.30% higher than ResNet+HF (AUC=96.10%), and 1.57% higher than ResNet($AUC = 95.83\%$) in SMC vs. MCI classification. This verifies that incorporating handcrafted features and CLIP-encoded demographic features into the proposed framework helps improve classification performance. A similar trend

Table 1. Results (%) of different methods in SMC progression prediction (i.e., SMC vs. MCI classification) with best results shown in boldface, as well as p-values via paired sample t-test between PML and each of the competing methods.

Method	AUC	ACC	SEN	SPE	F1	p-value
SVM [21]	88.90 ± 2.84	92.60 ± 2.16	81.00 ± 5.42	96.90 ± 2.29	85.40 ± 4.16	0.001
RF [2]	84.30 ± 4.04	90.30 ± 1.97	71.30 ± 8.87	97.20 ± 2.07	79.40 ± 5.67	0.001
XGBoost [24]	87.20 ± 5.08	92.40 ± 2.72	76.28 ± 10.42	98.20 ± 1.09	83.90 ± 6.42	0.001
ResNet [9]	95.83 ± 3.32	95.67 ± 3.25	**96.13 ± 3.35**	95.52 ± 2.14	92.81 ± 4.13	0.002
ViT [5]	91.87 ± 4.23	91.60 ± 5.69	92.35 ± 2.68	91.39 ± 7.42	85.78 ± 8.72	0.001
ResNet+HF	96.10 ± 3.08	96.17 ± 3.40	96.00 ± 2.67	96.21 ± 2.24	91.03 ± 2.61	0.010
ViT+HF	93.31 ± 2.84	93.92 ± 3.22	91.79 ± 7.10	94.83 ± 3.45	89.10 ± 4.94	0.001
ViT+HF+CLIP	94.42 ± 1.94	95.21 ± 1.33	92.61 ± 5.60	96.23 ± 2.80	91.16 ± 2.20	0.001
PML (Ours)	**97.40 ± 1.33**	**98.48 ± 1.01**	95.13 ± 2.47	**99.66 ± 0.16**	**96.89 ± 2.31**	--

Table 2. Results (%) of different methods in SMC progression prediction (i.e., PSCD vs. SSCD classification) with best results shown in boldface, as well as p-values via paired sample t-test between PML and each of the competing methods.

Method	AUC	ACC	SEN	SPE	F1	p-value
SVM [21]	51.56 ± 7.58	51.37 ± 7.92	53.11 ± 4.06	50.00 ± 15.49	53.05 ± 4.66	0.001
RF [2]	40.17 ± 10.28	39.26 ± 9.32	41.33 ± 11.66	39.00 ± 20.10	40.81 ± 9.07	0.001
XGBoost [24]	41.72 ± 3.13	40.37 ± 2.62	36.44 ± 12.83	47.00 ± 18.87	37.15 ± 6.87	0.001
ResNet [9]	53.00 ± 9.75	53.00 ± 9.75	56.00 ± 33.62	50.00 ± 33.17	48.96 ± 28.29	0.009
ViT [5]	51.56 ± 12.36	53.79 ± 12.73	54.11 ± 15.23	44.00 ± 31.77	54.01 ± 30.75	0.045
ResNet+HF	55.37 ± 9.00	56.05 ± 9.70	64.07 ± 25.20	46.67 ± 28.20	54.60 ± 17.57	0.018
ViT+HF	52.68 ± 10.34	51.32 ± 8.75	**63.82 ± 4.75**	43.76 ± 5.68	50.84 ± 7.69	0.002
ViT+HF+CLIP	53.83 ± 11.51	52.42 ± 9.79	63.67 ± 17.42	44.00 ± 32.86	**57.36 ± 7.55**	0.001
PML (Ours)	**61.44 ± 18.12**	**61.74 ± 18.35**	55.89 ± 35.29	**67.00 ± 21.17**	53.98 ± 42.71	--

appears in the task of PSCD vs. SSCD classification. On the other hand, it can be observed from Table 1 and Table 2 that ResNet-based methods (*i.e.*, ResNet, ResNet+HF, and PML) generally outperform their respective ViT-based counterparts (*i.e.*, ViT, ViT+HF, ViT+HF+CLIP). This implies that ResNet can extract more robust features from MRI compared to ViT, which may be due to the relatively limited dataset size, where convolutional architectures with strong inductive biases tend to generalize more effectively. In contrast, transformer-based models such as ViT typically require larger-scale data to fully exploit their capacity to model global spatial relationships. Furthermore, the proven transferability of ImageNet-pretrained ResNet models to medical imaging domains may also contribute to their robust performance in this context.

Feature Distribution Analysis. We visualize features from ResNet, ResNet+HF and PML in Fig. 4 using t-SNE [15]. The features in ResNet+HF are obtained by concatenating ResNet-extracted features with handcrafted features, while PML employs ResNet-extracted features, handcrafted features, and CLIP-encoded features. As shown in Fig. 4, the PML features are more separable between SMC and MCI categories compared with ResNet and ResNet+HF.

Table 3. Results (%) of PML and its four variants different methods in SMC progression prediction (i.e., SMC vs. MCI classification) with best results shown in boldface, as well as p-values via paired sample t-test between PML and each variant.

Method	AUC (%)	ACC (%)	SEN (%)	SPE (%)	F1 (%)	p-value
PMLw/oP	86.64 ± 11.23	92.36 ± 6.24	75.00 ± 13.30	98.28 ± 1.26	75.18 ± 12.49	0.001
PMLw/oCLIP	94.89 ± 4.94	97.20 ± 2.47	90.13 ± 5.69	99.66 ± 0.13	94.14 ± 5.56	0.002
PML-ADNI	95.70 ± 3.43	95.40 ± 3.76	**96.22 ± 2.47**	95.17 ± 2.87	92.85 ± 4.66	0.001
PML-CE	94.99 ± 3.89	96.95 ± 2.67	91.00 ± 3.45	98.98 ± 0.67	93.63 ± 5.08	0.001
PML (Ours)	**97.40 ± 1.83**	**98.48 ± 1.01**	95.13 ± 3.23	**99.66 ± 0.16**	**96.89 ± 2.41**	–

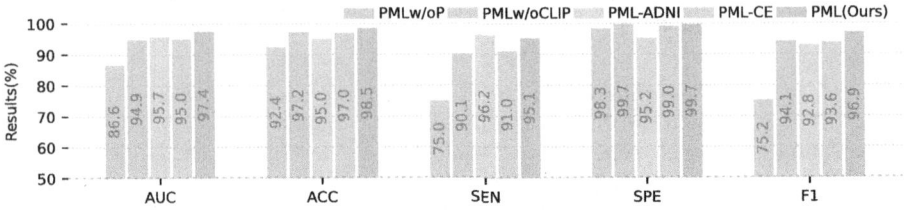

Fig. 3. Comparison between PML and its four variants in SMC vs. MCI classification.

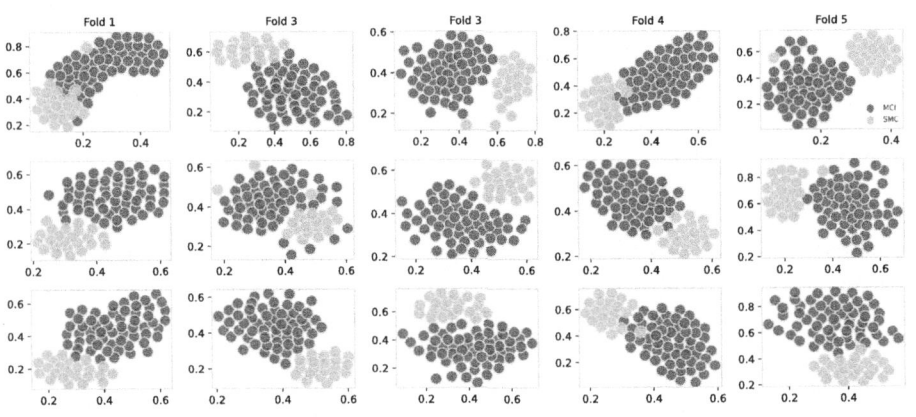

Fig. 4. T-SNE analysis of test data features from (top) ResNet, (middle) ResNet+HF, and (bottom) the proposed PML across 5 folds in SMC vs. MCI classification.

The features generated by ResNet+HF have better discriminative ability than ResNet, indicating the effectiveness of handcrafted MRI features in improving prediction performance.

Ablation Study. We compare PML with its four variants: (1) PMLw/oP without pretrained weights, (2) PMLw/oCLIP without CLIP-encoded demographic features, (3) PML-ADNI with weights pretrained in ADNI (AD vs. CN classification), and (4) PML-CE with cross-entropy loss. For a fair comparison, all four variants share the same architecture similar to PML. The results for SMC vs. MCI classification are presented in Table 3 and Fig. 3, which demonstrates that PML consistently outperforms the four variants, and the difference between PML and each of its four variants is statistically significant in most cases.

4 Conclusion and Future Work

This paper presents a prompt-driven multi-view learning (PML) framework for the clinical progression of significant memory concern (SMC) by integrating data-driven MRI features, handcrafted MRI features, and CLIP-encoded demographic features. Experimental evaluations conducted on 469 subjects demonstrate the effectiveness of the proposed PML in accurately predicting SMC progression. Our future work will focus on developing more advanced feature extraction techniques and exploring multimodal fusion strategies to integrate genetic data, cognitive assessments, and clinical information with imaging features. It is also interesting to employ hypergraph learning [6,14] to fuse these multi-view feature representations to further enhance the learning performance.

Acknowledgments. This work was completed during the period when C. Wang and Y. Gao were visiting the University of North Carolina at Chapel Hill. The research of C. Wang was supported in part by Shandong Agriculture and Engineering University Start-Up Fund for Talented Scholars (No. BSQJ202310).

Disclosure of Interests. The authors have no competing interests to declare that are relevant to the content of this article.

References

1. Barnes, L., Schneider, J., Boyle, P., Bienias, J., Bennett, D.: Memory complaints are related to Alzheimer disease pathology in older persons. Neurology **67**(9), 1581–1585 (2006)
2. Breiman, L.: Random forests. Mach. Learn. **45**, 5–32 (2001)
3. Brookmeyer, R., Abdalla, N., Kawas, C.H., Corrada, M.M.: Forecasting the prevalence of preclinical and clinical Alzheimer's disease in the united states. Alzheimer's Dementia **14**(2), 121–129 (2018)
4. Chen, S., Ma, K., Zheng, Y.: Med3D: Transfer learning for 3D medical image analysis. arXiv preprint arXiv:1904.00625 (2019)
5. Dosovitskiy, A., et al.: An image is worth 16 × 16 words: Transformers for image recognition at scale. In: International Conference on Learning Representations (2021)
6. Feng, Y., You, H., Zhang, Z., Ji, R., Gao, Y.: Hypergraph neural networks. In: AAAI, pp. 3558–3565 (2019)
7. Ferretti, M.T., et al.: Sex differences in Alzheimer disease-the gateway to precision medicine. Nat. Rev. Neurol. **14**(8), 457–469 (2018)
8. Fischl, B.: FreeSurfer. NeuroImage **62**(2), 774–781 (2012)
9. He, K., Zhang, X., Ren, S., Sun, J.: Deep residual learning for image recognition. In: CVPR, pp. 770–778 (2016)

10. Jack Jr, C.R., et al.: The Alzheimer's disease neuroimaging initiative (ADNI): MRI methods. J. Magn. Reson. Imaging: Official Journal of the International Society for Magnetic Resonance in Medicine **27**(4), 685–691 (2008)
11. Kunanbayev, K., Shen, V., Kim, D.S.: Training ViT with limited data for Alzheimer's disease classification: an empirical study. In: MICCAI, pp. 334–343. Springer (2024)
12. Leng, Y., et al.: Multimodal cross enhanced fusion network for diagnosis of Alzheimer's disease and subjective memory complaints. Comput. Biol. Med. **157**, 106788 (2023)
13. Lin, T.Y., Goyal, P., Girshick, R., He, K., Dollár, P.: Focal loss for dense object detection. In: ICCV, pp. 2980–2988 (2017)
14. Liu, M., Gao, Y., Yap, P.T., Shen, D.: Multi-hypergraph learning for incomplete multimodality data. IEEE J. Biomed. Health Inform. **22**(4), 1197–1208 (2017)
15. Van der Maaten, L., Hinton, G.: Visualizing data using t-SNE. J. Mach. Learn. Res. **9**(11) (2008)
16. Mitchell, A., Beaumont, H., Ferguson, D., Yadegarfar, M., Stubbs, B.: Risk of dementia and mild cognitive impairment in older people with subjective memory complaints: Meta-analysis. Acta Psychiatr. Scand. **130**(6), 439–451 (2014)
17. Pérez-García, F., Sparks, R., Ourselin, S.: TorchIO: a Python library for efficient loading, preprocessing, augmentation and patch-based sampling of medical images in deep learning. Comput. Methods Programs Biomed. **208**, 106236 (2021)
18. Peter, J., et al.: Gray matter atrophy pattern in elderly with subjective memory impairment. Alzheimer's Dementia **10**(1), 99–108 (2014)
19. Radford, A., et al.: Learning transferable visual models from natural language supervision. In: ICML, pp. 8748–8763. PmLR (2021)
20. Saykin, A., et al.: Older adults with cognitive complaints show brain atrophy similar to that of amnestic MCI. Neurology **67**(5), 834–842 (2006)
21. Steinwart, I., Christmann, A.: Support vector machines. Springer Science & Business Media (2008)
22. Stern, Y.: Cognitive reserve in ageing and Alzheimer's disease. Lancet Neurol. **11**(11), 1006–1012 (2012)
23. de Vos, F., et al.: Pre-trained MRI-based Alzheimer's disease classification models to classify memory clinic patients. NeuroImage: Clin. **27**, 102303 (2020)
24. Yi, F., et al.: XGBoost-SHAP-based interpretable diagnostic framework for Alzheimer's disease. BMC Med. Inform. Decis. Mak. **23**(1), 137 (2023)
25. Zhang, S., et al.: A multimodal biomedical foundation model trained from fifteen million image–text pairs. NEJM AI **2**(1), AIoa2400640 (2025)

Improving Late-Life Depression Analysis with Collaborative Domain Adaptation: Learning from Heterogeneous Structural MRI

Yuzhen Gao[1], Mengqi Wu[1], Li Wang[1], Lihong Wang[2], David C. Stephens[2], Guy G. Potter[3], and Mingxia Liu[1(✉)]

[1] Department of Radiology and BRIC, University of North Carolina at Chapel Hill, Chapel Hill, North Carolina 27599, USA
mingxia_liu@med.unc.edu
[2] Department of Psychiatry, University of Connecticut, Farmington, CT 06030, USA
[3] Department of Psychiatry and Behavioral Sciences, Duke University Medical Center, Durham, NC 27710, USA

Abstract. Accurate identification of late-life depression (LLD) based on brain MRI is crucial for monitoring its clinical progression over time. Existing learning-based LLD studies often suffer from limited (*e.g.*, tens) data, leading to unreliable model training. While using auxiliary data can enlarge the sample size, the inherent heterogeneity between auxiliary and target MRIs with different acquisition settings often hinders generalizability. To this end, we propose a collaborative domain adaptation (CDA) framework for LLD detection with T1-weighted MRIs, leveraging knowledge learned from large-scale auxiliary source domains to a small-sized target domain. The CDA contains a Vision Transformer (ViT) for capturing global MRI representation and a Convolutional Neural Network (CNN) for extracting local features, both pre-trained on 9,544 MRIs from a public cohort. Its training consists of three components: (1) *supervised training* on labeled source data, with ViT- and CNN-based encoders for feature extraction and two classifiers for prediction; (2) *fine-tuning with feature alignment* by minimizing the discrepancy between classifier outputs from two branches to make the categorical boundary clearer; and (3) *collaborative training* on unlabeled target MRIs, leveraging augmented MRI samples to ensure feature consistency. Extensive experiments on T1-weighted MRIs from a total of 238 subjects suggest that CDA outperforms several state-of-the-art approaches, achieving superior classification accuracy and improved cross-domain generalization.

Keywords: Late-life depression · Collaborative domain adaptation · Heterogeneous data · Structural MRI

1 Introduction

Late-life depression (LLD) is a prevalent neuropsychiatric disorder in older adults. Accurate identification of LLD with MRI is crucial to monitor its clinical progression over time and facilitate timely intervention. Early detection of LLD based on neurosurgical data such as brain MRI can significantly improve patient outcomes [1]. However, despite the promise of learning-based methods in neurosurgical-based diagnostics, LLD research is often constrained by limited data, and many studies rely on sample sizes of only a few dozen cases [2]. Small sample sizes hinder robust model training, often resulting in over-fitting and unreliable predictions, which in turn limit the generalizability of trained models [3].

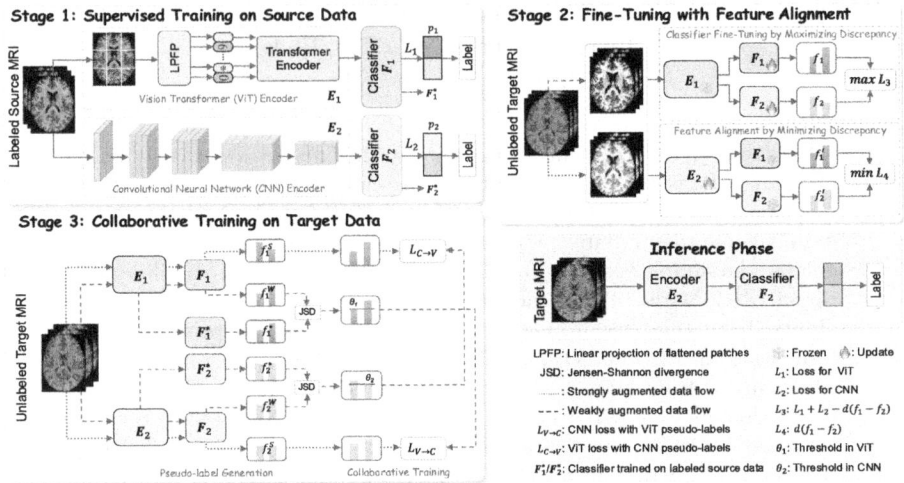

Fig. 1. Illustration of CDA, with (1) *supervised training* on labeled source data with ViT- and CNN-based branches for global and local feature learning, (2) *fine-tuning with feature alignment* to optimize differences between ViT and CNN predictions and improve decision boundaries, and (3) *collaborative training* on unlabeled target data with augmented samples, enhancing their consistency to facilitate cross-domain adaptation.

To mitigate this issue, incorporating auxiliary datasets has become a widely adopted strategy in medical image analysis [4]. By incorporating external MRI datasets, models can benefit from a larger and more diverse set of training images, potentially improving their robustness and generalizability. However, domain shift between auxiliary source and target MRI datasets can significantly hinder model training [5]. If these domain shifts are not adequately addressed, models trained on heterogeneous datasets can suffer performance degradation when applied to real-world clinical settings, ultimately limiting their clinical utility [6]. Therefore, developing robust domain adaptation strategies is essential to bridge the distribution gap between source and target data, ensuring improved cross-domain generalization for reliable LLD diagnosis [7].

To this end, we propose a collaborative domain adaptation (CDA) framework for MRI-based LLD identification. It integrates a Vision Transformer (ViT) [8] to capture global MRI representations and a Convolutional Neural Network (CNN) to extract local features. Both ViT and CNN are pretrained on 9,544 T1-weighted (T1w) MRIs from ADNI [9] through unsupervised image reconstruction, enhancing their feature extraction capability. As shown in Fig. 1, the CDA consists of *three key components*: (1) supervised training on a labeled source dataset; (2) model fine-tuning with feature alignment to optimize the difference between ViT and CNN predictions and improve decision boundaries; and (3) self-supervised training on unlabeled target data using strongly and weakly augmented data to enhance generalization. Experiments on 89 subjects from a target domain demonstrate that CDA outperforms several state-of-the-art (SOTA) approaches.

2 Methodology

Problem Formulation. Our goal is to identify patients with LLD from healthy controls (HCs) in a target domain with small data, by leveraging knowledge from relatively large-scale source data. N_S and N_T, respectively. We have a labeled source domain $D_S = \{(x_i^s, y_i^s)\}_{i=1}^{N_s}$, where each 3D MRI $x_i^s \in \mathbb{R}^{H \times W \times D}$ corresponds to a ground-truth category label $y_i^s \in \{0, 1\}$, and an unlabeled target domain $D_T = \{x_i^t\}_{i=1}^{N_t}$. Our CDA has two branches (ViT and CNN [10]) for global and local MRI feature learning, considering that ViT can effectively capture long-range dependencies and model comprehensive global representations and CNN is good at capturing local spatial hierarchies for robust feature learning. Both ViT and CNN encoders output 512-dimensional vectors, ensuring consistent input dimensionality for classifiers F_1 and F_2. To promote knowledge transfer from source to target domain, we develop 3 components in CDA, including *supervised training* on labeled source data, *fine-tuning with feature alignment*, and *collaborative training* on target data with augmented samples. To improve cross-domain generalization, we employ data augmentation in CDA, applying weak and strong augmentations [11] to enhance their feature consistency.

(1) Supervised Training on Labeled Source Data. As illustrated in the top left of Fig. 1, each input MRI is processed through two branches: ViT and CNN. The ViT tokenizes input MRI into patch embeddings and processes them using a Transformer encoder E_1 with multiple self-attention layers.

The encoded feature is then fed into a classifier F_1 with a multilayer perceptron (MLP) for prediction. CNN extracts MRI features using an encoder E_2 with stacked convolutional layers and residual blocks, followed by a classifier F_2 to generate logits for prediction. We train the two branches independently on labeled source data to facilitate effective MRI feature extraction. To address potential category imbalance in the source domain, we use a focal loss [12] for training to enhance the model's focus on samples from the minority class. The objective for each of the two branches is to minimize the following loss:

$$\mathcal{L}_1 = \frac{1}{N_s}\sum_{i=1}^{N_s} FL(y_i^s, p_1), \quad \mathcal{L}_2 = \frac{1}{N_s}\sum_{i=1}^{N_s} FL(y_i^s, p_2), \tag{1}$$

where $FL(\cdot)$ denotes the focal loss, the probability $p_1 = \sigma(F_1(E_1(x_i^s)))$ is from ViT and $p_2 = \sigma(F_2(E_2(x_i^s)))$ is from CNN for a source sample x_i^s, and $\sigma(\cdot)$ is the softmax function that converts logits into class probabilities. Finally, we obtain two initial classifiers F_1^* and F_2^* trained on labeled source data, which will be directly used in the 3rd stage (see Fig. 1). The trained ViT and CNN models will be further fine-tuned in the 2nd stage, with details introduced below.

(2) Fine-Tuning with Feature Alignment. In this stage, we focus on reducing the distribution discrepancy between source and target domains by aligning their encoded image representations. Specifically, we utilize ViT's properties to explicitly define class-specific decision boundaries by maximizing the difference between the two classifiers' outputs, identifying target samples distant from the source support. Conversely, the CNN encoder groups target features according to these pre-established boundaries by minimizing the difference between the classifiers' probability outputs. This process consists of a) *classifier fine-tuning* and b) *feature alignment*, as illustrated in the top right of Fig. 1.

a) Classifier Fine-Tuning. In this step, we freeze the ViT encoder E_1 and fine-tune classifiers F_1 and F_2 to maintain high consistency on labeled source data while maximizing their prediction differences on unlabeled target data. This encourages the classifiers to push their decision boundaries apart, highlighting misaligned target features and helping to detect target samples beyond the source distribution's support. Here, we employ a discrepancy loss $\mathcal{D}(\mathbf{a},\mathbf{b}) = \sum_{k=1}^{K}|a_k - b_k|/K$, by calculating the absolute difference between outputs of two classifiers $\mathbf{a}$ and $\mathbf{b}$ for each class, where a_k and b_k denote the probability outputs from the two classifiers corresponding to the k-th ($k = 1, \cdots, K$) class. We introduce a *classification discrepancy loss* $\mathcal{L}_3$ on target samples while incorporating supervised loss on labeled source data, formulated as:

$$\mathcal{L}_3 = -\frac{1}{N_t}\sum_{i=1}^{N_t} \mathcal{D}\Big(\sigma(F_1(E_1(x_i^t))), \sigma(F_2(E_1(x_i^t)))\Big) \\ + \frac{1}{N_s}\sum_{i=1}^{N_s} FL(y_i^s, \sigma(F_1(E_1(x_i^s)))) + \frac{1}{N_s}\sum_{i=1}^{N_s} FL(y_i^s, \sigma(F_2(E_2(x_i^s)))). \tag{2}$$

Our objective is to minimize $\mathcal{L}_3$ to improve class-specific boundaries. The enforced discrepancy between the two classifiers allows them to capture target samples that do not align with the source domain's distribution. We input labeled source data to ViT (E_1, F_1) and CNN (E_2, and F_2) for supervised learning to minimize the last two terms in Eq. (2), where only F_1 and F_2 are updated. This enables the classifiers to accurately distinguish between categories [13]. During training, we use the same number of source and target samples in each batch.

b) Feature Alignment. We then leverage the class-specific boundaries established by ViT encoder E_1 as a reference to guide the optimization of CNN encoder

E_2, while keeping classifiers F_1 and F_2 frozen. This is achieved by minimizing the discrepancy between outputs of F_1 and F_2, allowing features extracted by E_2 to be accurately classified, with the objective function defined as:

$$\mathcal{L}_4 = \frac{1}{N_t} \sum_{i=1}^{N_t} \mathcal{D}\left(f_1', f_2'\right) \quad (3)$$

where $f_1' = \sigma(F_1(E_2(x_i^t)))$ and $f_2' = \sigma(F_2(E_2(x_i^t)))$ denote the probability outputs of F_1 and F_2 for the unlabeled target sample x_i^t. The features extracted by encoder E_2 are clustered within the class-specific boundaries defined by E_1. This alignment is essential to ensure accurate classification of unlabeled target samples while maintaining robustness to changes in class-specific boundaries.

(3) Collaborative Training on Unlabeled Target Data. This stage focuses on reducing the gap between two branches by enabling mutual enhancement and leveraging each branch's strengths to generate pseudo-labels for unlabeled target samples. It also specifically boosts CNN's performance by utilizing ViT's ability to capture complex data patterns. To facilitate effective knowledge transfer between two branches, we use ViT to generate pseudo-labels of weakly augmented target samples $x^{t,W}$ to teach CNN branch with strongly augmented samples $x^{t,S}$. The reverse process is also applied as depicted in Fig. 1. We use random large-scale affine transformation and elastic deformation to generate strongly augmented samples, and random flip and slight random affine to obtain weakly augmented samples [11]. To improve pseudo-label quality, we use a pseudo-label optimization mechanism with Jensen-Shannon divergence (JSD) [14].

a) Pseudo-Label Generation. For each unlabeled target sample, we extract its features using encoders E_1 and E_2, followed by prediction via the classifiers F_1 and F_2 to generate pseudo-labels. The supervised training conducted in Stage 1 yields preliminary classifiers F_1^* and F_2^*, enabling consistency analysis on similar samples. The classification results for a set of weakly augmented samples $x^{t,W}$ include: (i) $q_1^* = F_1^*(E_1(x^{t,W}))$ and $q_1 = F_1(E_1(x^{t,W}))$ processed by ViT, where F_1 denotes the current classifier; and (ii) $q_2^* = F_2^*(E_2(x^{t,W}))$ and $q_2 = F_2(E_2(x^{t,W}))$ processed by CNN, where F_2 is the current classifier. The JSD is computed to assess the output consistency of preliminary and current classifiers for each weakly augmented sample within the same branch, defined as:

$$JSD(P_1, P_2) = \frac{1}{2} D_{KL}(P_1 \| M) + \frac{1}{2} D_{KL}(P_2 \| M), \quad (4)$$

where $M = \frac{1}{2}(P_1 + P_2)$ is the mean of two probability distributions and D_{KL} is Kullback-Leibler divergence [15]. We compute $JSD(q_1^*, q_1)$ for ViT and $JSD(q_2^*, q_2)$ for CNN.

When JSD divergence falls below a predefined parameter $\tau = 0.1$, it signifies a high degree of consistency between two classifiers' predictions on weakly augmented samples. Thus, these samples can be used to generate high-confidence pseudo-labels. For target samples that meet $\tau < 0.1$ in a batch, we can compute their pseudo-labels as follows:

$$\hat{y}_V^t = \frac{1}{2}(f_1^* + f_1^W), \hat{y}_C^t = \frac{1}{2}(f_2^* + f_2^W), \quad (5)$$

where $f_1^* = \sigma(q_1^*)$, $f_1^W = \sigma(q_1)$, $f_2^* = \sigma(q_2^*)$, and $f_2^W = \sigma(q_2)$.

b) Collaborative Training. To facilitate collaborative training of the two branches, we use pseudo-labels $\hat{y}_V^t$ generated by ViT for weakly augmented samples to teach CNN with strongly augmented ones $x^{t,S}$, by minimizing

$$L_{V \to C} = CE(F_2(E_2(x^{t,S})), \hat{y}_V^t) \cdot \mathbf{1}\left(\max(\hat{y}_V^t) > \theta_1\right), \tag{6}$$

where $\mathbf{1}(\cdot)$ masks the loss calculation to include only target samples with high-confidence pseudo-labels (above a confidence threshold θ_1), and $CE(\cdot)$ is cross-entropy. Similarly, we use pseudo-labels $\hat{y}_C^t$ generated by CNN to teach ViT as:

$$L_{C \to V} = CE(F_1(E_1(x^{t,S})), \hat{y}_C^t) \cdot \mathbf{1}\left(\max(\hat{y}_C^t) > \theta_2\right), \tag{7}$$

where the confidence threshold θ_2 is used in CNN to filter out those less reliable pseudo-labels. With Eqs. (6)–(7), we promote knowledge sharing between the ViT and CNN branches, improving their ability to capture global representations and local spatial hierarchies to enhance robust MRI feature learning. The CNN branch is used for inference due to its efficiency.

Implementation. We utilize a lightweight ViT design [16] and use CNN (*i.e.*, ResNet-34) from MONAI [11]. The ViT and CNN encoders are initially pre-trained on 9,544 MR images from ADNI [9] through autoencoder-based unsupervised image reconstruction. Each of F_1 and F_2 consists of 2 fully connected layers followed by a softmax function. For optimization, we use Stochastic Gradient Descent (SGD) with a momentum of 0.9 and a weight decay of 0.0005. The learning rates for ViT and CNN are set to 1×10^{-4}, with a batch size of 4. The confidence thresholds θ_1 and θ_2 for pseudo-label selection are set to 0.65.

3 Experiment and Discussion

Materials and Image Preprocessing. The target dataset from a local center consists of 56 LLD patients and 33 age- and gender-matched healthy controls (HC) with 3T MRIs. The source dataset includes 96 LLD patients and 53 HC patients with 3T T1w MRIs. MRIs are preprocessed using a standardized pipeline with FreeSurfer [17] and ANTs [18]. The preprocessing steps include N4 bias field correction, linear registration to the AAL3 [19] atlas, brain extraction (resulting in a skull-stripped MRI with $181 \times 217 \times 181$ dimension), motion correction, intensity normalization, and segmentation of regions-of-interest (ROIs) from the AAL3 atlas onto the registered MRI volumes. Additionally, 9,544 auxiliary T1w MRIs from ADNI [9] were used for ViT and CNN encoder pre-training.

Competing Methods. We compared our CDA with 9 competition methods for LLD vs. HC classification. These include 3 traditional domain adaptation methods using 1,373-dimensional handcrafted features extracted via FreeSurfer: **TCA** [20] that minimizes maximum mean discrepancy (MMD) to find a shared

Table 1. Results of different domain adaptation methods in LLD vs. HC classification.

Method	AUC (%)	ACC (%)	SEN (%)	SPE (%)	F1 (%)	BAC (%)	p-value
TCA [20]	49.62 ± 1.95	50.75 ± 1.70	63.88 ± 6.13	35.29 ± 8.32	39.18 ± 6.53	49.59 ± 5.17	0.0001
SCA [21]	51.95 ± 2.76	52.10 ± 1.66	58.26 ± 28.62	37.84 ± 27.41	32.11 ± 24.45	48.05 ± 19.81	0.0005
JDA [22]	54.15 ± 1.38	54.35 ± 1.12	62.03 ± 1.45	46.27 ± 4.01	50.54 ± 4.61	54.15 ± 2.13	0.0002
BNM [23]	56.26 ± 13.36	59.31 ± 5.52	65.56 ± 12.86	49.09 ± 16.86	46.78 ± 8.05	57.33 ± 10.60	0.0055
DAN [24]	54.32 ± 8.91	57.93 ± 5.51	68.89 ± 20.36	40.00 ± 22.71	39.01 ± 10.42	54.45 ± 15.25	0.0006
DSAN [25]	55.27 ± 17.62	54.48 ± 8.55	57.78 ± 14.74	49.09 ± 21.97	43.11 ± 14.31	53.44 ± 13.23	0.0322
DANN [26]	57.88 ± 9.46	57.93 ± 4.58	66.57 ± 7.43	46.16 ± 10.15	47.49 ± 8.60	56.37 ± 6.29	0.0003
DeepCoral [27]	66.87 ± 1.93	59.31 ± 3.38	**68.90 ± 5.67**	43.63 ± 3.64	44.88 ± 3.24	56.27 ± 3.37	0.0150
DAAN [28]	67.37 ± 10.98	65.28 ± 3.38	68.89 ± 7.54	42.73 ± 21.97	55.15 ± 16.13	55.81 ± 11.61	0.0327
CDA (Ours)	**74.18 ± 6.37**	**66.43 ± 4.29**	65.52 ± 9.13	**60.42 ± 9.50**	**61.18 ± 12.67**	**62.97 ± 6.59**	–

subspace between source and target domains, **SCA** [21] that projects original features into a reproducing kernel Hilbert space and performs domain adaptation, and **JDA** [22] that integrates MMD into principal component analysis to build a robust representation; and 6 deep domain adaptation methods with 3D MRI scans as input: **BNM** [23] that enhances feature discriminability by maximizing batch nuclear norm, **DAN** [24] that applies MMD to multiple layers of features in deep neural networks to align the distributions of source and target domains, **DSAN** [25] that minimizes distribution differences between the source and target domains across multiple subspaces through a subdomain matching strategy, **DANN** [26] that uses a Gradient Reversal Layer for adversarial training to learn domain-invariant features, **DeepCoral** [27] that minimizes the difference in means and covariance matrices between source and target features, and **DAAN** [28] that dynamically adjusts the feature alignment strategy via adversarial training to balance global alignment and class-conditional alignment.

For all methods, we randomly split the target data into 70/30 training and test sets, repeat this process five times, and record the average and standard deviation results. Six metrics are used: area under the ROC curve (AUC), accuracy (ACC), sensitivity (SEN), specificity (SPE), F1-score (F1), and balanced accuracy (BAC). Paired t-test (with a significance level of 0.05) is conducted to assess significant differences between CDA and each competing method.

Classification Results. Results of different methods in LLD vs. HC classification are reported in Table 1. One can see from Table 1 that our CDA consistently outperforms other approaches across most metrics. Notably, CDA achieves the highest AUC of 74.18%, surpassing the second-best method 6.81%, DAAN, by a significant margin. Deep learning models (*e.g.*, BNM, DANN, DeepCoral, DAAN and CDA) are generally superior to traditional domain adaptation methods (*i.e.*, TCA, SCA, and JDA) that rely on handcrafted features, validating the effectiveness of data-driven features in enhancing domain adaptation performance. Additionally, our CDA achieves better results in most metrics, compared with 6 deep methods. The possible reason is that CDA leverages large-scale MRI datasets for pre-training, enhancing model robustness and generalization to a new target cohort. While DeepCoral achieves the highest sensitivity (SEN) of 68.90%, its performance in other metrics such as SPE and F1 score is suboptimal. In contrast, CDA maintains a balanced performance across all metrics, effectively addressing the trade-off between sensitivity and specificity.

Table 2. Results of CDA and its three variants in LLD vs. HC classification.

Method	AUC (%)	ACC (%)	SEN (%)	SPE (%)	F1 (%)	BAC (%)	p-value
CDA-S_1	58.57 ± 2.68	58.33 ± 4.84	66.65 ± 4.07	44.49 ± 10.77	50.57 ± 7.41	55.57 ± 5.76	0.0006
CDA-$S_{1,2}$	60.00 ± 3.50	58.52 ± 8.21	63.10 ± 12.18	54.44 ± 8.38	50.64 ± 3.80	58.77 ± 7.39	0.0212
CDA-$S_{1,3}$	63.57 ± 8.92	60.09 ± 6.14	**68.57 ± 7.71**	45.99 ± 4.69	49.97 ± 5.45	57.28 ± 4.51	0.0116
CDA (Ours)	**74.18 ± 6.37**	**66.43 ± 4.29**	65.52 ± 9.13	**60.42 ± 9.50**	**61.18 ± 12.67**	**62.97 ± 6.59**	–

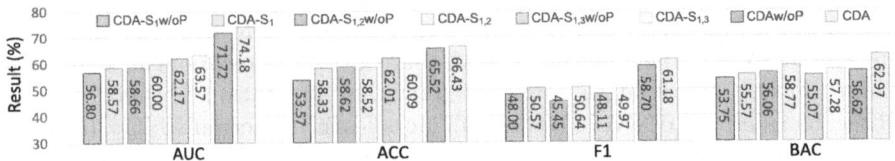

Fig. 2. Comparison of CDA and its 3 variants and their counterparts (*i.e.*, CDAw/oP, CDA-S_1w/oP, CDA-$S_{1,2}$w/oP, and CDA-$S_{1,3}$w/oP) without encoder pre-training.

Ablation Study. We compare the CDA with its three variants: 1) **CDA-S_1** with only Stage 1 (no fine-tuning and collaborative training), 2) **CDA-$S_{1,2}$** with Stage 1 and Stage 2 (no collaborative training), and 3) **CDA-$S_{1,3}$** with Stage 1 and Stage 3 (excluding fine-tuning), with results reported in Table 2. This table suggests that our CDA achieves an overall better performance than its three variants. This highlights the effectiveness of integrating all three stages. CDA-S_1 exhibits the poorest performance, implying that applying a source-trained model directly to target data leads to poor results. CDA-$S_{1,2}$ shows improvement over CDA-S_1, highlighting the importance of feature alignment in reducing domain discrepancy. Particularly, CDA-$S_{1,3}$ exhibits superior performance compared to CDA-$S_{1,2}$ in terms of AUC, ACC, and SEN, which suggests that pseudo-label guided collaborative learning enhances the model's discriminative ability.

Influence of Encoder Pre-training. We further study the influence of ViT and CNN encoder pre-training, with results given in Fig. 2. This figure shows that CDA outperforms its variant, CDA/woP, which lacks pre-training, in most cases. Similarly, CDA-S_1, CDA-$S_{1,2}$, and CDA-$S_{1,3}$ demonstrate superior performance compared to their respective versions without pre-training. This indicates that encoder pre-training plays a crucial role in boosting the performance of CDA.

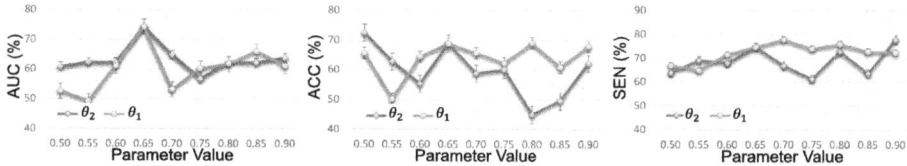

Fig. 3. Results of CDA using different values of confidence thresholds (θ_1 and θ_2).

Influence of Confidence Threshold. We vary the confidence thresholds (θ_1 and θ_2) and report the results of CDA in Fig. 3. We fix $\theta_1 = 0.65$ and vary θ_2 within $\{0.50, 0.55, \cdots, 0.90\}$. With fixed $\theta_2 = 0.65$, we vary θ_1 within the same range. This figure suggests that the choice of thresholds influences the performance of CDA. With fixed θ_1, the best performance is achieved with $\theta_2 = 0.65$. Lower θ_2 values (*e.g.*, 0.50, 0.55) result in moderate results, while higher θ_2 values (*e.g.*, 0.80, 0.85) reduce ACC and AUC but maintain high SEN. When fixing θ_2, CDA with a small θ_1 (*e.g.*, 0.55) yields worse performance. The likely reason is that a small θ_1 makes the model accept low-confidence pseudo-labels generated by ViT, hindering knowledge sharing between ViT and CNN.

4 Conclusion and Future Work

This paper proposes a Collaborative Domain Adaptation (CDA) framework for LLD detection using heterogeneous MRI data, and leverages ViT and CNN encoders to capture global and local features. Its training consists of supervised training on labeled source data, fine-tuning with feature alignment, and collaborative training on unlabeled target data. Experimental results demonstrate that CDA outperforms several state-of-the-art methods. In future work, we will apply CDA to longitudinal, multi-site datasets to further validate its generalizability. We will also explore adaptive thresholding strategies to dynamically adjust confidence thresholds during training to enhance robustness across domains. Additionally, we plan to investigate hypergraph learning [29,30] to model hierarchical brain structures.

Acknowledgments. This research work was supported in part by NIH grants (Nos. AG073297, AG082938).

Disclosure of Interests. The authors have no competing interests to declare that are relevant to the content of this article.

References

1. Hannon, K., Bijsterbosch, J.: Challenges in identifying individualized brain biomarkers of late life depression. Adv. Geriatr. Med. Res. **5**(4), e230010 (2024)
2. Shen, D., Wu, G., Suk, H.I.: Deep learning in medical image analysis. Annu. Rev. Biomed. Eng. **19**(1), 221–248 (2017)
3. Eitel, F., Schulz, M.A., Seiler, M., Walter, H., Ritter, K.: Promises and pitfalls of deep neural networks in neuroimaging-based psychiatric research. Exp. Neurol. **339**, 113608 (2021)
4. Kunanbayev, K., Shen, V., Kim, D.S.: Training ViT with limited data for Alzheimer s disease classification: an empirical study. In: International Conference on Medical Image Computing and Computer-Assisted Intervention, Springer, pp. 334–343 (2024)
5. Mårtensson, G., et al.: The reliability of a deep learning model in clinical out-of-distribution MRI data: a multicohort study. Med. Image Anal. **66**, 101714 (2020)
6. Babar, M., Qureshi, B., Koubaa, A.: Investigating the impact of data heterogeneity on the performance of federated learning algorithm using medical imaging. PLoS ONE **19**(5), e0302539 (2024)
7. Papoutsaki, M.V., et al.: Standardisation of prostate multiparametric MRI across a hospital network: a London experience. Insights Imaging **12**, 1–11 (2021)
8. Dosovitskiy, A., et al.: An image is worth 16 × 16 words: transformers for image recognition at scale. In: International Conference on Learning Representations (2020)
9. Jack Jr, C.R., et al.: The Alzheimer's disease neuroimaging initiative (ADNI): MRI methods. J. Magn. Reson. Imaging: An Official Journal of the International Society for Magnetic Resonance in Medicine **27**(4), 685–691 (2008)
10. He, K., Zhang, X., Ren, S., Sun, J.: Deep residual learning for image recognition. In: Proceedings of the IEEE Conference on Computer Vision and Pattern Recognition, pp. 770–778 (2016)
11. Cardoso, M.J., et al.: MONAI: An open-source framework for deep learning in healthcare. arXiv preprint arXiv:2211.02701 (2022)
12. Lin, T.Y., Goyal, P., Girshick, R., He, K., Dollár, P.: Focal loss for dense object detection. In: Proceedings of the IEEE International Conference on Computer Vision, pp. 2980–2988 (2017)
13. Saito, K., Watanabe, K., Ushiku, Y., Harada, T.: Maximum classifier discrepancy for unsupervised domain adaptation. In: Proceedings of the IEEE Conference on Computer Vision and Pattern Recognition, pp. 3723–3732 (2018)
14. Menéndez, M.L., Pardo, J., Pardo, L., Pardo, M.: The Jensen-Shannon divergence. J. Franklin Inst. **334**(2), 307–318 (1997)
15. Kullback, S., Leibler, R.A.: On information and sufficiency. Ann. Math. Stat. **22**(1), 79–86 (1951)
16. Touvron, H., Cord, M., Douze, M., Massa, F., Sablayrolles, A., Jégou, H.: Training data-efficient image transformers & distillation through attention. In: International Conference on Machine Learning, PMLR, pp. 10347–10357 (2021)
17. Jenkinson, M., Beckmann, C.F., Behrens, T.E., Woolrich, M.W., Smith, S.M.: FSL. NeuroImage **62**(2), 782–790 (2012)
18. Avants, B.B., Tustison, N., Song, G., et al.: Advanced normalization tools (ANTS). Insight J. **2**(365), 1–35 (2009)
19. Rolls, E.T., Huang, C.C., Lin, C.P., Feng, J., Joliot, M.: Automated anatomical labelling atlas 3. Neuroimage **206**, 116189 (2020)

20. Pan, S.J., Tsang, I.W., Kwok, J.T., Yang, Q.: Domain adaptation via transfer component analysis. IEEE Trans. Neural Netw. **22**(2), 199–210 (2010)
21. Ghifary, M., Balduzzi, D., Kleijn, W.B., Zhang, M.: Scatter component analysis: a unified framework for domain adaptation and domain generalization. IEEE Trans. Pattern Anal. Mach. Intell. **39**(7), 1414–1430 (2016)
22. Long, M., Wang, J., Ding, G., Sun, J., Yu, P.S.: Transfer feature learning with joint distribution adaptation. In: Proceedings of the IEEE International Conference on Computer Vision, pp. 2200–2207 (2013)
23. Cui, S., Wang, S., Zhuo, J., Li, L., Huang, Q., Tian, Q.: Towards discriminability and diversity: batch nuclear-norm maximization under label insufficient situations. In: Proceedings of the IEEE/CVF Conference on Computer Vision and Pattern Recognition, pp. 3941–3950 (2020)
24. Long, M., Cao, Y., Wang, J., Jordan, M.: Learning transferable features with deep adaptation networks. In: International Conference on Machine Learning, PMLR, pp. 97–105 (2015)
25. Zhu, Y., et al.: Deep subdomain adaptation network for image classification. IEEE Trans. Neural Netw. Learn. Syst. **32**(4), 1713–1722 (2020)
26. Ganin, Y., Lempitsky, V.: Unsupervised domain adaptation by backpropagation. In: International Conference on Machine Learning, PMLR, pp. 1180–1189 (2015)
27. Sun, B., Saenko, K.: Deep coral: Correlation alignment for deep domain adaptation. In: Computer Vision–ECCV 2016 Workshops: Amsterdam, The Netherlands, October 8-10 and 15-16, 2016, Proceedings, Part III 14, Springer, pp. 443–450 (2016)
28. Yu, C., Wang, J., Chen, Y., Huang, M.: Transfer learning with dynamic adversarial adaptation network. In: 2019 IEEE International Conference on Data Mining (ICDM), IEEE, pp. 778–786 (2019)
29. Liu, M., Gao, Y., Yap, P.T., Shen, D.: Multi-hypergraph learning for incomplete multimodality data. IEEE J. Biomed. Health Inform. **22**(4), 1197–1208 (2017)
30. Feng, Y., You, H., Zhang, Z., Ji, R., Gao, Y.: Hypergraph neural networks. Proc. AAAI Conf. Artif. Intell. **33**, 3558–3565 (2019)

Decoder-Free Supervoxel GNN for Accurate Brain-Tumor Localization in Multi-modal MRI

Andrea Protani[1,2(✉)], Marc Molina Van De Bosch[1,3], Lorenzo Giusti[1], Heloisa Barbosa Da Silva[1,4], Paolo Cacace[1,5], Albert Sund Aillet[1], Friedhelm Hummel[2], and Luigi Serio[1]

[1] European Organization for Nuclear Research, Geneva, Switzerland
[2] École Polytechnique Fédérale de Lausanne, Lausanne, Switzerland
andrea.protani@cern.ch
[3] Universitat Pompeu Fabra, Barcelona, Spain
[4] Universidade de Coimbra, Coimbra, Portugal
[5] Sapienza Università di Roma, Rome, Italy

Abstract. Modern vision backbones for 3D medical imaging typically process dense voxel grids through parameter-heavy encoder-decoder structures, a design that allocates a significant portion of its parameters to spatial reconstruction rather than feature learning. Our approach introduces **SVGFormer**, a decoder-free pipeline built upon a content-aware grouping stage that partitions the volume into a semantic graph of supervoxels. Its hierarchical encoder learns rich node representations by combining a patch-level Transformer with a supervoxel-level Graph Attention Network, jointly modeling fine-grained intra-region features and broader inter-regional dependencies. This design concentrates all learnable capacity on feature encoding and provides inherent, dual-scale explainability from the patch to the region level. To validate the framework's flexibility, we trained two specialized models on the BraTS dataset: one for node-level classification and one for tumor proportion regression. Both models achieved strong performance, with the classification model achieving a F1-score of 0.875 and the regression model a MAE of 0.028, confirming the encoder's ability to learn discriminative and localized features. Our results establish that a graph-based, encoder-only paradigm offers an accurate and inherently interpretable alternative for 3D medical image representation.

Keywords: Brain Tumor Localization · Graph Neural Networks · Multi-modal MRI · Supervoxel · Regression

1 Introduction

The automated characterization of brain tumors from multi-modal MRI has served as a key benchmark for advancing artificial intelligence in medical imaging, offering crucial support for diagnosis, surgical planning, and treatment

---

A. Protani and M. M. V. De Bosch—These authors contributed equally.

monitoring [2]. For years, convolutional neural networks (CNNs) [22], showed remarkable performance across a wide range of vision tasks [15]. The introduction of transformers motivated a shift towards attention-based architectures to overcome the limited receptive fields of CNNs and capture long-range dependencies [11,12,26]. However, both present challenges: architectural inefficiency and limited interpretability. They rely on parameter-heavy decoders, despite evidence suggesting that stronger encoders are the primary source of performance improvements [14,21,27]. This architectural inefficiency is compounded by a reliance on imprecise, post-hoc explainability methods like Gradient-weighted Class Activation Mapping (Grad-CAM) [23], which treats interpretability as an afterthought. Together, these challenges have motivated parallel trends towards more efficient models with lightweight decoders or even decoder-free architectures [3,16,20], and models with built-in, rather than post-hoc, explainability [13,24].

To address these challenges, we introduce **SuperVoxel Graph Transformer (SVGFormer)**, a new paradigm for image analysis that transforms dense volumetric data into a structured, hierarchical graph representation. Our approach begins by partitioning an image into anatomically coherent supervoxels, which serve as the fundamental nodes in our graph. Unlike traditional graph-based methods that rely on handcrafted statistical features for each node [8,17], SVG-Former introduces a novel, end-to-end hierarchical encoding scheme. At the finest level, a patch-level feature extractor learns deep, content-aware features directly from the raw multi-modal image patches within each supervoxel. These features are then aggregated to form a rich initial representation for each supervoxel node, which is subsequently refined by a Graph Attention Network (GAT) that explicitly models the spatial and contextual relationships between neighboring regions. We posit that this flexible graph-based backbone can learn a rich, hierarchical representation, and can be adapted to other downstream tasks, from region-based property prediction, which we benchmark here with tumor segmentation on the Brain Tumor Segmentation (BraTS) dataset, to whole-graph classification. This design offers two key advantages. First, by being decoder-free, it focuses the entire computational budget on creating a high-quality encoding. Second, it allows for inherent, multi-level interpretability studies, allowing any prediction to be traced back from graph-level interactions to the specific patch features that informed it. This work lays the foundation for a new family of graph-based models, opening avenues for future research in multi-task learning, fine-grained explainability, and novel graph-centric image data augmentation.

Our primary contributions are therefore: *(i)* a novel, end-to-end methodology for 3D medical image analysis that unifies voxel-level feature learning with semantic-level graph reasoning, enabling multi-level interpretability by design; *(ii)* a demonstration that accurate tumor localization can be achieved without complex decoders, shifting the paradigm from high-resolution reconstruction to efficient, region-based property prediction; and *(iii)* an empirical validation of a Transformer-supervoxel architecture on the BraTS dataset, establishing its competitive performance and proving its viability as a new approach for brain tumor analysis.

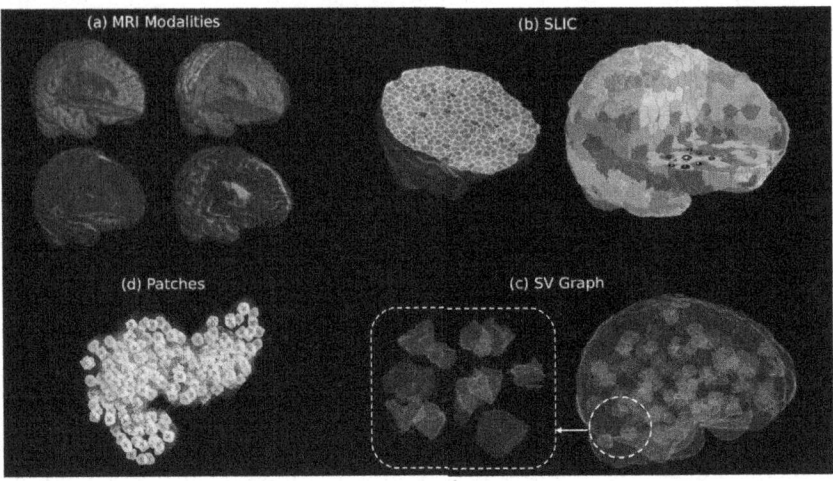

Fig. 1. Pipeline for supervoxel (SV) graph 3D-image encoding: (a) Initial MRI modalities; (b) Apply SLIC to T1-WI (SV grid shared across co-registered modalities); (c) Reconstructed graph of SVs as nodes connected to their 8 nearest neighbors; (d) Patchify method, relying on SV sampling of flattened voxel patches. While the graph is dense and packed, for visualization purposes a sparse graph is represented.

2 Background and Related Works

Deep Learning on Volumetric Brain MRI. Encoderdecoder CNNs such as U-Net [22] and 3D U-Net [29] remain predominant for volumetric brain MRI because their contracting path aggregates multi-scale context. Yet, the dense decoder inflates memory cost and distributes parameters across millions of upsampling weights that do not directly improve feature extraction.

Transformer-Based Encoders for Medical Images. Vision Transformers (ViT) showed that global self-attention can match or exceed convolutions on natural images [10]. Medical adaptations (TransUNet, Swin-UNet, etc.) combine ViT encoders with lightweight decoders, reducing the parameter count while capturing long-range context [5,6]. More recent hybrids move even further toward *encoder-only* designs, indicating a trend towards the investment of computational resources in feature learning rather than pixel-wise reconstruction.

Graph Neural Networks for Irregular Domains. Graph Neural Networks (GNNs) propagate information along the edges of a graph and are naturally suitable for anatomical structures [18]. GAT introduces masked self-attention for learnable neighbor weighting [25], and GATv2 refines this with dynamic attention coefficients that improve expressiveness without extra cost [4]. Recent work highlight their growing role in neuro-imaging for modelling long-range spatial relations and providing interpretable decision paths [19].

Supervoxel-Based Representations. Supervoxels group locally uniform voxels into anatomically meaningful regions. SLIC is the standard baseline for 2D and 3D over-segmentation [1], while energy-optimized variants such as 3D SEEDS [9] deliver order-of-magnitude speed-ups on large clinical datasets [28]. By transforming dense volumes into sparse region graphs, supervoxels make the application of GNNs and Attention mechanisms computationally feasible, especially in early encoding stages where the high dimensionality of the data would otherwise be prohibitive.

3 Supervoxel Graph Transformer

Pre-processing Pipeline. We adopt a graphbased representation of the BraTS 2025 multi-modal 3D MRI scans. Each patient volume comprises four modalities: T1-weighted (T1-WI), contrast-enhanced T1 (T1ce), T2-weighted (T2-WI), and T2-FLAIR (FLAIR) stored at native resolution, along with a voxel-level tumor segmentation mask. As illustrated in Fig. 1, our pre-processing pipeline converts the raw multi-modal MRI data into a compact supervoxel-based graph structure, encoding both spatial and feature-level information. The detailed processing steps are as follows:

i. *Supervoxel Generation:* To convert the continuous image domain into a graph structure, each normalized T1-WI volume is partitioned into a large number of small, locally uniform regions called supervoxels using 3D SLIC clustering. The number of supervoxels, n_{SV}, is a tunable hyper-parameter that controls the granularity of the graph (see Sect. 4). The T1-WI modality is chosen for its stable anatomical contrast, and the resulting supervoxel map is applied uniformly across all modalities to ensure anatomical consistency.

ii. *Dynamic Background Pruning:* To remove the background, supervoxels whose mean T1-WI intensity is below a data-driven threshold are discarded. We compute the largest gap in the sorted negative-mean distribution and set the cut-off $\theta = \frac{1}{2}(m_{(g)} + m_{(g+1)})$ halfway between the two means on either side of that gap. All subsequent processing is restricted to the retained index set $L = \{\ell \mid \bar{I}_\ell > \theta\}$.

iii. *Supervoxellevel tumor masks:* given the 4-class BraTS segmentation, we compute the fraction of voxels within each supervoxel belonging to any tumor class, obtaining a continuous regression target $y_{reg} \in [0, 1]$. In parallel, we define a classification ground truth by binarizing these proportions using a fixed threshold τ, thus assigning each supervoxel to a binary class (tumor vs. non-tumor).

iv. *Patch Extraction:* Within each retained supervoxel ℓ, we use k-means++ to select n_{patch} centroids [7]. Patches are formed from the flattened s nearest-neighbor voxels to each centroid, based on Euclidean distance in world coordinates. Each patch is augmented with its centroid coordinates, and features from all modalities are concatenated, resulting in a supervoxel tensor X_ℓ of shape $(n_{\text{patch}} \times n_{\text{modalities}}, s + 3)$.
v. *Graph Construction:* for each supervoxel node, we compute its centroid as the mean XYZ coordinate of its patches. We then link every node to its $k_{\text{NN}}=8$ nearest neighbors and record the connections in a symmetric adjacency matrix $A \in \{0,1\}^{|L| \times |L|}$, where $A_{ij} = 1$ iff nodes i and j are mutual k_{NN} neighbors.

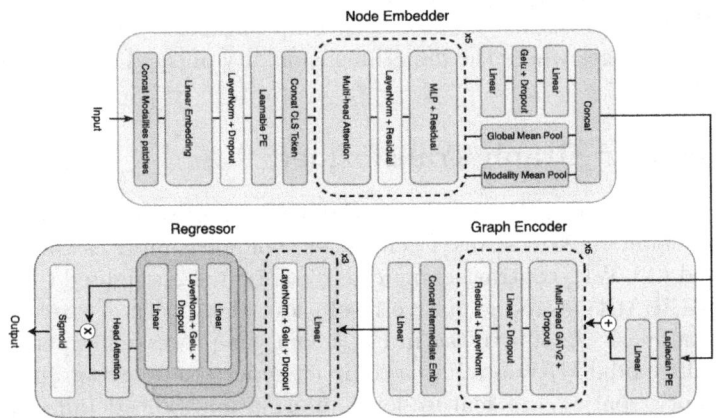

Fig. 2. Architecture of the SVGFormer; Hierarchical Encoder and Regressor. *(i) Node Embedder*: modality-aware patch tokens pass through a 5-layer Transformer to form node features. *(ii) Graph Encoder*: a 5-layer GATv2, augmented with Laplacian positional encodings, propagates context across supervoxels. *(iii) Regressor*: an attention-weighted ensemble of Multi-Layer Perceptron (MLP) heads predicts the tumor fraction for each node.

Patch-Graph Neural Network Architecture. The SVGFormer deep-learning module is the learnable core of the proposed end-to-end pipeline and is organized into three stages. As illustrated in Fig. 2, it begins with a **Node Embedder** that processes the patch tensor for each supervoxel (X_ℓ). Patches are linearly projected into a 256-dimensional space, summed with learnable modality embeddings, and prefixed with a [CLS] token. This sequence is then processed by a 5-layer, 8-head Transformer encoder. The final node representation is a rich descriptor formed by concatenating the output of the [CLS] token, the mean of all patch embeddings, and the per-modality means. Second, a **Graph Encoder** refines these features by modeling inter-supervoxel relationships. We employ a 5-layer GATv2, augmented with Laplacian positional encodings, to propagate context.

A multiscale fusion strategy combines outputs from all GNN layers to yield a final, context-rich embedding for each supervoxel. Finally, a **Regressor** head predicts the tumor proportion. These embeddings pass through a shared MLP into an ensemble of 8 parallel prediction heads. A dedicated attention module then computes a weighted average of these outputs to produce the final sigmoid-scaled prediction in $[0, 1]$, guided by an auxiliary loss that encourages diverse head specialization.

An advantage of the SVGFormer architecture is its built-in, dual-scale explainability. High-level predictions made on a supervoxel node by the Regressor can be traced back through the Graph Encoder to identify which neighboring supervoxels were most influential. Furthermore, the feature representation of any individual supervoxel can be analyzed by examining the attention weights within its patch-level Transformer encoder. This allows a direct link from a final prediction to the specific intra-supervoxel patch features that contributed most.

4 Experiments

Dataset. All experiments are conducted on the Pre-Treatment Glioma BraTS 2025 training release, which contains 1251 samples. Each sample includes four co-registered MRI modalities (T1-WI, T1ce, T2-WI, and T2-FLAIR) and a voxel-level ground-truth mask delineating enhancing core, non-enhancing core, and peritumoral edema. As detailed in Sect. 3, we normalize each volume and over-segment it into supervoxels at four distinct granularities, $n_{SV} \in \{1000, 2000, 3000, 4000\}$, to analyze the effect of graph resolution. For every supervoxel node, we then construct two distinct targets to facilitate the training of separate models: (i) a continuous *regression label*, $y_{\text{reg}} \in [0, 1]$, corresponding to the fraction of constituent voxels belonging to any tumor class, and (ii) a binary *classification label*, derived by thresholding y_{reg} at $\tau = 0.15$. Our evaluation protocol is based on a rigorous 5-fold cross-validation with patient-level splits to ensure that test sets are mutually exclusive and all scans from a single patient reside in the same fold to prevent data leakage. One split was reserved for an extensive hyper-parameter search.

Implementation Details. Models are implemented in PyTorch 2.2 and PyTorch Geometric 2.5, trained on a single NVIDIA A100 (40 GB) GPU. The architecture has 39M parameters (7M embedder, 28M graph encoder, 4M predictor). We use the AdamW optimizer (lr = 3×10^{-5}, weight-decay = 0.01) with cosine-annealing restarts ($T_0 = 100$ epochs, $\gamma = 0.5$). Each fold is trained for 72 hours with a batch size of 2 and accumulation of gradients in 8 steps. All random seeds are fixed per fold for reproducibility.

Evaluation Metrics. For **classification** we report F1-score and ROC-AUC. For **regression** we report MAE, R^2. Metrics are averaged over the four evaluation folds and expressed as mean ± standard deviation.

Experimental Results. To validate the flexibility of the SVGFormer encoding framework, we instantiated and trained two separate, specialized models: a **classification model** to assign a binary tumor label to each supervoxel, and a **regression model** to predict the precise tumor fraction within each supervoxel. Both models leverage the same core encoder but are optimized for their distinct tasks. We evaluated both models across four levels of graph granularity (n_{SV}) to assess their performance and robustness, with the results summarized in Tabel 1.

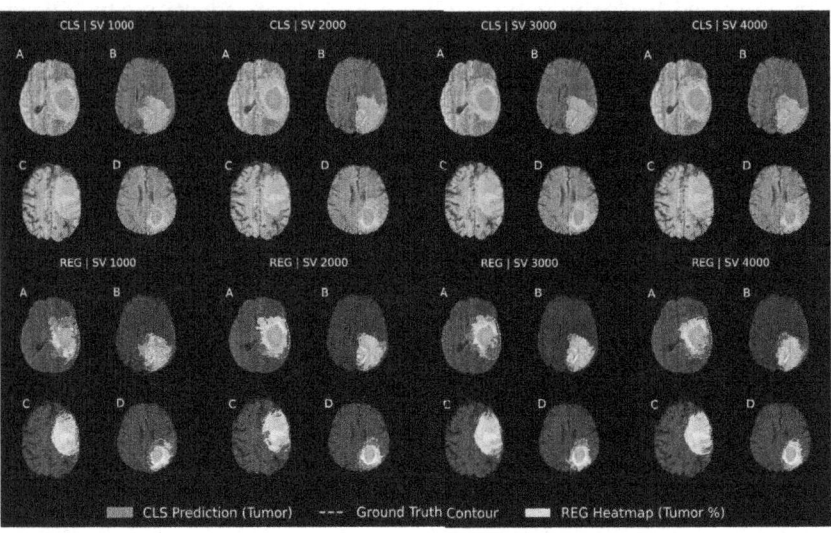

Fig. 3. Qualitative analysis of model predictions on four different patients. The top row shows results from binarized *Classification* models, while the bottom row shows predictions from the *Regression* models. From left to right, the columns correspond to models trained with *1000*, *2000*, *3000*, and *4000 supervoxels*. Each of these eight cells is itself a 2×2 quadrant, displaying a sagittal view of the model's prediction for each of the four patients (identified as A-D at the corner).

Model Performance. The classification model demonstrates consistently high performance across all granularities. With F1-scores remaining high (0.863–0.875) and ROC-AUC scores exceeding 0.97, the model can predict the presence or absence of the tumor in each supervoxel. This confirms that SVGFormer encoder learns discriminative features for localized binary classification objectives. Similarly, the regression model shows strong predictive capability. The low MAE, particularly at the 1000-SV level (**0.028**), and the high coefficient of determination ($R^2 = 0.793$) indicate that the encoder is capable of learning from a fine-grained target such as the localized tumor proportion. As granularity increases, node-level error slightly increases; this does not correspond to a clear degradation in the qualitative visualizations Fig. 3.

Table 1. Performance of two specialized models built on the SVGFormer encoder: a classification model and a regression model. Results show mean ± SD across four folds under varying graph granularities. These metrics evaluate the quality of node-level predictions for each task.

n_{SV}	Classification ↑		Regression	
	F1-score	ROC-AUC	MAE ↓	R^2 ↑
1000	**0.875 ± 0.006**	**0.976 ± 0.003**	**0.028 ± 0.001**	**0.793 ± 0.009**
2000	0.873 ± 0.006	**0.976 ± 0.003**	0.030 ± 0.001	0.749 ± 0.022
3000	0.863 ± 0.013	0.973 ± 0.006	0.032 ± 0.001	0.718 ± 0.014
4000	0.863 ± 0.003	0.973 ± 0.003	0.034 ± 0.001	0.697 ± 0.020

Framework Validation. Overall, these experiments validate our central hypothesis that a multi-stage graph-based supervoxel encoding pipeline serves as a powerful and versatile backbone for learning rich representations. It can be effectively adapted to distinct, node-level prediction tasks, both classification and regression, achieving high performance in both domains. Moreover, its applicability could be broadened by adding specific layers on top for tasks such as dense-segmentation or image reconstruction, all relying on a shared encoder backbone. To test the generalization power of our encoder, we evaluated its features on a downstream segmentation task without requiring any finetuning with a segmentation loss or adding any layer. Fixing a threshold over the output of the regression model slightly higher than the associated absolute error ($\tau = 0.04$), we achieve mean Dice scores ranging from $0.62 - 0.75$. Hence, this framework produces spatially coherent results for a task it was not trained on, proving robust region-specific representations.

5 Conclusion and Future Work

In this work, we introduced **SVGFormer**, a flexible, decoder-free framework that effectively bridges the gap between dense volumetric data and semantic graph processing for 3D medical image analysis. By converting multi-modal MRI volumes into supervoxel graphs, our end-to-end pipeline allocates its entire computational budget to a powerful Transformer-GAT encoding stack. Our experiments demonstrate that this approach produces high-quality, node-level representations for both classification and regression tasks.

The learnable graph-based image encoding pipeline serves as a backbone for general, multi-task models that could be used for finetuning specific model variations for predicting tumor severity, patient survival, and other clinical outcomes. A key advantage of this approach is its inherent multi-level explainability; the model's dual-scale attention mechanisms provide a clear path to trace high-level predictions back to the specific patch-level features that informed them, a crucial step for clinical translation. Furthermore, we plan to extend this architecture to perform fine-grained, decoder-free multi-class segmentation by regressing a vector of different tissue proportions for each supervoxel.

In conclusion, our work establishes that graph-based, encoder-only pipelines can be both performant and interpretable. By shifting the focus from voxel-level reconstruction to region-based graph encoding, SVGFormer provides a robust foundation for a new generation of multi-task, explainable models in medical imaging.

References

1. Achanta, R., Shaji, A., Smith, K., Lucchi, A., Fua, P., Susstrunk, S.: SLIC Superpixels compared to state-of-the-art Superpixel methods. IEEE Trans. Pattern Anal. Mach. Intell. **34**, 05 (2012)
2. Bakas, S., et al.: Identifying the best machine learning algorithms for brain tumor segmentation, progression assessment, and overall survival prediction in the brats challenge. arXiv preprint arXiv:1811.02629 (2018)
3. Brasé, G., Ošep, A., Leal-Taixé, L.: Native segmentation vision transformers (2025)
4. Brody, S., Alon, U., Yahav, E.: How attentive are graph attention networks? arXiv preprint arXiv:2105.14491 (2021)
5. Cao, H., et al.: Swin-unet: Unet-like pure transformer for medical image segmentation (2021)
6. Chen, J., et al.: TransUNet: Transformers make strong encoders for medical image segmentation (2021)
7. Choo, D., Grunau, C., Portmann, J., Rozhoň, V.: k-means++: few more steps yield constant approximation (2020)
8. Cosma, R.A., Knobel, L., der Linden, P.A.V., Knigge, D.M., Bekkers, E.J.: Geometric superpixel representations for efficient image classification with graph neural networks. In: 4th Visual Inductive Priors for Data-Efficient Deep Learning Workshop (2023)
9. Bergh, M.V., Boix, X., Roig, G., Gool, L.V.: Superpixels extracted via energy-driven sampling, Seeds (2013)
10. Dosovitskiy, A., et al.: An image is worth 16x16 words: Transformers for image recognition at scale (2021)
11. Hatamizadeh, A., Nath, V., Tang, Y., Yang, D., Roth, H.: Xu, D.: Swin UNETR: Swin transformers for semantic segmentation of brain tumors in MRI images (2022)
12. Hatamizadeh, A., et al.: UNETR: Transformers for 3D medical image segmentation (2021)
13. Hou, J., et al.: Self-explainable AI for medical image analysis: A survey and new outlooks (2024)
14. Ibtehaz, N., Rahman, M.S.: MultiResUNet: rethinking the U-Net architecture for multimodal biomedical image segmentation. Neural Netw. **121**, 74–87 (2020). Jan.
15. Isensee, F., Jaeger, P.F., Kohl, S.A.A., Petersen, J., Maier-Hein, K.H.: nnU-Net: a self-configuring method for deep learning-based biomedical image segmentation. Nat. Methods **18**(2), 203–211 (2020). Dec.
16. Kerssies, T., et al.: Your ViT is secretly an image segmentation model (2025)
17. Khatun, Z., et al.: Beyond pixel: Superpixel-based MRI segmentation through traditional machine learning and graph convolutional network. Comput. Methods Programs Biomed. **256**, 108398 (2024). Nov.

18. Kipf, T.N., Welling, M.: Semi-supervised classification with graph convolutional networks (2017)
19. Luo, X., et al.: Graph neural networks for brain graph learning: A survey (2024)
20. Ni, J., Mu, W., Pan, A., Chen, Z.: Rethinking the encoder–decoder structure in medical image segmentation from releasing decoder structure. J. Bionic Eng. **21**(3), 1511–1521 (2024). Apr.
21. Rahman, M.M., Munir, M., Marculescu, R.: EMCAD: efficient multi-scale convolutional attention decoding for medical image segmentation (2024)
22. Ronneberger, O., Fischer, P., Brox, T.: U-Net: Convolutional networks for biomedical image segmentation (2015)
23. Selvaraju, R.R., et al.: Grad-CAM: Why did you say that? Visual explanations from deep networks via gradient-based localization. CoRR, abs/1610.0239 (2016)
24. Sun, Q., Akman, A., Schuller, B.W.: Explainable artificial intelligence for medical applications: A review (2024)
25. Veličković, P., Cucurull, G., Casanova, A., Romero, A., Lio, P., Bengio, Y.: Graph attention networks. arXiv preprint arXiv:1710.10903 (2017)
26. Wald, T., et al.: Primus: Enforcing attention usage for 3D medical image segmentation (2025)
27. Xie, E., Wang, W., Yu, Z., Anandkumar, A., Alvarez, J.M., Luo, P.: SegFormer: Simple and efficient design for semantic segmentation with transformers (2021)
28. Zhao, C., Jiang, Y., Hollon, T.C.: Extending seeds to a supervoxel algorithm for medical image analysis (2025)
29. Çiçek, Ö., et al.: 3D U-Net: Learning dense volumetric segmentation from sparse annotation (2016)

Graph Conditioned Diffusion for Controllable Histopathology Image Generation

Sarah Cechnicka[1(✉)], Matthew Baugh[1], Weitong Zhang[1], Mischa Dombrowski[2], Zhe Li[2], Johannes C. Paetzold[3,4], Candice Roufosse[1,5], and Bernhard Kainz[1,2]

[1] Department of Computing, Imperial College London, London, UK
sc7718@imperial.ac.uk
[2] Department AIBE, FriedrichAlexander University ErlangenNürnberg, DE, Nürnberg, Germany
[3] Weill Cornell Medicine, New York City, NY, USA
[4] Cornell Tech, New York City, NY, USA
[5] Centre for Inflammatory Disease, Imperial College London, London, UK

Abstract. Recent advances in Diffusion Probabilistic Models (DPMs) have set new standards in high-quality image synthesis. Yet, controlled generation remains challenging particularly in sensitive areas such as medical imaging. Medical images feature inherent structure such as consistent spatial arrangement, shape or texture, all of which are critical for diagnosis. However, existing DPMs operate in noisy latent spaces that lack semantic structure and strong priors, making it difficult to ensure meaningful control over generated content. To address this, we propose graph-based object-level representations for Graph-Conditioned-Diffusion. Our approach generates graph nodes corresponding to each major structure in the image, encapsulating their individual features and relationships. These graph representations are processed by a transformer module and integrated into a diffusion model via the text-conditioning mechanism, enabling fine-grained control over generation. We evaluate this approach using a real-world histopathology use case, demonstrating that our generated data can reliably substitute for annotated patient data in downstream segmentation tasks. The code is available here.

Keywords: Diffusion · Histopathology · Graph conditioning

1 Introduction

Many medical fields are undergoing a transformation, moving from the traditional, manual examination of samples towards a fully digitized paradigm. In histopathology, this involves the high-resolution digital scanning of histologically prepared tissue samples, which are then processed into

Whole Slide Images (WSIs). This shift opens the door for a transition from primarily qualitative assessments to more quantifiable, automated, and feature-based analysis using Machine Learning (ML) algorithms. However, the vast size of WSIs, with resolutions upwards of 60,000 by 60,000 pixels per sample, containing thousands of structures, makes it a challenging task. This complexity, coupled with the difficulties in manual annotation for supervised ML and medical data sharing constraints [23], poses significant barriers to the clinical adoption of ML in digital pathology.

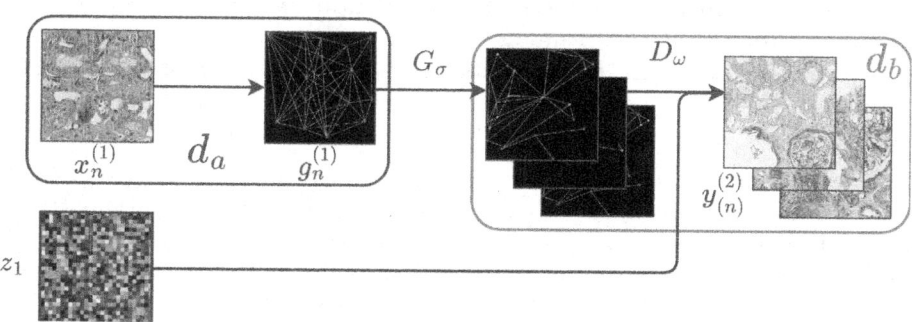

Fig. 1. Overview of our graph-based controllable synthesis pipeline. Graphs extracted from semantic image information (left) are processed by G_σ (e.g. node removal, node changes, node interpolations) to identify optimal graph representations. These representations condition D_ω's to generate synthetic histopathology images (right), creating datasets with enhanced utility for segmentation and diagnostic applications.

There has been a move towards generating synthetic samples to overcome the limitations of insufficient data in the face of privacy issues [5]. Expanding existing datasets with data augmentations has been explored [9], but fail to generate semantic variety by their nature as image operations. Techniques such as Generative Adversarial Networks (GANs) [14], variational autoencoders (VAEs) [12], and diffusion models [23] have shown promising results in learning the underlying image patterns and bridging domain gaps. However, despite synthetic images becoming increasingly indistinguishable from real ones [2,21], a shift in the distribution of sample diversity persists. This discrepancy impacts the efficacy of models trained solely on synthetic data, as they might not accurately represent the wide range of variations found in real-world samples. Part of the issue again lies in the data, as the distribution of data produced by a probabilistic generative model is as unbalanced as the data it was trained on. Without explicit control over feature distributions, the generation process may reinforce existing biases rather than remedy them. Thus models trained for downstream tasks on datasets enriched by such samples do not automatically have enhanced performance [3].

In this paper we address the diversity limitations of synthetic datasets and present a novel method based on graph proxy representations for conditional image generation. Our approach allows for control over the diversity of generated samples through Graph Conditioned Diffusion (GCD). Our contributions are: **(1)** We show that a histopathology image can be represented with a proxy graph, which can accurately guide a diffusion process. **(2)** To enable effective conditioning, we introduce a textual representation of graphs by providing an original architecture that allows for graph tokenization, replacing textual embeddings [26]; **(3)** We take advantage of these insights to investigate the causal relationships between objects in images by subtly altering their graph representations. **(4)** We achieve image generation quality en par with other generation methods while providing synthetic data that faithfully represents the training data distribution for downstream tasks such as image segmentation. Our method enables us to closely align the synthetic datasets with the statistical properties of real datasets, enhancing the utility of synthetic datasets for diagnostic applications and low-barrier data sharing. Our approach is outlined in Fig. 1.

Related Works: Diffusion models have demonstrated the ability to generate synthetic images more closely aligned with target dataset distributions [18,19, 22,27]. Recent applications of Conditional Diffusion Models (CDMs) [3,22,23] have significantly improved the integration of multi-scale information in medical imaging, advancing diagnostic precision and enhancing patient care [21]. However, research consistently shows that models trained exclusively on synthetic data underperform compared to those trained on real data when applied to downstream tasks [3,10], limiting their usefulness in practice. In the context of histopathology, especially within cancer research, graphs have become a of interest for navigating the challenges of WSI classifications, enabling detailed analysis beyond the capability of direct WSI processing [1,20,25]. However, the integration of graphs into image synthesis has not been explored yet.

2 Method

Synthetic Distribution Gap Analysis: Samples drawn from an unconditional diffusion model are not guaranteed to represent the underlying distribution of the original dataset $\tilde{\mathcal{X}}$. In fact, the diffusion model will mainly draw highly typical samples close to the distribution's mean $\mu(\tilde{\mathcal{X}})$. Conditioning of the diffusion model provides more precise control over the sampling process, but this only helps if the conditioning representation allows for targeted decisions to be made to balance datasets. Otherwise, there is no guarantee that the distribution's full range, including potential edge cases, will be captured. We hypothesize that graphs constructed from key features of the images in the original datasets provide such means of conditioning, as they offer a structured and interpretable encoding of the data, allowing for explicit decision-making regarding which structures should appear in specific contexts, and enabling more balanced and diverse sampling while maintaining control over the model's outputs.

Graph Construction: Ground Truth (GT) graphs are generated using a defined protocol. The pixel-level Center of Mass (COM) is calculated for all segmented objects of each image from the GT label masks. Different classes are marked with distinct labels, and the vertex set is connected according to the following criterion: for vertices v_i and v_j with centers of mass C_i and C_j, an edge (i, j) exists if and only if the line segment from C_i to C_j does not intersect any other labeled region, i.e.,

$$\forall t \in [0,1], (1-t)C_i + tC_j \notin \bigcup_{k \neq i,j} R_k.$$

Several examples demonstrating this can be seen in the "images" row of Fig. 2.

Graph Textual Embedding: To condition a diffusion model on a graph structure, we introduce a transformer with modified attention masks, specifically designed to replace the standard auto-regressive text-conditioning mechanism. This involves the use of adjacency matrices $\mathbf{A}$ as the basis for constructing attention matrices, effectively embedding graph structural information into the model. This means our attention mechanism is defined as softmax $\left(\frac{\mathbf{Q}\mathbf{K}^T}{\sqrt{d_k}} \odot \mathbf{A} \right)$, where $\mathbf{Q}$ and $\mathbf{K}$ are the query and key matrices of dimensions $N \times d_k$, derived from the graph nodes' feature embeddings, and d_k is the dimensionality of the key vectors, ensuring the scaled dot-product attention incorporates the adjacency matrix $\mathbf{A}$. The feature embeddings $\mathbf{F}$ of nodes are designed to replace traditional textual embeddings and are defined as $\mathbf{F} = \mathbf{E}_{class} \oplus \mathbf{E}_{BYOL} \oplus \mathbf{E}_{pos}$, where $\oplus$ denotes concatenation. The first component $\mathbf{E}_{class}$ is a one-hot vector encoding the nodes class. The second component $\mathbf{E}_{BYOL}$ is the result of applying a binary mask to the original image such that only pixels belonging to the target object are kept, followed by obtaining a vector embedding using a convolutional neural network trained with BYOL [11] on such masked images. Finally the third component $\mathbf{E}_{pos}$ is the positional encoding generated using a sinusoidal function from vision transformers. This configuration enables the diffusion model to be conditioned on the graph by embedding rich, graph-structured information into the model's learning process.

Graph Conditioned Diffusion (GCD): Following graph construction, we leverage these graphs G to condition an SDE-based diffusion process. The forward diffusion process follows the standard stochastic formulation; gradually adding noise to an image $\mathbf{x_0}$ over time $t \in [0, 1]$ to produce images with varying degrees of noise $\mathbf{x_t}$. This process transforms a structured image into pure noise, enabling the learning of a reverse process that can generate realistic images from noise. For graph-conditioned generation, the backward process aims to reverse the process conditioned on the constructed G. To facilitate this we treat each graph as a high-dimensional entity characterized by $\mathbf{G} \in \mathbb{R}^{N \times F} \times \mathbb{R}^{N \times N}$, where N represents the number of nodes with F dimensional features, alongside an adjacency matrix $\mathbf{A}$ for edge attributes. After incorporating this as additional conditioning for the neural model, the sample during reverse process $\boldsymbol{x}_s$ is defined as:

$$\boldsymbol{x}_s = \tilde{\boldsymbol{\mu}}_{s|t}(\boldsymbol{x}_t, \hat{\boldsymbol{x}}_\theta(\boldsymbol{x}_t, \lambda_t, \boldsymbol{g})) + \sqrt{(\tilde{\sigma}_{s|t}^2)^{1-\gamma}(\sigma_{t|s}^2)^\gamma} \epsilon, \tag{1}$$

where $\lambda_t = \log\left[\frac{\alpha_t^2}{\sigma_t^2}\right]$ represents the noise schedules. The entire GCD model adheres to score-matching principles and incorporates a weighted variational lower bound within the graph embedding framework:

$$\mathbb{E}_{t,x_0 \sim p(x), \epsilon \sim \mathcal{N}(0,I), g}[w(\lambda_t) \| \hat{x}_\theta(x_t, \lambda_t, g) - x \|_2^2]. \quad (2)$$

We incorporate this graph conditioning into a cascaded diffusion model [22], initially generating images at a resolution of 64×64 followed by two super-resolution models upsampling it first to 256×256 and then 1024×1024. All three models are augmented with our graph conditioning.

Graph Interventions: Following our flexible sampling in Eq. 1, we propose a graph augmentation mechanism. For each graph G, this process begins with minimal modifications, such as the removal of a single node, denoted as G^{-v}, or the change of a node's class from an initial class c to a new class c', represented as $G^{v:c \to c'}$. More generally, any modification to a specific node v, such as alterations to its features or attributes, can be expressed as $G^{v,\sigma}$, where σ captures the nature of the modification.

These targeted interventions form the foundation for more extensive intervention mechanisms as it has been explored for molecular modelling [13] and image augmentation techniques like Cut-Paste [7]. This process begins by identifying segments within graphs by locating lone bridges-edges whose removal results in the creation of subgraphs. These identified subgraphs are then randomly mixed and matched with other subgraphs from different graphs in the training set, emulating a Cut-Paste methodology adapted for graph structures. For each graph G in the training set, we identify edges e that, when removed, partition G into two distinct subgraphs, G_1 and G_2. Then we randomly select pairs of subgraphs (G_i, G_j) from the pool of generated subgraphs across the training set and combine the selected pairs to form new graphs $G_{new} = G_i \oplus G_j$, where $\oplus$ denotes the operation of mixing and matching subgraph structures.

In cases where direct mixing and matching are infeasible or result in raph structures that violate the defigned design constraints, we employ a linear interpolation strategy between two graphs, G_a and G_b, to synthesize intermediate graph structures. This option serves as a mechanism to generate variations between existing graph structures, enhancing the diversity of the augmented dataset.

3 Experiments

Data Preprocessing: We use an in-house Kidney Transplant Pathology WSI dataset (courtesy of Charing Cross Hospital) containing 334 patients, similar in construction to [17,24]. Diagnosis of kidney pathology images is based predominantly on the appearance and configuration of structures such as tubules and glomeruli, meaning correctly replicating the distribution of such objects is vital. We randomly select 6 WSIs for testing on downstream tasks, and use the remaining to train the diffusion models. As each WSI has a resolution of $40,000 \times 40,000$

we split them into 1024×1024 pixel sections and discard any non-tissue or empty patches. The remaining 1654 sections are annotated for kidney cortex features by an expert pathologist and an assistant, of which 68 belong to the test images. In order to train each stage of our cascaded diffusion model we make copies of each patch at both 64 × 64 and 256 × 256 resolution, applying simple data augmentations such as shifting and flipping dynamically during training.

Training Setup: All models are trained on a compute node with eight Nvidia A100 GPUs for an average training time of 72 h.

Metrics: To evaluate generative models, we employ several complementary metrics. The Fréchet Inception Distance (FID) evaluates generated image quality and diversity by comparing two Gaussian distributions $\mathcal{N}(\boldsymbol{\mu}_r, \boldsymbol{\Sigma}_r)$ and $\mathcal{N}(\boldsymbol{\mu}_g, \boldsymbol{\Sigma}_g)$, where r indicates real images and g represents generated samples. The mean $\boldsymbol{\mu}$ and covariance $\boldsymbol{\Sigma}$ are computed from latent-space feature vectors extracted from both real and generated images. Though FID serves as a well-established metric for evaluating distributional similarities between real and generated data, additional metrics provide deeper insights. Improved Precision (IP) quantifies how well the generated samples correspond to the real data manifold, while Improved Recall (IR) measures how effectively the synthetic data manifold encompasses the real data distribution, indicating the model's coverage of data diversity. Both metrics utilize k-Nearest Neighbors distances to create non-parametric approximations of these data manifolds. Together, IR and IP provide insights into a generative model's capability to produce both diverse and high-quality samples that accurately reflect the real data characteristics. We provide additional experiments on a downstream segmentation task that are evaluated on real WSIs to verify the utility of our generated images in a setting closely resembling real-world usage, using Dice Similarity Coefficient (Dice) [4] and Aggregated Jaccard Index (AJI) [15] as evaluation metrics.

4 Results

We compare our GCD against other state-of-the-art diffusion-based methods with results shown in Table 1. Since the primary focus of this work is on enhancing diffusion models' capabilities through graph-based representations, rather than a strict evaluation of their generative abilities, the comparisons are based on existing diffusion-based approaches. As such, an unconditional diffusion model and a mask-conditioned diffusion model were chosen (text-conditioned diffusion was not possible as the dataset has no text labels). Additionally, besides common fidelity metrics (IP, IR, FID) downstream segmentation tasks were performed to better assess the practical utility of the generated images, evaluating not only their visual quality but also their effectiveness in improving performance on task-specific objectives. Our method achieves higher IP and IR scores, indicating that GCD captures the diversity of the real image distribution more effectively. Despite this, the images generated with GCD have a lower FID, although this is not too surprising as previous work has shown that FID does not effectively

Table 1. Comparison of dataset diversity and fidelity for various diffusion model methods. Comparison of the original dataset for generated images on downstream segmentation tasks.

Method	IP↑	IR↑	FID↓	Dice(%)↑	AJI(%)↑
Original images	n.a	n.a	n.a	88.01	62.05
Unconditional Diffusion	0.82	0.57	**10.35**	**90.44**	**66.80**
Mask Conditioned Diffusion	0.36	0.07	162.43	82.00	42.40
Graph Image Conditioned Diffusion	**0.90**	0.30	79.11	89.85	66.60
Graph Text Conditioned Diffusion	0.77	**0.64**	39.78	86.05	59.45

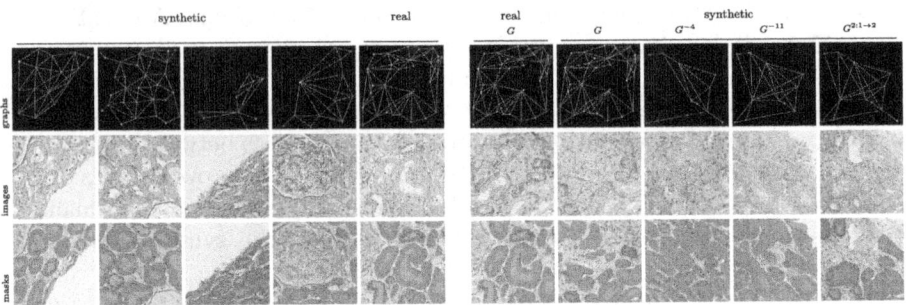

Fig. 2. Left group: synthetic data and a real example. Right group: causally inspired interventions starting with real graph G and synthetic modifications (G^{-4}, G^{-11}, $G^{2:1 \to 2}$).

measure image diversity [6,16]. Looking at the downstream tasks, the models trained on data produced by GCD achieve performance on par with a pure diffusion model while providing the added benefit of control over the image content, whereas the masked diffusion approach performs significantly worse. Figure 2 left illustrates the effectiveness of the generated synthetic data for downstream segmentation tasks.

Ablation Study: To understand the effect of each graph modification, we conducted an ablation study examining different combinations of graph generation and graph conditioning methods, as seen in Table 2. Further, an example of changes for an individual graph is illustrated in Fig. 2 right.

Evaluation: As shown in Tables 1 and 2 and Fig. 2, the GCD models perform effectively across various tasks, consistently producing images that faithfully represent the graphs they are conditioned on both in image and textual form. They generate a more diverse set of samples compared to other methods while maintaining high accuracy, as evidenced by the IP and IR scores in Table 1. Interestingly, occasionally less faithful images that are noisier in appearance lead to higher Dice and AJI scores, as shown in Table 2, on downstream tasks. This finding suggests that diversity in generated samples may be more critical for down-

Table 2. Different methods of feature extractions used as node vector encodings. The method names are the graph intervention followed by the graph encoding. 'Real' uses real graphs from the train set, 'Cut-Paste' and 'Cut-Paste(short)' are examples of $G_{new} = G_i \oplus G_j$ where short restricts the number of subgraphs to 2, and 'Interpolated' linearly interpolates between two graphs. 'Image' stands for Graph-Image-Conditioned-Model, 'Extracted' follows the graph generation models using $\mathbf{F} = \mathbf{E}_{class} \oplus \mathbf{E}_{BYOL} \oplus \mathbf{E}_{pos}$, and 'Manual' relies on manual graph extraction using a centroid, average areas and bounding boxes for each node.

Method	IP↑	IR↑	FID↓	Dice(%)↑	AJI(%)↑
Real + Manual	0.08	0.00	322.7	21.59	14.11
Real + Extracted	0.77	0.64	**39.78**	86.05	59.45
Real + Image	0.81	0.49	55.65	87.35	61.48
Cut-Paste + Manual	0.05	0.00	326.8	33.64	16.93
Cut-Paste + Extracted	0.70	0.60	186.35	37.18	22.32
Cut-Paste + Image	0.88	0.41	83.19	82.68	52.73
Cut-Paste(short) + Manual	0.10	0.00	326.6	87.39	61.59
Cut-Paste(short) + Extracted	0.71	**0.75**	199.2	58.87	31.08
Cut-Paste(short) + Image	0.89	0.37	77.93	79.27	48.36
Interpolated + Manual	0.10	0.00	327.7	36.20	21.34
Interpolated + Extracted	0.65	0.35	201.1	89.35	65.31
Interpolated + Image	**0.90**	0.30	79.11	**89.85**	**66.60**

stream tasks than the realism of individual images. The ablation study results (Table 2) indicate that linear interpolations may be more effective than more complex structural manipulations, such as Cut-Paste. This is likely due to the individual node representations already capturing critical information encoded in the $\mathbf{E}_{BYOL}$, $\mathbf{E}_{pos}$, or the adjacency matrix $\mathbf{A}$, which in turn may result in conflicting signals being transmitted to the model. This is further underlined by simpler changes in Cut-Paste(short) giving better results then Cut-Paste. It is important to note that all these manipulations were used as replacements to train new segmentation models, rather than as data augmentations. As a result, the Dice scores reflect performance on purely synthetic images, not augmented ones. However, as per [3], strong synthetic performance often leads to improved results when combining synthetic and real data.

Discussion: The results show that simple, rule-based interventions lead to better outcomes than more complex transformations. Although this may seem counter-intuitive, it aligns with the sparse representation theory [8] in image reconstruction, which suggests that regularized, simplified structures capture essential features more effectively, while excessive complexity can disrupt these features. The GCD model can be seen as a 'sparsity-guided diffusion' paradigm, where its interventions focus on connectivity and relational patterns, generating clinically relevant diversity without introducing unnecessary complexity.

5 Conclusion

In this work, we introduced a novel approach to generating synthetic histopathological images by introducing GCD models. Our approach leverages graph-based representations to enhance diversity, fidelity, and control in synthetic image generation. By explicitly encoding spatial relationships and anatomical structures, our method preserves critical structural consistency, addressing longstanding challenges in synthetic medical imaging. Furthermore, targeted interventions enrich dataset diversity in clinically meaningful ways. Our results demonstrate that GCD not only improves image diversity metrics but also achieves comparable or superior performance in downstream segmentation tasks compared to traditional methods.

Acknowledgments. S. Cechnicka is supported by the UKRI Centre for Doctoral Training AI4Health (EP / S023283/1). Support was also received from the ERC project MIA-NORMAL 101083647, the State of Bavaria (HTA) and DFG 512819079. HPC resources were provided by NHR@FAU of FAU Erlangen-Nürnberg under the NHR project b180dc. NHR@FAU hardware is partially funded by the DFG 440719683. Dr. Roufosse is supported by the National Institute for Health Research (NIHR) Biomedical Research Centre based at Imperial College Healthcare NHS Trust and Imperial College London (ICL). The views expressed are those of the authors and not necessarily those of the NHS, the NIHR or the Department of Health. Dr Roufosse's research activity is made possible with generous support from Sidharth and Indira Burman. Human samples used in this research project were obtained from the Imperial College Healthcare Tissue & Biobank (ICHTB). ICHTB is supported by NIHR Biomedical Research Centre based at Imperial College Healthcare NHS Trust and ICL. ICHTB is approved by Wales REC3 to release human material for research (22/WA/2836)

Disclosure of Interests. The authors have no competing interests to declare that are relevant to the content of this article.

Appendix

Visual representations of Structural Interventions

Figure 3 shows that simple removal or exchange of a node leads to significantly better visual when inspected. Cut+Paste (G_{cp}) and linear interpolations (G_{int}) likely lead to graphs that are too far out of the distribution of graphs the diffusion model trained and changes of nodes no longer align with adjacency matrixes as dictated by graph generation rules. The graphs are also likely too complex for the system as can be seen with a slightly better handle on (G_{cps}) which reinforces some simplicity in Cut+Paste graphs limiting them to be made up of a maximum of 3 subgraphs. Interestingly, despite this poor visual performance Table 2 shows that segmentation models trained even on visually incorrect data give the model enough context to perform reasonably well on real data. This is likely due to large similarities in most tissue images, showcasing how small differences in performance scores signify much larger differences in relative performance.

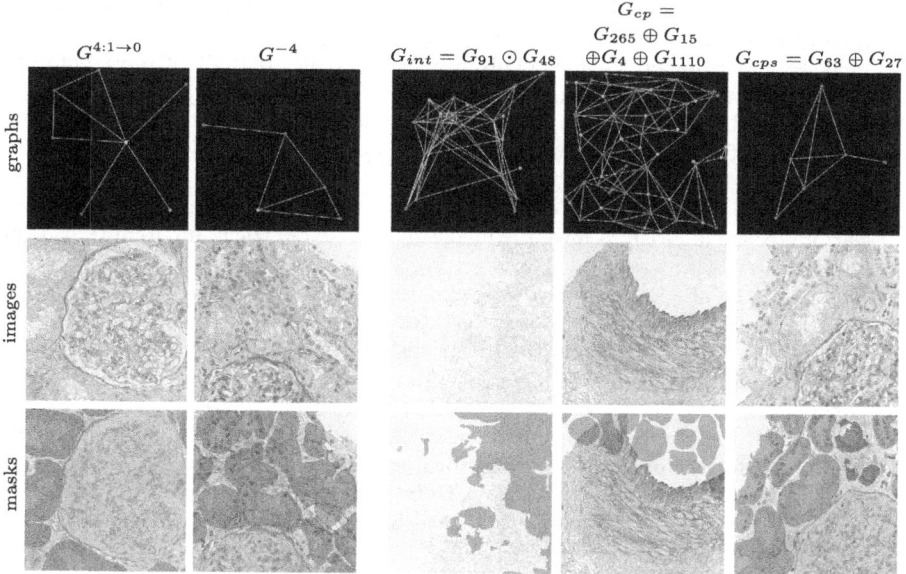

Fig. 3. Effects of both rule-based and complex interventions. Columns depict a graph after the removal of node 4 a graph after the node 4 changed classes from 'tubules' to 'glomerulus' as well as a graph interpolated between graphs 91 and 48, and 2 graphs that had sections of graphs 265, 15, 4 and 1110 and 63 and 27 Cut-Paste to generate them

References

1. Adnan, M., Kalra, S., Tizhoosh, H.R.: Representation learning of histopathology images using graph neural networks. In: 2020 IEEE/CVF Conference on Computer Vision and Pattern Recognition Workshops (CVPRW) (2020)
2. Cechnicka, S., et al.: URCDM: ultra-resolution image synthesis in histopathology. In: Medical Image Computing and Computer Assisted Intervention (MICCAI), pp. 535–545. Springer, Cham (2024)
3. Cechnicka, S., Ball, J., Reynaud, H., Arthurs, C., Roufosse, C., Kainz, B.: Realistic data enrichment for robust image segmentation in histopathology. In: Koch, L., et al. (eds.) Domain Adaptation and Representation Transfer. Springer, Cham (2024)
4. Dice, L.R.: Measures of the amount of ecologic association between species. Ecology **26**(3), 297–302 (1945)
5. Dombrowski, M., Kainz, B.: Quantifying sample anonymity in score-based generative models with adversarial fingerprinting. arXiv preprint arXiv:2306.01363 (2023)
6. Dombrowski, M., Zhang, W., Cechnicka, S., Reynaud, H., Kainz, B.: Image generation diversity issues and how to tame them. arXiv preprint arXiv:2411.16171 (2024)

7. Dwibedi, D., Misra, I., Hebert, M.: Cut, paste and learn: surprisingly easy synthesis for instance detection. In: Proceedings of the IEEE International Conference on Computer Vision (2017)
8. Elad, M., Aharon, M.: Image denoising via sparse and redundant representations over learned dictionaries. IEEE Trans. Image Process. **15**(12), 3736–3745 (2006)
9. Faryna, K., Laak, J., Litjens, G.: Automatic data augmentation to improve generalization of deep learning in H&E stained histopathology. Comput. Biol. Med. **170**, 108018 (2024)
10. Gao, C., et al.: Synthetic data accelerates the development of generalizable learning-based algorithms for x-ray image analysis. Nat. Mach. Intell. **5**(3), 294–308 (2023)
11. Grill, J.B., et al.: Bootstrap your own latent-a new approach to self-supervised learning. Adv. Neural Inf. Process. Syst. **33** (2020)
12. Guleria, H., et al.: Enhancing the breast histopathology image analysis for cancer detection using variational autoencoder. Int. J. Environ. Res. Public Health **20**(5) (2023)
13. Jensen, J.H.: A graph-based genetic algorithm and generative model/Monte Carlo tree search for the exploration of chemical space. Chem. Sci. **10**(12) (2019)
14. Jose, L., Liu, S., Russo, C., Nadort, A., Di Ieva, A.: Generative adversarial networks in digital pathology and histopathological image processing: a review. J. Pathol. Inf. (2021)
15. Kumar, N., Verma, R., Sharma, S., Bhargava, S., Vahadane, A., Sethi, A.: A dataset and a technique for generalized nuclear segmentation for computational pathology. IEEE Trans. Med. Imag. **36**(7), 1550–1560 (2017)
16. Kynkäänniemi, T., Karras, T., Laine, S., Lehtinen, J., Aila, T.: Improved precision and recall metric for assessing generative models. arXiv preprint arXiv:1904.06991 (2019)
17. Loupy, A., Mengel, M., Haas, M.: Thirty years of the international Banff classification for allograft pathology: the past, present, and future of kidney transplant diagnostics. Kidney Int. **101**(4), 678–691 (2022)
18. Meng, C., Rombach, R., Gao, R., Kingma, D.P., Ermon, S., Ho, J., Salimans, T.: On distillation of guided diffusion models (2023)
19. Ramesh, A., Pavlov, M., Goh, G., Gray, S., Voss, C., Radford, A., et al.: Zero-shot text-to-image generation. arXiv preprint arXiv:2102.12092 (2021)
20. Reisenbüchler, D., Wagner, S.J., Boxberg, M., Peng, T.: Local attention graph-based transformer for multi-target genetic alteration prediction. In: MICCAI 2022 (2022)
21. Reynaud, H., et al.: Feature-conditioned cascaded video diffusion models for precise echocardiogram synthesis. In: Greenspan, H., et al. (eds.) MICCAI 2023. Springer, Cham (2023)
22. Saharia, C., et al.: Photorealistic text-to-image diffusion models with deep language understanding. In: Oh, A.H., Agarwal, A., Belgrave, D., Cho, K. (eds.) Advances in Neural Information Processing Systems (2022)
23. Shrivastava, A., Fletcher, P.T.: NASDM: nuclei-aware semantic histopathology image generation using diffusion models. In: Greenspan, H., et al. (eds.) MICCAI 2023. Springer, Cham (2023)
24. Studer, L., van Midden, D., Ayatollahi, F., Hilbrands, L., Kers, J., van der Laak, J.: Monkey dataset (2024). https://monkey.grand-challenge.org/dataset/
25. Weers, A., Berger, A.H., Lux, L., Schüffler, P., Rueckert, D., Paetzold, J.C.: From pixels to histopathology: a graph-based framework for interpretable whole slide image analysis. arXiv preprint arXiv:2503.11846 (2025)

26. Ye, J., et al.: Synthetic sample selection via reinforcement learning. In: MICCAI 2020, Part I 23 (2020)
27. Zhang, W., Zang, C., Li, L., Cechnicka, S., Ouyang, C., Kainz, B.: Stability and generalizability in SDE diffusion models with measure-preserving dynamics. In: NeurIPS (2024)

X-Node: Self-explanation is All We Need

Prajit Sengupta and Islem Rekik(✉)

BASIRA Lab, Imperial-X (I-X) and Department of Computing,
Imperial College London, London, UK
i.rekik@imperial.ac.uk
http://basira-lab.com

Abstract. Graph neural networks (GNNs) have achieved state-of-the-art results in computer vision and medical image classification tasks by capturing structural dependencies across data instances. However, their decision-making remains largely opaque limiting their trustworthiness in high-stakes clinical applications, where interpretability is essential. Existing explainability techniques for GNNs are typically post-hoc and global, offering limited insight into individual node decisions or local reasoning. We introduce **X-Node**, a self- explaining GNN framework in which each node generates its own explanation as part of the prediction process. For every node, we construct a structured *context vector* encoding interpretable cues, such as degree, centrality, clustering, feature saliency, and label agreement within its local topology. A lightweight *Reasoner* module maps this context into a compact *explanation vector*, which serves three purposes: (1) reconstructing the node's latent embedding via a Decoder to enforce faithfulness, (2) generating a natural language explanation using a pre-trained LLM (e.g., Grok or Gemini), and (3) guiding the GNN itself via a "text-injection" mechanism that feeds explanations back into the message-passing pipeline. We evaluate **X-Node** on two graph datasets derived from *MedMNIST* and *MorphoMNIST*, integrating it with GCN, GAT, and GIN backbones. Our results show that X-Node maintains competitive classification accuracy while producing faithful, per-node explanations. https://github.com/basiralab/X-Node.

Keywords: Graph Neural Network · Graph Topological Measures · Interpretable · Explainable AI · Natural Language Processing

1 Introduction

Graph neural networks (GNNs) have emerged as powerful tools for modeling structured medical data, such as cellular interactions in histopathology, organ topologies in medical imaging, and anatomical relationships in population-level brain graphs [1,4,15]. By learning over nodes and edges, GNNs naturally capture both local and global structure in such data, and have shown state-of-the-art performance in various diagnostic tasks [7,11]. However, **interpretability remains a key bottleneck**: in high-stakes clinical environments, it is insufficient for models to be merely accurate—they must also explain

---

GitHub: https://github.com/basiralab/X-Node.

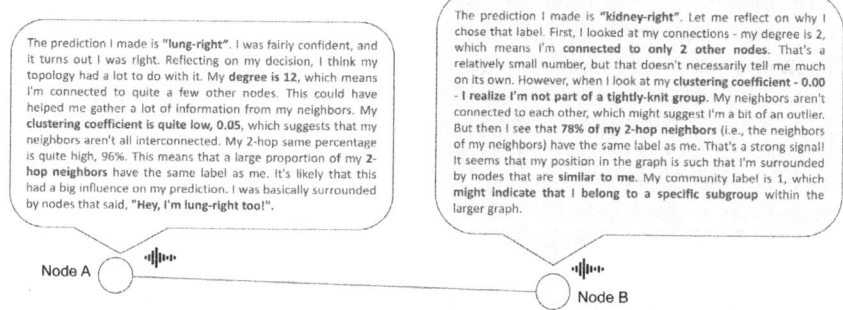

Fig. 1. Two neighbouring nodes self-explaining

their decisions in a way that is faithful, transparent, and verifiable [8, 16]. Despite growing interest in graph explainability [13, 22, 23], current approaches face several critical limitations: **First**, most existing explainability methods for GNNs are *post-hoc* and *non-intrinsic*. Tools such as GNNExplainer [22] or PGExplainer [14] identify subgraphs or features after training, with no guarantee that these explanations reflect the model's actual reasoning process [17]. Such post-hoc explanations are prone to instability and adversarial inconsistency, especially in sparse medical graphs where subtle topological changes can shift model predictions without warning [3]. **Second**, existing models do not offer *localized, node-level reasoning*. A GNN's decision for a given node emerges from hidden message passing over the graph, but rarely can the node itself articulate why it received a certain label. In contrast, clinical reasoning is often local and explainable: radiologists, for instance, interpret findings based on visual features and regional context. Node-level explainability—where each node reasons about its own state as shown in Fig. 1—is notably absent from current architectures. **Finally**, current GNN pipelines treat explanation as disconnected from learning. That is, even if explanations are available, they are not used to guide or constrain training. This decoupling limits the utility of explanations in improving robustness or trust. Worse, explanations can be optimized separately to "look plausible", but still diverge from the actual decision pathway—a phenomenon known as *rationalization over reasoning* [9]. To bridge this gap between decision accuracy and faithful interpretability, we introduce **X-Node**, a fully novel graph learning framework that equips each node with a self-explainable capability. By modeling nodes as introspective agents—capable of constructing contextual explanations, reflecting on their local topology and feature space, and injecting this explainability reasoning back into the network—**X-Node** offers an interpretable and performance-aligned solution to node classification as summarized below:

1. *On a methodological level:* X-Node introduces *self-explainable graph nodes* by embedding explanation-aware learning directly into the GNN. Each node builds a local context, generates an explanation vector, and reinjects it into the network— enabling faithful, intrinsic interpretability beyond post-hoc approaches.

2. *On a clinical level:* X-Node allows each node—representing a patient, organ, or region—to justify its prediction using topological cues, label agreement, and optimization signals. This supports transparent, clinician-aligned decision-making.
3. *On a generic level:* Though evaluated on MedMNIST and MorphoMNIST, X-Node can augment any GNN (e.g., GCN, GAT, GIN) with self-explaining capabilities, offering a modular interpretability layer across graph learning tasks.

2 Related Work

Understanding the landscape of explainability techniques in GNNs is essential for positioning our contribution. Figure 2 summarizes this taxonomy, categorizing existing methods as either post-hoc or ante-hoc, and highlighting where our approach, **X-Node**, fits within this space.

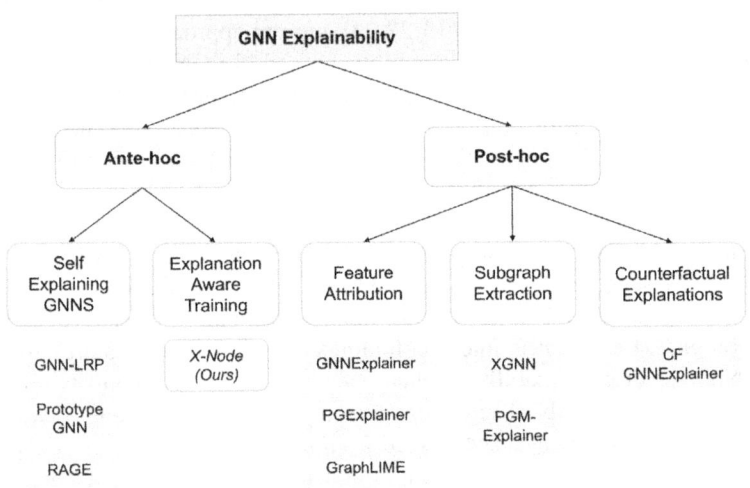

Fig. 2. Types of GNN explainability methods

Graph Neural Networks. Our work builds on the rich literature of GNNs for node classification. Kipf and Welling introduced GCNs, which perform spectral convolutions on graphs [10]. GAT adds attention weights over neighbors [18], while [20] proposed GIN (Graph Isomorphism Networks) showed how certain aggregation functions can make GNNs as powerful as the Weisfeiler-Lehman test. We use these classic architectures (GCN, GAT, GIN) as baselines in our experiments. However, unlike standard GNNs, our architecture includes *reasoning modules* that explicitly compute explanations at every node.

Explainable AI and GNN Explanation. Interpretability has long been recognized as essential in critical fields like healthcare [2]. Many methods seek to explain deep models post-hoc (e.g. LIME, SHAP in vision/NLP), but these do not guarantee faithful introspection [12]. In graphs, GNNExplainer [22] was among the first general, model-agnostic explainers: it finds a subgraph and feature mask that maximize the mutual information with a given prediction, identifying a compact rationalizing substructure. Other approaches (PGExplainer, XGNN) also produce edge/feature importance masks. While useful, these methods are post-hoc and have been shown to be brittle: even small graph perturbations can drastically change their output without affecting the model's prediction [12]. In contrast, X-Node embeds explainability intrinsically into the model. Our approach is inspired by the framework of self-explaining neural networks, where explanations are generated by the model itself as part of the forward pass.

Self-explaining Models. [2] introduced SENN, a class of networks where predictions come with explicit contributions and concepts. Similarly, modular or capsule networks use interpretable sub-components. Our Self-Explainable Nodes generalize this idea to graph domains: each node uses its local graph-based context to self-reason about its label. This is akin to building a hybrid deep learning with symbolic reasoning at the node level. The use of language models for explanations also resonates with recent work on chain-of-thought reasoning in LLMs [19] (Table 1).

Table 1. Limitations of current GNN-based medical pipelines and proposed solutions.

Limitations	Proposed Solutions
Opaque predictions: Standard GNNs yield high accuracy but provide little insight into *why* a decision was made. In medicine this lack of interpretability reduces clinician trust [8,16,22].	**Self-Explainable Nodes:** Each node produces an explanation vector reflecting explicit reasoning (e.g., graph topology and 2-hop label patterns). This makes the models logic transparent.
Post-hoc explanations can mislead: Common explainers (e.g., Grad-CAM, LIME) are post-hoc and may not reflect the model's true reasoning. Clinicians cannot rely on them in high-stakes cases [12,17,22].	**Integrated Explainability:** We incorporate explanation into the training loop (self-explainable AI). Because the Reasoner is part of the model, its explanations are inherently aligned with decision-making, improving faithfulness by aligning symbolic and neural features.
No feedback from explanations: Conventional pipelines do not use explanations to improve learning. The GNN and the explainer are independent, so explanatory signals do not shape the model [2,9].	**Feedback Loop:** The explanation embeddings are *injected* back into the GNN layers. This adaptive reinjection lets the model refine its representations based on its own reasoning, effectively using explanation loss as auxiliary supervision.

LLM-Augmented Graph Explanation Models. Recent advancements have explored the integration of Large Language Models (LLMs) to enhance the interpretability of Graph Neural Networks (GNNs). For instance, GraphXAIN [6] employs LLMs to translate technical outputs, such as subgraphs and feature importance scores, into coherent natural language narratives, thereby improving the understandability of GNN predictions for non-expert users. Similarly, LLMExplainer [24] integrates LLMs as Bayesian inference modules within GNN explanation networks to mitigate learning biases and generate more robust explanations. Our approach diverges from these by utilizing an LLM as an explanation decoder within the GNN framework. In our model, each node constructs a local context vector, which is then transformed into a natural language explanation by the LLM. This explanation is subsequently reintegrated into the

GNN through a process we term "text-injection", guiding further message passing and enhancing interpretability. Unlike methods that rely on symbolic reasoning components, our framework leverages the generative capabilities of LLMs to produce human-readable justifications. Throughout this paper, we are formulating **3 hypotheses**:

H1: Node-level explainability improves interpretability *Each node can produce a faithful, self-contained explanation by summarizing local graph structure and topology.*

H2: Reasoning as regularization improves learning *Injecting explanation vectors into GNNs provides inductive bias, improving generalization and aligning embeddings.*

H3: LLMs enhance explanation *A pre-trained LLM can map structured context into fluent, human-readable text that captures reasoning aligned with clinical logic.*

3 Methodology

Our proposed framework, **X-Node**, introduces self-explainable nodes that classify and explain decisions based on graph structure, features, and topological characteristics during training. The overall pipeline is shown in Fig. 3.

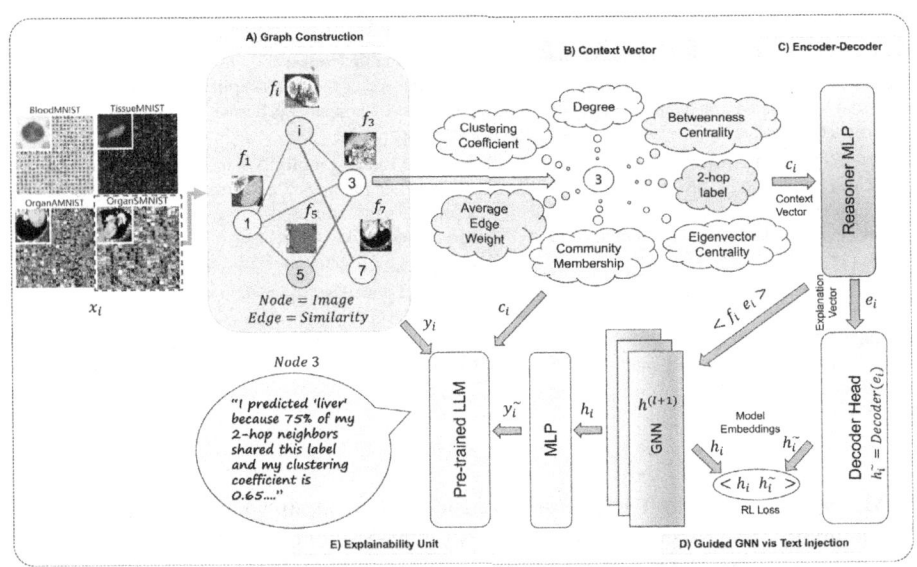

Fig. 3. Proposed X-Node Architecture

A. Problem Setup and Graph Construction. Let $\mathcal{D} = (x_i, y_i)_{i=1}^{N}$ denote a dataset of N medical images with class labels $y_i \in \mathcal{Y}$. Each image x_i is encoded into a feature vector $f_i \in \mathbb{R}^d$ using a pre-trained CNN encoder F, i.e., $f_i = F(x_i)$. We then build a k-nearest neighbor (k-NN) graph $\mathcal{G} = (\mathcal{V}, \mathcal{E}, \mathbf{X})$, $\mathcal{V}$: Nodes representing images.

$\mathbf{X} = [f_1, \ldots, f_N]^\top \in \mathbb{R}^{N \times d}$: Node features. $\mathcal{E}$: Edges based on cosine similarity encoded into the adjacency matrix: $A_{ij} = \cos(f_i, f_j)$ if $f_j \in$ Top-$k(f_i)$; 0 otherwise. This graph encodes both visual and structural similarity across instances as in (Fig. 3-A).

B. Context Vector Extraction from Graph Topology. In graph-based medical datasets, a node's local topology often encodes clinically relevant patterns—such as label homophily (e.g., organs of the same type clustering together), centrality (e.g., typical vs. atypical organs), or community structure as shown in (Fig. 3-B). To capture such interpretable patterns, we construct a compact *context vector* $\mathbf{c}_i \in \mathbb{R}^{d_c}$ for each node v_i based on topological and label-aware descriptors. These serve as reasoning cues for explanation generation and auxiliary supervision. While this vector is numerical for model input, it is preserved as labeled key-value pairs (e.g. "degree": 3, "2-hop agreement": 0.85) where each key is an interpretable feature.

$$\mathbf{c}_i = \text{Concat}(d_i, cc_i, \rho_i^{(2)}, ec_i, bc_i, \bar{w}_i, c_i) \tag{1}$$

The following features are selected for their empirical and theoretical relevance to graph classification, especially in sparse or hierarchical domains like medical imaging:

- **Degree** (d_i): Node connectivity. High degree may indicate prototypical or redundant nodes.
- **Clustering Coefficient** (cc_i): Reflects local cohesion. High values suggest dense neighborhoods.
- **2-hop Label Agreement** ($\rho_i^{(2)}$): Measures semantic consistency in the extended neighborhood:

$$\rho_i^{(2)} = \frac{\text{\# same-label nodes in 2-hop}}{\text{\# 2-hop neighbors}}$$

- **Eigenvector Centrality** (ec_i): Importance in global graph flow. Nodes connected to other important nodes receive high scores.

- **Betweenness Centrality** (bc_i): Captures "bridge" roles. Important for detecting outliers or misclassified nodes.
- **Average Edge Weight** ($\bar{w}_i$): Indicates confidence in neighborhood similarity:

$$\bar{w}_i = \frac{1}{|\mathcal{N}(i)|} \sum_{j \in \mathcal{N}(i)} w_{ij}$$

- **Community Membership** (c_i): Structural cluster ID, indicating coarse graph-level partitioning.

C. Explanation Vector Generation via Reasoner. To interpret a node's behavior, its context vector $\mathbf{c}_i$ is passed through a MLP Reasoner parameterized by trainable weights W_1, W_2 and biases b_1, b_2 to produce a low-dimensional explanation vector $\mathbf{e}_i \in \mathbb{R}^{d_e}$:

$$\mathbf{e}_i = \text{Reasoner}\,\phi(\mathbf{c}_i) = \sigma(W_2 \cdot \text{ReLU}(W_1 \cdot \mathbf{c}_i + b_1) + b_2) \tag{2}$$

D. Embedding Reconstruction via Decoder. To ensure faithfulness, $\mathbf{e}_i$ is decoded to reconstruct the node's latent GNN embedding $\hat{\mathbf{h}}_i \in \mathbb{R}^{d_h}$: $\hat{\mathbf{h}}_i = \text{Decoder}\,\psi(\mathbf{e}_i)$ (Fig. 3-C). This enforces alignment between $\hat{\mathbf{h}}_i$ and the actual embedding $\mathbf{h}_i$ from the GNN.

E. Textual Explanation via LLM. To generate interpretable explanations (Fig. 3-E), each node v_i uses its structured *context vector* $\mathbf{c}_i \in \mathbb{R}^{d_c}$ and its predicted label $\hat{y}_i$ (and optionally, true label y_i) as input to a pre-trained large language model (LLM), such as Grok's `llama-4-scout-17b-16e-instruct` or Google's Gemini 2.5 Pro:

$$\mathcal{T}_i = \text{LLM}_\psi (\text{prompt}(\mathbf{c}_i, \hat{y}_i, y_i)) \tag{3}$$

The LLM serves as a natural language decoder, converting structured node-level statistics into faithful textual rationales $\mathcal{T}_i$ supporting **H3**. The prompt is formatted as follows:

LLM Prompt for Explanation

You are a node in a medical graph.
Your topological context is: `<context_vector>`
Your predicted label: `<predicted_label>`. True label: `<true_label>`
Explain in natural language why you predicted `<predicted_label>`. If incorrect, describe what might have misled you based on your structure, features, and neighbors.

This design allows the node to *self-narrate* its decision logic—either validating a correct classification or introspecting its own failure.

F. Explanation-Guided GNN via Text Injection. The explanation vector $\mathbf{e}_i$ is concatenated with the GNN embedding $\mathbf{h}_i$ for final classification, allowing reasoning signals to directly inform prediction, creating a feedback loop as shown in (Fig. 3-D):

$$\mathbf{z}_i = \text{Concat}(\mathbf{h}_i, \mathbf{e}_i), \quad \hat{y}_i = \text{MLPclass}(\mathbf{z}_i) \tag{4}$$

G. Loss Function and Joint Training. The model is trained by jointly minimizing classification, alignment, and reconstruction losses:

$$\mathcal{L} = \sum_{i=1}^{N} \left[\underbrace{\text{CE}(\hat{y}_i, y_i)}_{\text{Classification}} + \alpha \underbrace{\|\mathbf{e}_i - \mathbf{c}_i\|^2}_{\text{Alignment}} + \beta \underbrace{\|\hat{\mathbf{h}}_i - \mathbf{h}_i\|^2}_{\text{Reconstruction}} \right] \tag{5}$$

4 Results and Discussion

We assess X-Node on six image-derived graph datasets—five from MedMNIST (OrganCMNIST, OrganAMNIST, OrganSMNIST, TissueMNIST, BloodMNIST) [21] and one synthetic benchmark (Morpho-MNIST) [5]. Each dataset is converted into a k-NN graph using pretrained image embeddings. X-Node is evaluated not only for classification accuracy but also for its ability to generate faithful per-node explanations. Results are averaged over **3-fold cross-validation** with seeds 42, 43, and 44 (9 experiments).

Dataset and Experimentation. As shown in Table 2 the datasets span a range of medical domains, with node counts from 17K (BloodMNIST) to 236K (TissueMNIST), and label spaces from 4 to 11 classes. Experiments ran on an Apple M2 Air with 16GB RAM and MPS acceleration GPU, using an 80/20 train-validation split in a 3-fold CV setup. Adding *reasoner* increased epoch time and memory moderately as in Table 3.

Classification Performance. X-Node consistently improves over baseline GNNs across all datasets. For example, on OrganAMNIST, it raises F1 from 91.19% to 93.16%

Table 2. Graph Dataset Details created from Images

	OrganCMNIST	OrganAMNIST	OrganSMNIST	TissueMNIST	BloodMNIST	Morpho-MNIST
# of Nodes	23583	58830	25211	236386	17092	280000
# of Edges	82315	205404	88951	826714	60116	970699
# of Features	512	512	512	512	512	512
# of Labels	11	11	11	8	8	4
Task Type	Multi-class	Multi-class	Multi-class	Multi-class	Multi-class	Multi-class
Training Type	Inductive	Inductive	Inductive	Inductive	Inductive	Inductive
Training Nodes	12975	34561	13932	165466	11959	216000
Validation Nodes	2392	6491	2452	23640	1712	24000
Test Nodes	8216	17778	8827	47280	3421	40000

Table 3. Comparison of **Average Epoch Time (s)** and **Peak Memory (MB)** across datasets.

Method	OrganCMNIST		OrganAMNIST		OrganSMNIST		TissueMNIST		BloodMNIST		MoprhoMNIST	
	Time	Memory	Time	Memory	Time	Memory	Time	Memory	Time	Memory	Time	Memory
GCN	0.48	758.95	1.15	1234.64	0.49	1024.92	2.76	2472.47	0.35	557.36	4.20	2648.61
GCN + Reasoner	1.90	944.25	1.78	1439.06	1.25	889.30	4.76	1781.39	0.38	718.44	4.71	1717.73

and sensitivity from 91.18% to 94.07% as in Table 4. Gains are especially notable in sensitivity—crucial in medical diagnosis showing up to 35% improvement. On large-scale graphs like TissueMNIST and MorphoMNIST, although overall accuracy slightly drops, X-Node improves F1 and sensitivity metrics which is critical in medical domain.

Interpretability (Per-Node Explanations). Beyond classification accuracy, X-Node generates faithful, introspective justifications for each node's prediction. For example, Node 3 from Fig. 3, which produces the following self-explanation based on its context vector c_i and prediction $\hat{y}_i$ supporting **H1**. This reasoning aligns closely with graph-theoretic descriptors like neighborhood sparsity, edge strength, and label agreement:

> Node 3 (True = Kidney-Right, Pred = Femur-Left) Explanation
>
> *"I predicted that I was representing the 'femur-left' class, but my true label is 'kidney-right'. Let's reflect on this. I have a degree of 4, meaning I'm moderately connected. My clustering coefficient is 0.00, which shows my neighbors are not well connected to each other. One potential cause for the error could be my high average edge weight (0.929), indicating strong connections that might have biased my prediction. While my top feature (F[117] = 10.00) may be distinctive for kidney-right, it was likely overridden by misleading neighborhood influence. My community label and low centrality further suggest I'm not centrally embedded in the kidney-right cluster. This misclassification reveals how structural signals can dominate node identity when feature signals are ambiguous."*

The explanation is both structured and contextualized, allowing users to **trace misclassification** as in this example back to interpretable topological features. This per-node transparency highlights the value of explanation-aware training.

Table 4. Comparison of GCN variants on six inductive datasets (best results in green).

	OrganCMNIST				OrganAMNIST			
Method	ACC	F1	Sensitivity	ROC-AUC	ACC	F1	Sensitivity	ROC-AUC
GCN	88.20±0.61	86.03±0.89	86.13±0.06	99.09±0.08	91.85±0.30	91.19±0.33	91.18±0.03	99.51±0.03
GCN + Reasoner	89.22±0.79	87.46±0.87	87.93±0.08	99.18±0.07	93.64±0.21	93.16±0.19	93.36±0.02	99.64±0.03
GAT	90.31±0.28	88.47±0.34	88.51±0.03	99.38±0.05	93.69±0.36	93.26±0.28	93.36±0.04	99.69±0.02
GAT+ Reasoner	90.75±0.54	89.11±0.60	89.29±0.55	99.37±0.05	94.17±0.20	93.85±0.16	94.07±0.13	99.69±0.02
GIN	87.96±0.59	85.61±0.75	85.56±0.75	98.86±0.09	91.54±0.71	90.45±0.73	90.50±0.69	99.41±0.08
GIN + Reasoner	89.61±0.24	87.78±0.24	87.88±0.21	99.09±0.08	93.24±0.27	92.75±0.27	92.98±0.22	99.61±0.02

	OrganSMNIST				TissueMNIST			
Method	ACC	F1	Sensitivity	ROC-AUC	ACC	F1	Sensitivity	ROC-AUC
GCN	78.62±0.82	73.74±0.99	73.85±0.08	97.80±0.11	50.90±0.32	32.61±0.79	32.51±0.07	81.98±0.31
GCN + Reasoner	79.34±1.36	74.81±1.51	75.23±0.14	97.94±0.17	51.51±0.36	34.30±0.30	34.21±0.15	82.66±0.11
GAT	81.80±0.68	77.22±0.73	77.16±0.07	98.39±0.09	51.53±0.35	33.10±0.56	33.06±0.07	83.11±0.12
GAT + Reasoner	82.08±0.59	77.70±0.69	77.99±0.72	98.36±0.09	43.98±0.41	37.59±0.40	41.02±0.42	82.51±0.29
GIN	77.23±0.62	71.65±0.65	71.77±0.76	97.36±0.10	50.51±1.09	30.31±3.70	32.50±2.05	81.72±0.85
GIN + Reasoner	80.29±0.60	75.70±0.74	75.99±0.63	98.00±0.10	43.26±0.54	36.91±0.46	40.53±0.41	82.29±0.19

	BloodMNIST				MorphoMNIST			
Method	ACC	F1	Sensitivity	ROC-AUC	ACC	F1	Sensitivity	ROC-AUC
GCN	80.49±0.66	77.46±0.96	77.15±0.09	96.56±0.18	90.89±0.05	90.82±0.02	90.73±0.02	98.88±0.03
GCN + Reasoner	80.32±0.55	78.01±0.62	78.18±0.07	96.77±0.10	90.78±0.48	90.97±0.35	90.78±0.53	98.90±0.20
GAT	82.02±0.31	79.38±0.40	79.18±0.04	97.45±0.08	91.50±0.15	91.48±0.16	91.33±0.05	98.73±0.04
GAT+ Reasoner	80.59±0.62	78.29±0.70	79.31±0.56	97.18±0.11	91.69±0.32	91.65±0.20	91.66±0.18	98.79±0.02
GIN	80.30±0.60	77.21±0.91	76.44±0.95	96.58±0.21	91.60±0.19	91.59±0.18	91.60±0.19	98.75±0.05
GIN + Reasoner	80.56±0.37	78.19±0.35	79.18±0.32	96.89±0.07	91.55±0.85	91.64±0.55	91.63±0.26	98.74±0.06

Conclusion. This work introduces **X-Node**, a self-explainable GNN architecture that integrates per-node explanation into training. Beyond accuracy, it enables interpretable decision-making through learned context vectors and LLM-based textual rationales. *Grok 3* outperforms *Gemini 2.5 Pro* in clarity of explanation, and future work can explore the effect of prompt variations. X-Node offers a transferable framework for faithful, explanation-aware learning marking a step toward trustworthy graph intelligence.

References

1. Ahmedt-Aristizabal, D., Armin, M.A., Denman, S., Fookes, C., Petersson, L.: Graph-based deep learning for medical diagnosis and analysis: past, present and future. Sensors **21**(14), 4758 (2021). https://doi.org/10.3390/s21144758
2. Alvarez Melis, D., Jaakkola, T.: Towards robust interpretability with self-explaining neural networks. In: Bengio, S., Wallach, H., Larochelle, H., Grauman, K., Cesa-Bianchi, N., Garnett, R. (eds.) Advances in Neural Information Processing Systems, vol. 31. Curran Associates, Inc. (2018). https://proceedings.neurips.cc/paper_files/paper/2018/file/3e9f0fc9b2f89e043bc6233994dfcf76-Paper.pdf

3. Bordt, S., Finck, M., Raidl, E., von Luxburg, U.: Post-hoc explanations fail to achieve their purpose in adversarial contexts. In: Proceedings of the 2022 ACM Conference on Fairness, Accountability, and Transparency, pp. 891–905. FAccT '22, Association for Computing Machinery, New York, NY, USA (2022). https://doi.org/10.1145/3531146.3533153
4. Bronstein, M.M., Bruna, J., LeCun, Y., Szlam, A., Vandergheynst, P.: Geometric deep learning: going beyond Euclidean data. IEEE Sig. Process. Mag. **34**(4), 18–42 (2017). https://doi.org/10.1109/MSP.2017.2693418
5. Coelho de Castro, D., Tan, J., Kainz, B., Glocker, B.: Morpho-MNIST: quantitative assessment and diagnostics for representation learning. J. Mach. Learn. Res. **20** (2019)
6. Cedro, M., Martens, D.: GraphXAIN: narratives to explain graph neural networks. arXiv preprint arXiv:2411.02540 (2024)
7. Gao, Y., Yang, H., Chen, Y., Wu, J., Zhang, P., Wang, H.: LLM4GNAS: a large language model based toolkit for graph neural architecture search. arXiv preprint arXiv:2502.10459 (2025)
8. Holzinger, A., Biemann, C., Pattichis, C., Kell, D.: What do we need to build explainable ai systems for the medical domain? arXiv preprint arXiv:1712.09923 (2017)
9. Jacovi, A., Goldberg, Y.: Towards faithfully interpretable NLP systems: how should we define and evaluate faithfulness? In: Jurafsky, D., Chai, J., Schluter, N., Tetreault, J. (eds.) Proceedings of the 58th Annual Meeting of the Association for Computational Linguistics, pp. 4198–4205. Association for Computational Linguistics, Online (2020). https://doi.org/10.18653/v1/2020.acl-main.386
10. Kipf, T.N., Welling, M.: Semi-supervised classification with graph convolutional networks. In: International Conference on Learning Representations (2017). https://openreview.net/forum?id=SJU4ayYgl
11. Li, X., Zhao, H., Han, L., Tong, Y., Tan, S., Yang, K.: Gated fully fusion for semantic segmentation. Proc. AAAI Conf. Artif. Intell. **34**, 11418–11425 (2020). https://doi.org/10.1609/aaai.v34i07.6805
12. Li, Z., Geisler, S., Wang, Y., Günnemann, S., Leeuwen, M.: Explainable graph neural networks under fire. arXiv preprint arXiv:2406.06417 (2024)
13. Luo, D., et al.: Parameterized explainer for graph neural network. In: Larochelle, H., Ranzato, M., Hadsell, R., Balcan, M., Lin, H. (eds.) Advances in Neural Information Processing Systems, vol. 33, pp. 19620–19631. Curran Associates, Inc. (2020). https://proceedings.neurips.cc/paper_files/paper/2020/file/e37b08dd3015330dcbb5d6663667b8b8-Paper.pdf
14. Luo, D., et al.: Parameterized explainer for graph neural network. In: Proceedings of the 34th International Conference on Neural Information Processing Systems. NIPS '20, Curran Associates Inc., Red Hook, NY, USA (2020)
15. Parisot, S., Ktena, S.I., Lee, M., Guerrero, R., Glocker, B., Rueckert, D.: Disease prediction using graph convolutional networks: application to Autism spectrum disorder and Alzheimer's disease. Med. Image Anal. **48** (2018). https://doi.org/10.1016/j.media.2018.06.001
16. Samek, W., Wiegand, T., Müller, K.R.: Explainable artificial intelligence: Understanding, visualizing and interpreting deep learning models. ITU J. ICT Discov. Spec. Issue 1 Impact AI Commun. Netw. Serv. **1**, 1–10 (2017). https://doi.org/10.48550/arXiv.1708.08296
17. Slack, D., Hilgard, S., Jia, E., Singh, S., Lakkaraju, H.: Fooling lime and shap: adversarial attacks on post hoc explanation methods. In: Proceedings of the AAAI/ACM Conference on AI, Ethics, and Society, pp. 180–186. AIES '20, Association for Computing Machinery, New York, NY, USA (2020). https://doi.org/10.1145/3375627.3375830
18. Veličković, P., Cucurull, G., Casanova, A., Romero, A., Liò, P., Bengio, Y.: Graph attention networks. In: International Conference on Learning Representations (2018). https://openreview.net/forum?id=rJXMpikCZ

19. Wei, J., et al.: Chain-of-thought prompting elicits reasoning in large language models. In: Proceedings of the 36th International Conference on Neural Information Processing Systems. NIPS '22, Curran Associates Inc., Red Hook, NY, USA (2022)
20. Xu, K., Hu, W., Leskovec, J., Jegelka, S.: How powerful are graph neural networks? In: International Conference on Learning Representations (2019). https://openreview.net/forum?id=ryGs6iA5Km
21. Yang, J., Shi, R., Wei, D., et al.: MedMNIST v2 - a large-scale lightweight benchmark for 2D and 3D biomedical image classification. Sci. Data **10**, 41 (2023). https://doi.org/10.1038/s41597-022-01721-8
22. Ying, Z., Bourgeois, D., You, J., Zitnik, M., Leskovec, J.: GNNExplainer: generating explanations for graph neural networks. In: Wallach, H., Larochelle, H., Beygelzimer, A., d'Alché-Buc, F., Fox, E., Garnett, R. (eds.) Advances in Neural Information Processing Systems, vol. 32. Curran Associates, Inc. (2019). https://proceedings.neurips.cc/paper_files/paper/2019/file/d80b7040b773199015de6d3b4293c8ff-Paper.pdf
23. Yuan, H., Tang, J., Hu, X., Ji, S.: XGNN: towards model-level explanations of graph neural networks, pp. 430–438 (2020). https://doi.org/10.1145/3394486.3403085
24. Zhang, J., Liu, J., Luo, D., Neville, J., Wei, H.: LLMExplainer: large language model based Bayesian inference for graph explanation generation. arXiv preprint arXiv:2407.15351 (2024)

Population-Graph Post-hoc Correction of Survival Predictions for Improved Risk Stratification

Oriane Thiery[1], Mira Rizkallah[1], Hakima Laribi[2], Martin Vallières[2], Thomas Carlier[3,4], and Diana Mateus[1]

[1] Nantes Université, École Centrale Nantes, CNRS, LS2N, UMR 6004, Nantes, France
diana.mateus@ec-nantes.fr
[2] Department of Computer Science, Université de Sherbrooke, Sherbrooke, Canada
[3] Nuclear Medicine Department, University Hospital, Nantes, France
[4] Nantes Université, INSERM, CNRS, Université d'Angers, CRCI2NA, Nantes, France

Abstract. Accurate risk stratification of cancer patients is essential to propose a relevant treatment for each patient. In the clinical field, risk stratification is usually based on models or scoring systems relying on known biomarkers for the pathology. Yet, learning-based methods have been proven to be able to improve stratification compared to such empirically-defined indicators. These methods usually either learn directly to classify the patients into binary or multiclass risk groups [21,22], or they perform a survival analysis from which they derive such risk groups, an approach strongly privileged by clinicians (as in [2]). However, these methods only use the individual information of each patient to make a prediction, without taking advantage of the information of individuals who have a profile similar to that of the patient considered.

Keywords: Population graph · Error passing network · Survival analysis · Risk stratification

Accurately stratifying cancer patients into low- and high-risk groups is essential for guiding treatment. Although approaches based on known clinical biomarkers [4] or machine learning [2,21,22] exist, these methods usually only use individual patient data and do not leverage information from patients with similar profiles. Learning-based stratification can be done by classification of the patients [22] or by a survival analysis [2,14], an approach preferred by clinicians.

Population graphs can leverage the similarity between subjects, for example to improve the quality of individual predictions [1,24], or to correct prediction errors [9,18,23]. Such error-correction methods have been applied to regression [18], classification [9,23], or language modeling tasks [23]. However, the application of these graph-based post-hoc correction methods on survival predictions has not yet been explored, to the best of our knowledge.

Designing a graph-based error propagation method for survival predictions presents several challenges. Unlike a regression or classification problem, a survival model must compare a scalar label (event time) and a survival curve prediction. A second challenge is to maintain, after correction, the specific properties of survival curves (*e.g.* bounded and monotonically decreasing). The third challenge is dealing with censoring, *i.e.* having patients for whom we only know the time at which they left the study, not the true event time. Finally, the definition of the population graph on which the residuals are propagated is not straightforward as feature-based similarities alone may not reflect times-to-event closeness.

To address the above challenges, we propose a population-graph error correction method that improves a survival baseline model's ability to stratify. Our proposition draws inspiration from the Error Passing Network (EPN) [18], designed to correct regression errors. We extend this model in the following ways:

- We adapt the framework to address survival predictions and censoring,
- We propose explicitly introducing the patients' event times in the population graph construction,
- We introduce a post-processing step to ensure that the corrected survival curves maintain their probabilistic nature with similar performances,
- And we target the clinically crucial problem of risk stratification.

Moreover, regarding the population graph construction, we draw inspiration from [20], but replace the matrix completion with a transformer [19] as a last contribution to propagate the known labels and patient features towards predicting pseudo time-to-event labels, enabling building a time-to-event-aware graph.

An experimental validation on the METABRIC dataset shows a recurrent improvement of the stratification after the correction, with the possibility of enforcing the properties of a conventional survival curve upon the corrections.

1 Proposed Method

Problem Statement. Consider a population $\mathcal{P}$ of N_{pop} patients. Let $\mathbf{x}^{(n)} \in \mathbb{R}^{D_{\text{feat}}}$ be the D_{feat} dimensional feature vector associated with a patient $n \in \{1, \ldots, N_{\text{pop}}\}$, $\delta^{(n)}$ its censorship indicator equal to 0 if the patient is censored and 1 otherwise, and $t_c^{(n)}$ the continuous label corresponding to the patient's event time $E^{(n)}$ when uncensored ($\delta^{(n)} = 1$) and to its censoring time $C^{(n)}$ otherwise ($\delta^{(n)} = 0$). Let $\mathcal{P}_L$ be the subgroup of N_L individuals for whom we have access to both features and true target values, *i.e.* $\mathcal{P}_L = \{\mathbf{x}^{(j)}, t_c^{(j)}, \delta^{(j)}\}_{j=1}^{N_L}$, and $\mathcal{P}_U$ a subpopulation with access only to the features, *i.e.* $\mathcal{P}_U = \{\mathbf{x}^{(i)}\}_{i=N_L+1}^{N_{\text{pop}}}$.

Also consider a base model $f_\Theta(\cdot)$ parameterized by Θ, which takes as input the feature vector $\mathbf{x}^{(n)}$ of a patient n and predicts a continuous survival curve $\hat{S}_{BM}^{(n)}(t) = f_\Theta(\mathbf{x}^{(n)}, t)$[1]. Based on predicted survival curves, the population can be stratified into two risk groups $\mathcal{P}_{low}$ and $\mathcal{P}_{high}$. In this paper, given the populations $\mathcal{P}_L$ and $\mathcal{P}_U$ and the baseline survival analysis model f_Θ, we seek to design a post-hoc correction method to improve the stratification of the population $\mathcal{P}$.

---
[1] A survival curve $S^{(n)}$ is defined as $S^{(n)}(t, \mathbf{x}^{(n)}) = \text{P}(E^{(n)} > t | \mathbf{x}^{(n)})$ for $t > 0$.

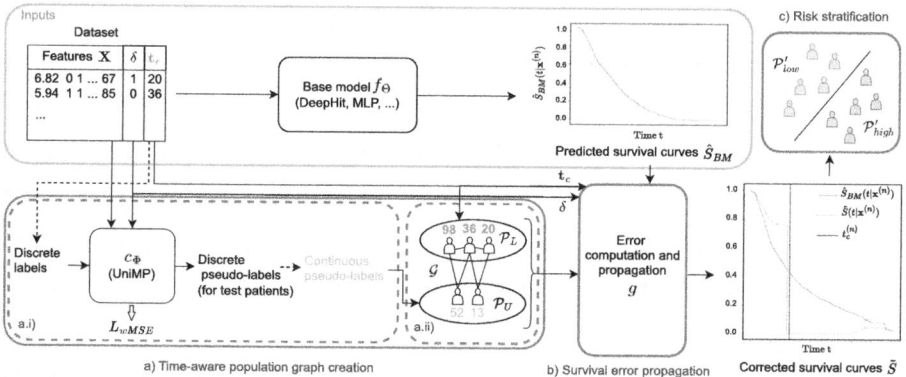

Fig. 1. The survEPN framework takes as input the feature matrix $\mathbf{X}$ describing all patients, their labels $\{t_c^{(n)}\}$ and censorship values $\{\delta^{(n)}\}$ and their survival curves $\{\hat{S}_{BM}^{(n)}\}$ predicted by the base model f_Θ we want to improve. It consists of three stages: (a) the construction of a time-aware population graph $\mathcal{G}$ by learning to predict pseudolabels for patients without labels with a model c_Φ, (b) an error correction module g which propagates the prediction error of the $\hat{S}_{BM}$ curves over the population graph and (c) the stratification of the patients into low- $\mathcal{P}'_{low}$ and high-risk $\mathcal{P}'_{high}$ groups based on their corrected survival curve $\tilde{S}$.

Proposed survEPN Framework. Building upon the original EPN [18] described in Appendix A, we introduce the survEPN method, tailored for survival analysis. Our model takes as input the concatenated features $\mathbf{X} \in \mathbb{R}^{N_{\text{pop}} \times D_{\text{feat}}}$ of all the patients, their labels $\{t_c^{(n)}\}$ and censorship values $\{\delta^{(n)}\}$ when available, and all the survival curves $\{\hat{S}_{BM}^{(n)}\}$ predicted by f_Θ. The proposed framework consists of three steps as illustrated in Fig. 1: a) the construction of a time-aware population graph, b) the propagation and correction of errors, and c) the revised stratification of the population. The output is an improved risk group separation $\mathcal{P} = \{\mathcal{P}'_{low} \cup \mathcal{P}'_{high}\}$, where $\mathcal{P}'_{low}$ and $\mathcal{P}'_{high}$ stand for the low and high risk groups respectively obtained after the survEPN correction step. We can also retrieve, if necessary, the corrected survival curves $\{\tilde{S}^{(n)}\}$.

a.i) *Building time-aware connections:* We first build an attributed population graph $\mathcal{G} = \{\mathcal{V}, \mathcal{E}\}$, where each node n in $\mathcal{V}$ represents a patient in $\mathcal{P}$ described by his feature vector $\mathbf{x}^{(n)}$ and edges establish time-aware connections between patients, such that patients with similar time-to-events are connected. Since during inference the patient's time-to-event label is unknown, we first design and train a model c_Φ to predict pseudolabels and find the edges and their weights.

In practice, the pseudolabel predictor model c_Φ is a transformer-based UniMP (Unified Message Passing) [19], capable of propagating both features and available labels, to predict pseudolabels for the unlabeled patients. For the training of c_Φ (Sec. a.i)), we partition $\mathcal{P}_L$ into two distinct subsets: $\mathcal{P}_{L^\bullet}$ with access to the true labels, and $\mathcal{P}_{L^\circ}$, whose labels are deliberately masked (not used as inputs) during training and validation. The number of patients considered in each of the two sets is denoted $N_{L^\bullet}$ and N_{L°, respectively. Formally, c_Φ predicts a pseudolabel $\hat{t}_c^{(j^\circ)}$ for every masked patient $j^\circ \in \mathcal{P}_{L^\circ}$, taking into account all the features as well as the unmasked labels; that is, $\hat{t}_c^{(j^\circ)} = c_\Phi(\{\mathbf{x}^{(n)}\}_{n=1}^{N_{\text{pop}}}, \{t_c^{(j^\bullet)}\}_{j^\bullet=1}^{N_{L^\bullet}}, \{\delta^{(j^\bullet)}\}_{j^\bullet=1}^{N_{L^\bullet}})$. See Appendix B for details on the UniMP architecture. In particular, c_Φ adds the unmasked labels to the input feature matrix, building a time-aware feature matrix. To enable the effective combination of features and labels, the continuous labels are first discretized into N_{cl} classes with quantile-based time intervals, ensuring a similar number of training patients per class, to get $t_d^{(n)}$. Then, the feature matrix $\mathbf{H}_0$ entering the transformer blocks results from adding the original patients' features $\mathbf{X}$ and the unmasked discrete labels, converted into one-hot vectors $\mathbf{t}_{oh}^{(j^\bullet)} \in \mathbb{R}^{N_{cl}}$, projected by a learned matrix $\mathbf{W}_T \in \mathbb{R}^{N_{cl} \times D_{\text{feat}}}$ and combined as: $\mathbf{H}_0 = \mathbf{X} + \bar{\mathbf{T}}_d$, with both $\mathbf{H}_0, \bar{\mathbf{T}}_d \in \mathbb{R}^{N_L \times D_{\text{feat}}}$, and $\bar{\mathbf{t}}_d^{(j)}$ the j-th row of $\bar{\mathbf{t}}_d^{(j)}$, computed as $\bar{\mathbf{t}}_d^{(j)} = \mathbf{t}_{oh}^{(j)} \mathbf{W}_T$ if $j \in \mathcal{P}_{L^\bullet}$ and $\delta^{(j)} = 1$ or kept null otherwise. The output of the UniMP's last transformer layer is the N_{cl} one-hot pseudolabel prediction $\hat{\mathbf{t}}_{oh}^{(j)}$ for patient j. The final pseudolabel is its argmax converted back to continuous time, $\hat{t}_c^{(j)}$.

We train the UniMP with a multi-class weighted Mean Square Error (MSE) loss penalizing the predictions far from the true class in an ordinal way. We also avoid considering the predictions of classes after their censoring time. Formally:

$$L_{wMSE} = \frac{1}{N_{L^\circ} N_{cl}} \sum_{j^\circ=1}^{N_{L^\circ}} \sum_{k=1}^{N_{cl}} \Big[\mu_k^{(j^\circ)} \cdot \overbrace{(\mathbf{t}_{oh,k}^{(j^\circ)} - \text{softmax}(\hat{\mathbf{t}}_{oh}^{(j^\circ)})_k)^2}^{\text{Squared error per class k}} \Big], \quad (1)$$

$$\text{with } \mu_k^{(j^\circ)} = \underbrace{\left(\delta^{(j^\circ)}|t_d^{(j^\circ)} - k| + (1-\delta^{(j^\circ)})\max((t_d^{(j^\circ)}-k),0)\right)}_{\text{Censoring-dependent ordinal penalization}}$$

with $\mathbf{t}_{oh,k}^{(i)}$ being the kth element (so the element of the kth class) of the one-hot vector encoding the discrete label of patient i; and $\text{softmax}(\hat{\mathbf{t}}_{oh}^{(j^\circ)})_k$ the kth element/class of the output vector of the UniMP after a softmax operation. An example of computation of this loss is presented in Appendix C. During test time, c_Φ is applied to the unseen population $\mathcal{P}_U$ to predict pseudolabels $\hat{t}_c^{(i)}$.

a.ii) *Time-aware population graph creation:* With the predicted pseudolabels $\{\hat{t}_c^{(i)}\}$, we build the edge set of our population graph. To this end, we first define similarities between pairs of patients based on their event times t_1 and t_2:

$$\text{sim}(t_1, t_2) = (|t_1 - t_2| + \omega)^{-1}, \quad (2)$$

with $|\cdot|$ the absolute value and ω a small value to avoid dividing by 0. In practice, we define the similarities between an unlabeled and a labeled patient as $\text{sim}(\hat{t}_c^{(p)}, t_c^{(q)})$ and between two labeled patients as $\text{sim}(t_c^{(p)}, t_c^{(q)})$.

Based on the observed similarities, the underlying connectivity $\mathcal{E}$ of the population graph is defined using a K-Nearest Neighbors (K-NN) approach.

Specifically, for each patient, only the K most similar labeled neighbors are retained, resulting in a directed graph structure. The edge weights $\alpha^{(p,q)}$ are computed for each patient p based on its neighbourhood $\mathcal{N}(p)$ as the softmax of its similarities:

$$\alpha^{(p,q)} = \text{softmax}(\{\tau \cdot \text{sim}(t_c^{(p)}, t_c^{(q)})\}_{q|q \in \mathcal{N}(p)}), \quad (3)$$

for patients $p \in \mathcal{P}_L$, and if $p \in \mathcal{P}_U$ we replace $t_c^{(p)}$ by $\hat{t}_c^{(p)}$. The temperature hyperparameter τ controls the distribution's sharpness.

b) *Computation and correction of the survival errors:* Inspired by [18], we refine the baseline survival predictions with an error propagation module, denoted hereafter g. The latter takes as input the times to event and censorships of the labeled subjects $\{t_c^{(j)}, \delta^{(j)}\}_{j \in \mathcal{P}_L}$, the predicted survival curves of the baseline model on the entire population $\{\hat{S}_{BM}^{(n)}\}_{n \in \mathcal{P}}$, and the time-aware population graph $\mathcal{G}$, and predicts a corrected version $\tilde{S}^{(n)}$ of each survival curve:

$$\tilde{S}^{(n)} = g(\{t_c^{(j)}, \delta^{(j)}\}_{j \in \mathcal{P}_L}, \{\hat{S}_{BM}^{(n)}\}_{n \in \mathcal{P}}, \mathcal{G}) \quad (4)$$

More precisely, we approximate the true survival function $S^{(j)}$ of patient j with a step-function dropping at the event time $t_c^{(j)}$ [8]. To avoid propagating uncertain errors, $S^{(j)}$ is considered equal to the predicted curve on the censored intervals. Then, g adjusts the baseline survival curves by computing the errors on the neighboring patients and propagating the respective corrections to obtain the corrected survival curve for patient n:

$$\tilde{S}^{(n)} = \left(\sum_{j | j \in \mathcal{N}(n)} \alpha^{(n,j)} \left(S^{(j)} - \hat{S}_{BM}^{(j)} \right) \right) + \hat{S}_{BM}^{(n)}, \quad (5)$$

Note that this model does not learn additional parameters. Since the corrected curves may not respect the expected properties of survival curves e.g. monotonically decreasing and bounded between 0 and 1, the resultant survival curves $\tilde{S}^{(n)}$ are first clamped to the $[0,1]$ interval. When beyond the stratification of the population, proper survival curves are required per patient, we propose to optionally fit a sigmoid to the corrected curves as follows:

$$\tilde{S}_{sig}^{(n)}(t) = 1 - \frac{A_{sig}}{1 + \exp(\frac{t - t_{sig}}{s_{sig}})}, \quad \text{with } A_{sig} \in [0, 1], \, t_{sig} \geq 0, \, s_{sig} \leq 0, \quad (6)$$

where A_{sig} controls the amplitude of the sigmoid; t_{sig} is related to the time at which the curve decreases, and s_{sig} corresponds to the slope of the function. The added constraints enforce both monotony and boundaries on the curve.

c) *Risk stratification:* Finally, we use the corrected curves $\{\tilde{S}^{(n)}\}_{n \in \mathcal{P}}$ (or $\{\tilde{S}_{sig}^{(n)}\}_{n \in \mathcal{P}}$) to divide the patients into low- and high-risk groups by using the Area Under their Survival Curve (AUSC) or their Estimated Event Time (EET), the latter computed as the time at which the survival curve of a patient equals 0.5. In the two cases, we take the median AUSC or EET value of the training patients $\mathcal{P}_L$'s corrected curves and use it as a threshold to split the $\mathcal{P}_U$ test patients into $\mathcal{P}'_{low}$ and $\mathcal{P}'_{high}$ from their corrected curves $\{\tilde{S}^{(i)}\}_{i \in \mathcal{P}_U}$.

2 Experimental Validation

2.1 Dataset and Experimental Setup

To evaluate our method, we rely on the METABRIC dataset [3], consisting of gene and protein expression profiles (with $D_{feat} = 9$ features described in [12]) used to determine new breast cancer subgroups. It is composed of 1904 patients, and the events corresponding to the patients' deaths have a censoring rate of 42%. The numerical covariates are standardized by removing their mean and scaling their variance to unity, while the binary ones are left unchanged.

In the next experiments, the base model f_Θ is a 2-layer Multi-Layer Perceptron (MLP) with ReLU activations, trained with a CoxTime loss [15]. Details on the MLP and UniMP modules, along with the population graph construction, are provided in Appendix D. Accounting for random uncertainty in the data, we report results over five random train-validation/test splits. For each test set $\mathcal{P}_U$, we perform ten loops of training and validation, using random splits of the remaining patients. Training, validation and test sets comprise 80%, 10% and 10% of the data respectively. We force the proportion of censored patients to be the same in all sets, and for a fair comparison, we use the same sets to train the base model. The training and validation steps correspond to the training of the UniMP module, with the validation patients used for early stopping: during these steps, we respectively compute on the masked patients the L_{wMSE} loss and the cAMAE metric defined below. The test step corresponds to building the population graph, then performing the error propagation and the risk stratification. We illustrate the labeling handling for these three steps in Appendix E.

The results presented in this paper correspond to the test performance of the model with the best validation cAMAE (for censoring-aware Average Mean Absolute Error) of the loop, computed as ($\#\mathcal{P}$ being the cardinal of set $\mathcal{P}$):

$$\text{cAMAE} = \frac{1}{N_{cl}} \sum_{k=1}^{N_{cl}} \sum_{j^\circ | t_d^{(j^\circ)} = k} \frac{\delta^{(j^\circ)} |t_d^{(j^\circ)} - \hat{t}_d^{(j^\circ)}| + (1 - \delta^{(j^\circ)}) \max(t_d^{(j^\circ)} - \hat{t}_d^{(j^\circ)}, 0)}{\#\{j^\circ | t_d^{(j^\circ)} = k\}}$$

To assess the quality of the stratification, we compute the p-value after a log-rank test [6] with the hypothesis that the Kaplan-Meier curves [11] obtained from the two risk groups separated from the survival curves are identical. Additionally, we evaluate the quality of the curves with 2 metrics: (1) the Integrated Brier Score (IBS), characterizing both the discrimination and the calibration of the prediction [8], and (2) the c-index [10] linked to the ordering of the curves of the patients. We report the average IBS and the c-index over all training loops. For the p-values, we report the median, first and third quartiles of the 50 test values obtained. Finally, the whole framework was coded in Python using the `PyTorch` [16] `PyCox` [7], `scikit-survival` [17] and `pytorch-geometric` [5] modules.

Table 1. Stratification quality and survival performances of the base model, and of the base model corrected by our survEPN method applied on either random graphs; a feature graph; and our time-aware graph presented in part 1.

	Quantile	Base model	Random graph survEPN	Feature graph survEPN	Time-aware graph survEPN
p-value EET ($\downarrow$)	25%	1.86×10^{-4}	4.17×10^{-5}	3.53×10^{-6}	1.15×10^{-6}
	50%	2.78×10^{-3}	1.35×10^{-3}	6.86×10^{-5}	1.96×10^{-5}
	75%	1.03×10^{-2}	9.47×10^{-3}	6.44×10^{-4}	1.69×10^{-4}
p-value AUSC ($\downarrow$)	25%	5.14×10^{-5}	1.54×10^{-5}	1.69×10^{-5}	9.45×10^{-8}
	50%	8.40×10^{-4}	6.26×10^{-4}	1.48×10^{-4}	1.70×10^{-5}
	75%	4.10×10^{-3}	5.18×10^{-3}	2.79×10^{-3}	5.02×10^{-4}
IBS ($\downarrow$)		0.163 ± 0.004	0.168 ± 0.005	0.165 ± 0.004	0.190 ± 0.007
c-index ($\uparrow$)		0.66 ± 0.03	0.63 ± 0.03	0.65 ± 0.03	0.64 ± 0.03

2.2 Results and Analysis

Quality of the Correction. Table 1 reports the stratification performance of the base model's output, and after correction with our survEPN (first and last columns). We observe from the p-value decrease an important improvement in the separation between low- and high-risk patients. The stratification refinement can also be observed in Fig. 2a where there is an increased difference between the Kaplan-Meier curves of the two risk groups before and after our correction.

Ablation Study. In order to show the value of the time-aware graph, we further compare our survEPN against the error propagation applied on two other types of graphs: (1) random graphs[2] (100 different graphs per training/validation/test loop for robustness); (2) a feature graph with a cosine similarity between the patient's features. In both cases, $K = 50$ and $\tau = 0.1$ are chosen empirically for the population graph construction. The results are shown in Table 1. We observe that applying the survEPN on random graphs gives equivalent or slightly better stratification results than the base model. However, as soon as the graph becomes informative (as with the feature graph), applying the survEPN on the base model significantly improves its stratification ability, especially when splitting the risk groups based on the estimated event times (EET) of the patients. Finally, explicitly introducing the time information in the graph construction further improves the stratification, even when the split is based on the AUSC.

Survival Curves Quality. Concerning the survival curves' quality, the IBS worsening is mainly due to a few late-event patients predicted too early by UniMP. Because late-event classes are wide, a small prediction error can connect a patient to others with earlier events, resulting in unrealistically low survival

---

[2] A normal random similarity value is assigned to each edge and used in Eq. 3.

Table 2. Quality of the stratification and the survival curves after applying our time-aware survEPN and refining survival curves with sigmoid fitting.

p-value EET (↓)			p-value AUSC (↓)			IBS (↓)	c-index (↑)
25%	50%	75%	25%	50%	75%		
4.65×10^{-7}	1.91×10^{-5}	2.98×10^{-4}	1.04×10^{-7}	1.70×10^{-5}	4.87×10^{-4}	0.190 ± 0.007	0.63 ± 0.04

predictions that the IBS strongly penalizes. A more balanced dataset, with more late-event patients, would likely reduce IBS degradation. In contrast, p-values improve when patients shift risk groups, regardless of the survival curve accuracy.

Post-refinement of Survival Curves. We observe from Fig. 2b that the corrected survival curve (in orange) may not monotonically decrease as expected, while the refined curve obtained via Eq. 6 (in green) does. The results after refinement in Table 2 show the stratification improvement is maintained. This optional step is thus recommended when using our framework to improve risk stratification while maintaining the clinical interpretation of survival curves.

3 Conclusion and Perspectives

In summary, we addressed the problem of improving the risk stratification of a given survival model in a post-hoc manner. To this end, we proposed a semi-supervised framework, Survival Error Passing Network (survEPN), relying on a time-aware population graph and an error correction strategy. Experimental results showed the effectiveness of the proposed method, including when ensuring

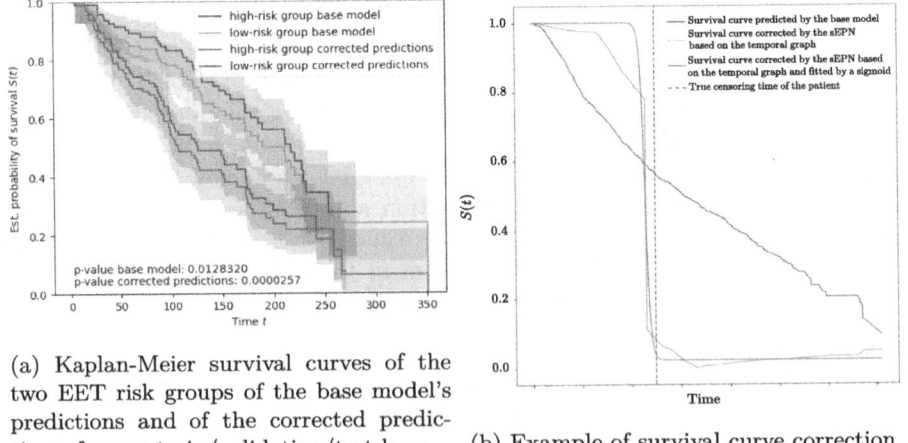

(a) Kaplan-Meier survival curves of the two EET risk groups of the base model's predictions and of the corrected predictions, for one train/validation/test loop.

(b) Example of survival curve correction.

Fig. 2. Qualitative evaluation of the Kaplan Meier and the refined survival curves.

that the corrections produce proper survival curves. Some perspectives include exploring ways of integrating the censored labels in the UniMP module, and making the survEPN process learnable end-to-end, with a joint formulation of the population graph creation and error propagation processes.

Acknowledgements and Disclosure of Interests. This work has been funded by the AIby4 project (Centrale Nantes-Project ANR-20-THIA-0011), INCa-DGOS-INSERM-ITMO Cancer 18011 (SIRIC ILIAD) with the support from the Pays de la Loire region (GCS IRECAN 220729). The authors have no competing interests to declare that are relevant to the content of this article.

Appendix

A Original EPN

Formally, the original EPN [18] addresses prediction correction in the context of a regression problem. In such case, the target value u of the base model $f_\Theta(\cdot)$, and the corresponding predictions $\hat{u}^{(n)} = f_\Theta(\mathbf{x}^{(n)})$ are both scalar values. The EPN seeks to improve the quality of the initial predictions $\hat{u}^{(n)}$ by computing the residuals $\{\epsilon^{(j)}\}_{j=1}^{N_L}$ for the labeled patients in $\mathcal{P}_L$ and propagating them towards the whole population $\mathcal{P}$ to obtain improved predictions $\{\tilde{u}^{(n)}\}_{n=1}^{N_{\text{pop}}}$.

a) *Building the population graph:* The propagation operation is performed over an attributed population graph $\mathcal{G}_o = \{\mathcal{V}_o, \mathcal{E}_o\}$, where each node n in $\mathcal{V}_o$ represents a patient of the population $\mathcal{P}$, described by his feature vector $\mathbf{x}^{(n)}$. The edges $e^{(n,j)}$ connect any patient $n \in \mathcal{P}$ to all labeled individuals $j \in \mathcal{P}_L$ (with no self-loops), defining a neighbourhood $\mathcal{N}_o(n) = \{j | e^{(n,j)} \in \mathcal{E}_o\}$. Learned edge weights $\alpha^{(n,j)}$ are associated with these connections, computed as:

$$\alpha_{EPN}^{(n,j)} = \text{softmax}(\{s_{EPN}^{(n,j)}\}_{j|j \in \mathcal{N}_o(n)}), \text{ with} \tag{7}$$

$$s_{EPN}^{(n,j)} = \frac{(\mathbf{W}_{EPN,Q}\mathbf{x}^{(n)})^\top \mathbf{W}_{EPN,K}\mathbf{x}^{(j)}}{\sqrt{D_{\text{feat}}}}, \tag{8}$$

and where $\mathbf{W}_{EPN,Q}$ and $\mathbf{W}_{EPN,K}$ are two projection matrices, learned with a Mean Square Error (MSE) loss.

b) *Propagating the errors and correcting predictions:* The residuals, for the labeled patients $j \in \mathcal{P}_L$, are computed as $\epsilon^{(j)} = u^{(j)} - f_\Theta(\mathbf{x}^{(j)})$. The corresponding corrections are then propagated towards the nodes $n \in \mathcal{P}$ on the population graph as follows:

$$\tilde{u}^{(n)} = \Big(\sum_{j|j \in \mathcal{N}_o(n)} \alpha_{EPN}^{(n,j)} (u^{(j)} - f_\Theta(\mathbf{x}^{(j)})) \Big) + f_\Theta(\mathbf{x}^{(n)}). \tag{9}$$

B UniMP Description

Here, we describe the details of the UniMP module used for predicting the pseudolabels of unlabeled patients, necessary for the time-aware population graph construction. Overall, it first learns to project the information from labels to combine them with the features, and then relies on several multi-head transformer blocks to compute the attention between patients and propagate this information, to finally predict a class per patient (as presented in Fig. 3).

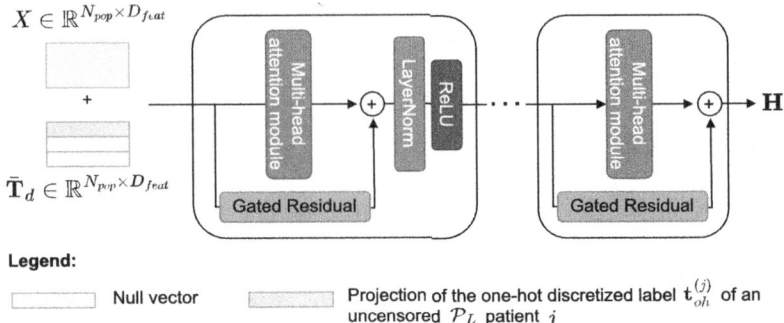

Fig. 3. UniMP module overview: the patients' features $\mathbf{X}$ are summed to the projection of the partially observed labels $\bar{\mathbf{T}}_d$, then updated via L transformer layers, which finally returns a discrete time-to-event prediction for all patients $\mathbf{H}_L \in \mathbb{R}^{N_{\text{pop}} \times N_{cl}}$.

First, we create the initial feature matrix $\mathbf{H}_0$ by adding to the original patients' features $\mathbf{X}$ the uncensored discrete labels available, converted into one-hot encoding $\mathbf{t}_{oh}^{(n)} \in \mathbb{R}^{N_{cl}}$ and projected by a learned matrix $\mathbf{W}_T \in \mathbb{R}^{N_{cl} \times D_{\text{feat}}}$ in the same space as the features:

$$\mathbf{H}_0 = \mathbf{X} + \bar{\mathbf{T}}_d, \tag{10}$$

with both $\mathbf{H}_0, \bar{\mathbf{T}}_d \in \mathbb{R}^{N_{\text{pop}} \times D_{\text{feat}}}$, and $\bar{\mathbf{t}}_d^{(n)}$ the n-th row of $\bar{\mathbf{T}}_d$:

$$\bar{\mathbf{t}}_d^{(n)} = \begin{cases} \mathbf{t}_{oh}^{(n)} \mathbf{W}_T & \text{if } n \in \mathcal{P}_L \text{ and } \delta^{(n)} = 1 \\ 0_{D_{\text{feat}}} & \text{otherwise.} \end{cases} \tag{11}$$

Two types of multi-head transformer blocks are then used: one for intermediate layers and one for the output, the difference lying in the aggregation and activation steps. In both cases, a self-attention module per head is applied on $\mathbf{H}_l$, transforming the features of the lth layer according to learned projection matrices and biases. For the transformer at layer l of the UniMP having M attention heads, the self-attention provides an intermediate representation $\tilde{\mathbf{h}}_{l+1}^{(n)}$ for each patient n:

$$\tilde{\mathbf{h}}_{l+1}^{(n)} = \text{self-attention}(\mathbf{H}_l, \{\mathbf{W}_{l,m,q}, \mathbf{b}_{l,m,q}, \mathbf{W}_{l,m,k}, \mathbf{b}_{l,m,k}, \mathbf{W}_{l,m,v}, \mathbf{b}_{l,m,v}\}_{m=1}^M). \tag{12}$$

This representation $\tilde{\mathbf{h}}_{l+1}^{(n)}$ is obtained by computing a common query-key-value operation per node $n \in \mathcal{P}$, based on the representation $\mathbf{h}_l^{(n)}$ of all patients, and relying on trainable query ($\mathbf{W}_{l,m,q}$, $\mathbf{b}_{l,m,q}$), key ($\mathbf{W}_{l,m,k}$, $\mathbf{b}_{l,m,k}$) and value ($\mathbf{W}_{l,m,v}$, $\mathbf{b}_{l,m,v}$) parameters per attention head m[3].

A gated residual connection is introduced between layers to prevent the model from oversmoothing, computing an adaptive parameter $\beta_l^{(n)}$ for each patient and layer from the learned parameters $\mathbf{W}_{l,r}$, $\mathbf{b}_{l,r}$ and $\mathbf{W}_{l,g}$, as follows:

$$\mathbf{r}_l^{(n)} = \mathbf{W}_{l,r} \mathbf{h}_l^{(n)} + \mathbf{b}_{l,r}, \tag{18}$$

$$\beta_l^{(n)} = \text{sigmoid}(\mathbf{W}_{l,g}[\tilde{\mathbf{h}}_{l+1}^{(n)} || \mathbf{r}_l^{(n)} || \tilde{\mathbf{h}}_{l+1}^{(n)} - \mathbf{r}_l^{(n)}]). \tag{19}$$

This parameter is tailored to balance the weight of the previous and newly learned representations of the node. As such, the final output of the intermediate transformer block after normalization and activation is computed as:

$$\mathbf{h}_{l+1}^{(n)} = \text{ReLU}(\text{LayerNorm}((1 - \beta_l^{(n)})\tilde{\mathbf{h}}_{l+1}^{(n)} + \beta_l^{(n)} \mathbf{r}_l^{(n)})). \tag{20}$$

For the last output layer (layer L), the transformer block instead employs averaging over the multi-head outputs and removes the non-linear transformation as follows:

$$\tilde{\mathbf{h}}_L^{(n)} = \frac{1}{M} \sum_{m=1}^{M} \left[\sum_{d=1}^{N_{\text{pop}}} a_{L-1,m}^{(nd)} \mathbf{v}_{L-1,m}^{(d)} \right], \tag{21}$$

$$\mathbf{h}_L^{(n)} = (1 - \beta_{L-1}^{(n)})\tilde{\mathbf{h}}_L^{(n)} + \beta_{L-1}^{(n)} \mathbf{r}_{L-1}^{(n)}. \tag{22}$$

---

[3] For each pair of patient, the mth self-attention head first transforms the source node feature $\mathbf{h}_l^{(s)}$ and distant node feature $\mathbf{h}_l^{(d)}$ into query vector $\mathbf{q}_{l,m}^{(s)} \in \mathbb{R}^{d_h}$ and key vector $\mathbf{k}_{l,m}^{(d)} \in \mathbb{R}^{d_h}$ respectively, with d_h the hidden size of each head:

$$\mathbf{q}_{l,m}^{(s)} = \mathbf{W}_{l,m,q} \mathbf{h}_l^{(s)} + \mathbf{b}_{l,m,q}, \tag{13}$$

$$\mathbf{k}_{l,m}^{(d)} = \mathbf{W}_{l,m,k} \mathbf{h}_l^{(d)} + \mathbf{b}_{l,m,k}. \tag{14}$$

The attention between this pair of patients is then computed for each head:

$$a_{l,m}^{(sd)} = \text{softmax}\left(\frac{\mathbf{q}_{l,m}^{(s)\top} \mathbf{k}_{l,m}^{(d)}}{\sqrt{d_h}}\right). \tag{15}$$

Then, for each of the M attention heads we do the message aggregation from the N_{pop} distant patients d to the source s, and aggregate them by a concatenation operation:

$$\mathbf{v}_{l,m}^{(d)} = \mathbf{W}_{l,m,v} \mathbf{h}_l^{(d)} + \mathbf{b}_{l,m,v}, \tag{16}$$

$$\tilde{\mathbf{h}}_{l+1}^{(s)} = ||_{m=1}^{M} \left[\sum_{d=1}^{N_{\text{pop}}} a_{l,m}^{(sd)} \mathbf{v}_{l,m}^{(d)} \right]. \tag{17}$$

This $\mathbf{h}_L^{(n)}$, corresponding to $\hat{\mathbf{t}}_{oh}^{(n)}$ in the paper, is of dimension N_{cl}. To get a continuous pseudo-label for each $\mathcal{P}_U$ patient i, we first take the class corresponding to the argmax of $\hat{\mathbf{t}}_{oh}^{(i)}$, named $\hat{t}_d^{(i)}$. We then convert it back to continuous time by taking the middle value of this class's time interval. This corresponds to the final continuous pseudo-label $\hat{t}_c^{(i)}$ for patient i.

C Example of Computation of the MSE Loss

For an example of computation of our L_{wMSE} loss and its comparison against a basic MSE, let's consider a situation with $N_{cl} = 3$ classes and two patients p and q with labels $t_d^{(p)} = 2$, $\delta^{(p)} = 1$, $t_d^{(q)} = 1$ and $\delta^{(q)} = 0$. For a UniMP model whose output predictions are softmax($\hat{\mathbf{t}}_{oh}^{(p)}$) = $\{0.3, 0.5, 0.2\}$ and softmax($\hat{\mathbf{t}}_{oh}^{(q)}$) = $\{0.25, 0.4, 0.35\}$, we get:

$$L_{wMSE}^{(p)} = (0-0.3)^2 \cdot 2 + (0-0.5)^2 \cdot 1 + (1-0.2)^2 \cdot 0, \tag{23}$$

$$L_{wMSE}^{(q)} = (0-0.25)^2 \cdot 0 + (1-0.4)^2 \cdot 0 + (0-0.35)^2 \cdot 0, \tag{24}$$

while for a normal MSE loss:

$$L_{MSE}^{(p)} = (0-0.3)^2 + (0-0.5)^2 + (1-0.2)^2, \tag{25}$$

$$L_{MSE}^{(q)} = (0-0.25)^2 + (1-0.4)^2 + (0-0.35)^2. \tag{26}$$

Thus, our method avoids enforcing predictions after censoring time, and penalizes more strongly wrong predictions from classes far from the true class in an ordinal way.

D Models and Training Configuration

The MLP baseline model is trained with a batch size of 256 on 100 epochs (with early stopping if the loss value does not improve for 10 epochs), and relies on the CoxTime survival loss function [15]. We selected its configuration by a grid search on its hidden size (16 from $\{16, 32, 64, 128\}$) and learning rate (0.001 from $\{0.1, 0.01, 0.001, 0.0001\}$). As for the UniMP module, it is trained on 400 epochs (and stopped early if its cAMAE does not improve for 200 epochs) with a dropout rate of 0.3. The difference in the number of epochs between the base model and the UniMP module comes from the difference in their speed of convergence, as the MLP does not improve when trained on more than 100 epochs. An exploration of the hyperparameters of the UniMP led to considering $L = 2$ layers, $M = 2$ attention heads, a hidden size d_h of 32, using a learning rate of 0.1 and discretizing the continuous labels in $N_{cl} = 9$ classes. As for the graph construction, after performing experiments to find the optimal settings, the time-aware graph is defined as a K-NN with K= 400 and $\tau = 0.1$. During the training step of UniMP, we consider as in [19] a label rate of 0.625, corresponding to the proportion of unmasked patients $\mathcal{P}_{L^\bullet}$ randomly selected from $\mathcal{P}_L$ for training. Both models are optimized via Adam [13] with $\beta_1 = 0.9$, $\beta_2 = 0.999$, and $\epsilon = 10^{-8}$, including a weight decay of 0.0005 for the UniMP.

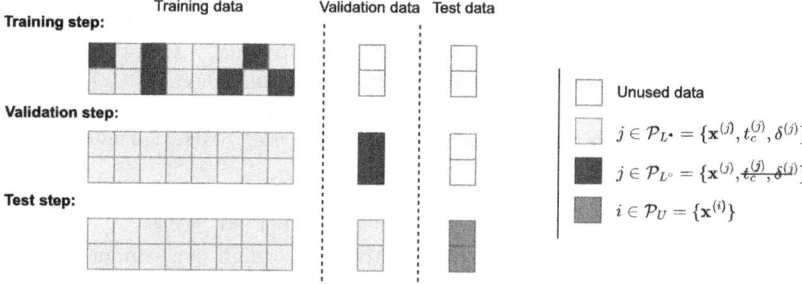

Fig. 4. For a given training/validation/test split, definition of the $\mathcal{P}_{L^\bullet}$, $\mathcal{P}_{L^\circ}$ and $\mathcal{P}_U$ sets. Training and validation only concern the UniMP (Sec. a.i)).

E Masking Process

We describe briefly in Fig. 4 how the $\mathcal{P}_{L^\bullet}$, $\mathcal{P}_{L^\circ}$ and $\mathcal{P}_U$ sets are defined for a given training/validation/test split:

References

1. Bintsi, K.M., Baltatzis, V., Potamias, R.A., Hammers, A., Rueckert, D.: Multimodal brain age estimation using interpretable adaptive population-graph learning. In: International Conference on Medical Image Computing and Computer Assisted Interventions (MICCAI), pp. 195–204 (2023)
2. Carlier, T., et al.: Prognostic value of 18 F-FDG PET radiomics features at baseline in PET-guided consolidation strategy in diffuse large B-cell lymphoma: a machine-learning analysis from the GAINED study. J. Nucl. Med. (2024)
3. Curtis, C., et al.: The genomic and transcriptomic architecture of 2,000 breast tumours reveals novel subgroups. Nature **486**, 346–352 (2012)
4. Farha, M.W., Salami, S.S.: Biomarkers for prostate cancer detection and risk stratification. Ther. Adv. Urol. **14** (2022)
5. Fey, M., Lenssen, J.E.: Fast graph representation learning with PyTorch Geometric. In: ICLR Workshop on Representation Learning on Graphs and Manifolds (2019)
6. Fleming, T.R., Harrington, D.P.: A class of hypothesis tests for one and two sample censored survival data. Commun. Stat. Theory Methods **10**(8), 763–794 (1981)
7. Geck, M.: PyCox: computing with (finite) Coxeter groups and Iwahori-Hecke algebras. LMS J. Comput. Math. **15**, 231–256 (2012)
8. Graf, E., Schmoor, C., Sauerbrei, W., Schumacher, M.: Assessment and comparison of prognostic classification schemes for survival data. Stat. Med. **18**(17–18), 2529–2545 (1999)
9. Huang, Q., He, H., Singh, A., Lim, S.N., Benson, A.R.: Combining label propagation and simple models out-performs graph neural networks. In: International Conference on Learning Representations (ICLR) (2020)
10. Ishwaran, H., Kogalur, U.B., Blackstone, E.H., Lauer, M.S.: Random survival forests. Ann. Appl. Stat. **2**, 841–860 (2008)
11. Kaplan, E.L., Meier, P.: Nonparametric estimation from incomplete observations. J. Am. Stat. Assoc. **53**(282), 457–481 (1958)

12. Katzman, J.L., Shaham, U., Cloninger, A., Bates, J., Jiang, T., Kluger, Y.: DeepSurv: personalized treatment recommender system using a Cox proportional hazards deep neural network. BMC Med. Res. Methodol. **18**(1), 24 (2018)
13. Kingma, D.P., Ba, J.: Adam: a method for stochastic optimization (2014)
14. Klein, J.P., Moeschberger, M.L.: Survival Analysis. SBH, Springer, New York (2003). https://doi.org/10.1007/b97377
15. Kvamme, H., Borgan, O., Scheel, I.: Time-to-event prediction with neural networks and Cox regression. J. Mach. Learn. Res. **20**(129), 1–30 (2019)
16. Paszke, A., et al.: PyTorch: an imperative style, high-performance deep learning library. In: Neural Information Processing Systems (NeurIPS), pp. 8024–8035 (2019)
17. Pölsterl, S.: scikit-survival: a library for time-to-event analysis built on top of scikit-learn. J. Mach. Learn. Res. **21**(212), 1–6 (2020)
18. Raymond, N., et al.: Development of error passing network for optimizing the prediction of VO_2 peak in childhood acute leukemia survivors. In: Proceedings of the fifth Conference on Health, Inference, and Learning (2024)
19. Shi, Y., Huang, Z., Feng, S., Zhong, H., Wang, W., Sun, Y.: Masked label prediction: unified message passing model for semi-supervised classification. In: Electronic Proceedings of IJCAI 2021, vol. 2, pp. 1548–1554 (2021)
20. Taherkhani, F., Kazemi, H., Nasrabadi, N.M.: Matrix completion for graph-based deep semi-supervised learning. In: Proceedings of the Thirty-Third AAAI Conference on Artificial Intelligence and Thirty-First Innovative Applications of Artificial Intelligence Conference and Ninth AAAI Symposium on Educational Advances in Artificial Intelligence, pp. 5058–5065 (2019)
21. Thiery, O., et al.: PET-based lesion graphs meet clinical data: an interpretable cross-attention framework for DLBCL treatment response prediction. Comput. Med. Imag. Graph. **120**, 102481 (2025)
22. Yala, A., Lehman, C., Schuster, T., Portnoi, T., Barzilay, R.: A deep learning mammography-based model for improved breast cancer risk prediction. Radiology **292**(1), 60–66 (2019)
23. Yang, Z., et al.: ResMem: learn what you can and memorize the rest. Adv. Neural. Inf. Process. Syst. **36**, 60768–60790 (2023)
24. Zhang, H., et al.: Classification of brain disorders in rs-fMRI via local-to-global graph neural networks. IEEE TMI **42**(2), 444–455 (2023)

Spectral Graph Autoregressive Modeling for Conditional Brain Network Augmentation

Hayoung Ahn[1], Seungjoo Lee[1], Jaeyoon Sim[1], Yechan Hwang[1], Hyuna Cho[1], Guorong Wu[2], and Won Hwa Kim[1(✉)]

[1] Pohang University of Science and Technology (POSTECH), Pohang, South Korea
{ahnha,wonhwa}@postech.ac.kr
[2] University of North Carolina at Chapel Hill, Chapel Hill, USA

Abstract. We present Spectral Graph AutoRegressive (SGAR) model, a novel conditional node signal synthesis method for brain networks. Unlike conventional generative models, SGAR employs a coarse-to-fine graph generation strategy in the spectral space: it first predicts low-frequency components that capture the global graph structure and then progressively refines high-frequency details to encode local feature dependencies. SGAR leverages Graph Fourier Transform (GFT) to decompose graph signals in the spectral domain and utilizes a conditional autoregressive transformer to generate spectral components based on disease stage labels. The continuous node signals are subsequently reconstructed via Inverse Graph Fourier Transform (IGFT), preserving the overall network topology. Applied to the Alzheimer's Disease Neuroimaging Initiative (ADNI) dataset, our framework effectively addresses the challenges of data scarcity and label imbalance by augmenting brain networks with realistic, structured node features. Experimental results demonstrate that SGAR improves downstream AD classification performance while maintaining the global structure of brain networks.

Keywords: Data Augmentation in Neuroimaging · Alzheimer's Disease · Spectral Graph Analysis

1 Introduction

Early diagnosis of neurodegenerative diseases such as Alzheimer's disease (AD) is crucial for timely treatment planning, given their irreversible nature. To improve understanding of degenerative structural patterns in the white-matter connectome, recent studies have employed brain network analyses that model the brain as a graph structure of interconnected regions [5,29]. Leveraging the graph structure provides a more comprehensive understanding of disease mechanisms beyond regional analyses [6,21,34]. However, the analysis faces challenges from the dataset curation stage with reliable brain network data, as it requires both brain regional features (i.e., node signals) and connectivity information (i.e., edges) that map all pairs of brain regions. Due to the high costs for data acquisition, data scarcity, class imbalance, and even privacy issues, these challenges limit the robustness of network-based brain analyses.

To address these issues, graph generation methods [14,33] have been studied and shown successful results in synthesizing realistic graph data. However, existing models are computationally expensive [3], as the complexity of modeling edges grows quadratically with respect to the number of nodes, and they are often unconditional. Thus, they only can be restrictively applied to small graphs with short sequences, making them unsuitable for larger brain networks with larger sequences of nodes. Autoregressive models [22,35] are able to handle larger graphs with longer sequences by sequentially generating nodes and edges, where each generative step is conditioned on previously generated elements. Yet, this sequential dependency introduces local biases, as new nodes or edges depend only on a subset of the generated graph so far. Moreover, the generation order of graph signals is often arbitrary, which leads to inconsistencies in graph quality and reliability of the generated graph signals.

To address these limitations, we propose a graph (node) signal synthesis model for brain network analysis, dubbed **S**pectral **G**raph **A**uto-**R**egressive (SGAR). Unlike existing autoregressive works that generate individual graph components (e.g., a node or an edge) sequentially in the graph space, our method gradually synthesizes global-to-local graph features in the graph spectral space. This is realized by decomposing graphs into its spectral components using Graph Fourier Transform (GFT), which allows the model to first generate low-frequency global structural patterns and progressively impose high-frequency local details based on the global characteristics generated so far. This hierarchical graph generation scheme with spectral decomposition enables more coherent and globally consistent graph signal generation, mitigating local biases inherent in existing autoregressive approaches.

Given that the structural connectivity remains relatively stable over time, whereas regional measures substantially change with disease progression, our method utilizes fixed structural connectivity and dynamic node signals across multiple time points. Notably, when obtained labels and graph signals are sparse and limited within a sequence, SGAR exploits structural connectivity of observed time point to generate missing node signals with labels. This strategy is especially promising for neurological disorder studies–such as in Alzheimer's disease research–where ample data is essential for a deeper understanding of the longitudinal degenerative progression but securing sufficient data is challenging.

To evaluate the effectiveness of SGAR, we conduct extensive experiments on the Alzheimer's Disease Neuroimaging Initiative (ADNI) [25] dataset, demonstrating: (1) the plausibility of generated graphs compared to real data, and (2) improved downstream AD classification performance by accurately capturing disease characteristics throughout its progression.

2 Background

Spectral Graph Analysis. Spectral graph analysis has emerged as a powerful tool for understanding and processing graph-structured data [7,9,20,28]. The Graph Fourier Transform (GFT) converts node features into the spectral domain to analyze the frequency components inherent in the data. This spectral representation effectively separates global low-frequency information and local high-frequency details, which can be exploited in subsequent processing stages.

Autoregressive Modeling via Next-Scale Prediction. Autoregressive models are effective in various domains such as natural language processing [1,2,4,32] and visual modeling [11,23,24]. Next-scale prediction, originally proposed in visual autoregressive modeling [30], involves setting an appropriate scale unit (i.e., resolution) and applying autoregression at each scale, a technique shown to be critical for high-quality results. This approach allows models to first capture the coarse structure and then refine finer details, a strategy that can be adapted for graph data to improve the generation of node features.

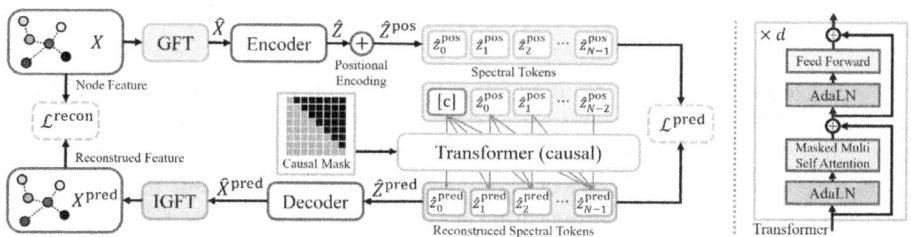

Fig. 1. Overview of SGAR. From node signals X, a spectral encoder produces N token maps $[\hat{z}_0^{pos}, \hat{z}_1^{pos}, ..., \hat{z}_{N-1}^{pos}]$. Our transformer is trained to predict next-frequency signals: it takes $[\texttt{[c]}, \hat{z}_0^{pos}, \hat{z}_1^{pos}, ..., \hat{z}_{N-2}^{pos}]$ as input to predict $[\hat{z}_0^{pred}, \hat{z}_1^{pred}, ..., \hat{z}_{N-1}^{pred}]$. A causal attention mask ensures each token can only refer to its prefix (i.e., lower frequency signals). Both $\mathcal{L}^{pred}$ and $\mathcal{L}^{recon}$ are used to reconstruct node feature X^{pred}.

3 Method

Let an undirected graph $G = (V, E)$, where V denotes the node set with $|V| = N$, and E represents the edge set. The connectivity between nodes is encoded by a symmetric adjacency matrix $A \in \mathbb{R}^{N \times N}$, where each element A_{pq} indicates the connection between node p and node q. Given the diagonal degree matrix $D \in \mathbb{R}^{N \times N}$, where $D_{pp} = \sum_q A_{pq}$, the graph Laplacian L is defined as $L = D - A$, and a normalized one is defined as $\tilde{L} = D^{-1/2} L D^{-1/2}$. Since $\tilde{L}$ is real and positive semi-definite, it has orthonormal eigenvectors $U = [u_0, u_1, ..., u_{N-1}]$ and corresponding real and non-negative eigenvalues $\Lambda = diag(\lambda_0, \lambda_1, ..., \lambda_{N-1})$, where

$0 = \lambda_0 \leq \lambda_1 \leq ... \leq \lambda_{N-1}$. In this spectral decomposition, smaller eigenvalues correspond to eigenvectors capturing global, low-frequency information of a graph, while larger eigenvalues capture local, high-frequency details.

Our objective is to generate continuous node features $X \in \mathbb{R}^{N \times d}$ with feature dimension d, given a graph G and a target label. To achieve this, our approach consists of three components: 1) Aligning each graph in a common coordinate system via graph Laplacian projection, 2) Autoregressive spectral token generation using a label-conditional SGAR Transformer, and 3) Training objective. Overall framework is illustrated in Fig. 1.

3.1 Aligning Individual Graphs with Laplacian Projection

To ensure a fair comparison and consistent generation across different subjects, individual graphs are aligned within a common spectral coordinate system. Given a set of M subjects, this alignment is performed by projecting each graph Laplacian onto a shared spectral space derived from the mean graph Laplacian $\bar{L}$. For this, the $\bar{L}$ across all subjects with eigen-decomposition is computed as:

$$\bar{L} = \frac{1}{M} \sum_{i=1}^{M} \tilde{L}_i = \bar{U}\bar{\Lambda}\bar{U}^\top, \tag{1}$$

where $\tilde{L}_i$ represents the normalized graph Laplacian matrix of the i-th subject. Here, $\bar{U} \in \mathbb{R}^{N \times N}$ is an eigenvector matrix of $\bar{L}$, forming a common orthonormal basis, and $\bar{\Lambda} \in \mathbb{R}^{N \times N}$ is a diagonal matrix of the corresponding eigenvalues.

For each subject i, we project the individual graph Laplacian $\tilde{L}_i$ onto $\bar{U}$, and then perform an eigendecomposition on this projected Laplacian $\tilde{L}_i^{\text{proj}}$ as:

$$\tilde{L}_i^{\text{proj}} = \bar{U}^\top \tilde{L}_i \bar{U} = U_i \Lambda_i U_i^\top, \quad i = 1, \ldots, M, \tag{2}$$

where U_i and Λ_i denote the eigenvector and eigenvalue of $\tilde{L}_i^{\text{proj}}$, respectively.

3.2 Autoregressive Spectral Token Generation

In this section, we describe the key process of autoregressively generating spectral tokens incorporating label condition and positional information of eigenvalues. This autoencoder architecture involves the eigenvalues and eigenvectors of $\tilde{L}_i^{\text{proj}}$ described as Λ_i and U_i in Eq. 2. For simplicity, we drop the index i and denote them as $\Sigma = \Lambda_i$ and $Q = U_i$ for a single subject.

Given a node signal X and eigenvector Q, using GFT, the spectral representation of the node feature $\hat{X}$ is computed as:

$$\hat{X} = Q^\top X = [\hat{x}_0, \hat{x}_1, \ldots, \hat{x}_{N-1}], \tag{3}$$

where each element $\hat{x}_n \in \mathbb{R}^d$ represents the coefficient indicating the contribution of the n-th eigenvector to the original feature X.

Subsequently, we aim to generate a prediction $\hat{X}^{\text{pred}}$, given $\hat{X}$ and a label as a condition, using a conditional autoregressive transformer. First, we obtain a h-dimensional latent representation $\hat{Z} \in \mathbb{R}^{N \times h}$ for $\hat{X}$ using an encoder $\mathcal{E}_X$:

$$\hat{Z} = \mathcal{E}_X(\hat{X}) = [\hat{z}_0, \hat{z}_1, ..., \hat{z}_{N-1}], \tag{4}$$

where $\mathcal{E}_X$ is a MLP with 2 layers. Since eigenvalues represent the magnitude of influence for their corresponding eigenvectors in describing the structure, to incorporate positional information, we compute a h-dimensional spectral positional encoding $p_n \in \mathbb{R}^h$ for each eigenvalue σ_n using an encoder $\mathcal{E}_\Sigma$ as:

$$p_n = \mathcal{E}_\Sigma(\sigma_n), \tag{5}$$

where $\mathcal{E}_\Sigma$ is also composed of a 2-layer MLP. These positional encodings $P = [p_0, p_1, ..., p_{N-1}]$ are added to $\hat{Z}$ to form the token representation $\hat{Z}^{\text{pos}} = \hat{Z} + P = [\hat{z}_0^{\text{pos}}, \hat{z}_1^{\text{pos}}, ..., \hat{z}_{N-1}^{\text{pos}}] \in \mathbb{R}^{N \times h}$ for training an autoregressive module.

We adopt an autoregressive transformer architecture from Visual Autoregressive Transformer (VAR) [30] and revise it for autoregressive graph generation. Specifically, our transformer generates $\hat{Z}^{\text{pred}}$ from $\hat{Z}^{\text{pos}}$. Given the previous n tokens, i.e., $\hat{z}_{0:n-1}^{\text{pos}}$, with the label condition c, the transformer generates the next token $\hat{z}_n^{\text{pred}}$ as follows:

$$\hat{z}_n^{\text{pred}} = \text{Transformer}(c, \hat{z}_{0:n-1}^{\text{pos}}), \tag{6}$$

where the label c is processed by a learnable embedding layer and conditioned into the model. For the Transformer, we adopt the architecture of standard decoder-only transformer akin to GPT-2 [27] with adaptive normalization (AdaLN), which is a common approach in many generative models [8,16–19]. We modify the standard architecture by incorporating an additional embedding layer for positional encoding of eigenvalues for each spectral token and replacing the CE loss with MSE loss to better suit continuous data generation.

During training, ground truth tokens $\hat{z}_{0:n-1}^{\text{pos}}$ are used as input to the model for stable learning, whereas during inference, the predicted tokens $\hat{z}_{0:n-1}^{\text{pred}}$ are fed back as input for the next token prediction. The generated token sequence $\hat{Z}^{\text{pred}} = [\hat{z}_0^{\text{pred}}, \hat{z}_1^{\text{pred}}, ..., \hat{z}_{N-1}^{\text{pred}}]$ is then decoded using a decoder $\mathcal{D}_X$ to obtain the predicted spectral representation $\hat{X}^{\text{pred}} \in \mathbb{R}^{N \times d}$ as:

$$\hat{X}^{\text{pred}} = \mathcal{D}_X(\hat{Z}^{\text{pred}}) = [\hat{x}_0^{\text{pred}}, \hat{x}_1^{\text{pred}}, ..., \hat{x}_{N-1}^{\text{pred}}]. \tag{7}$$

Finally, using the Inverse Graph Fourier Transform (IGFT), the predicted node feature X^{pred} is computed as $X^{\text{pred}} = Q\hat{X}^{\text{pred}}$, reconstructing the original X.

Training Objective. Given X_i, X_i^{pred}, $\hat{Z}_i^{\text{pos}}$, and $\hat{Z}_i^{\text{pred}}$ for subject i, the overall objective $\mathcal{L}$ is the summation of $\mathcal{L}^{\text{pred}}$ and $\mathcal{L}^{\text{recon}}$ as:

$$\mathcal{L} = \mathcal{L}^{\text{pred}} + \mathcal{L}^{\text{recon}} = \frac{1}{M} \sum_{i=1}^{M} \left\| \hat{Z}_i^{\text{pos}} - \hat{Z}_i^{\text{pred}} \right\|^2 + \frac{1}{M} \sum_{i=1}^{M} \left\| X_i - X_i^{\text{pred}} \right\|^2, \tag{8}$$

where it ensures accurate spectral prediction via the SGAR transformer through $\mathcal{L}^{\text{pred}}$ and faithful reconstruction of node signals by joint training of the encoder and decoder through $\mathcal{L}^{\text{recon}}$, while preserving the graph structure.

4 Experiment

4.1 Dataset and Experimental Setup

Dataset. From the ADNI dataset [25], we use two AD-specific biomarkers parcellated to 148 cortical and 12 subcortical regions based on Destrieux atlas [10]. Cortical thickness (CT) was obtained from MRI and measures on subcortical regions are replaced with gray matter volume. While CT and standardized uptake value ratio of fluorodeoxyglucose (FDG) from PET are used as nodal features each, the white matter fiber tracts between these nodes serves as edges of graphs. Five diagnostic labels were used to categorize the subjects: Control (CN), Significant Memory Concern (SMC), Early/Late Mild Cognitive Impairment (EMCI/LMCI), and AD, and their demographics are detailed in Table 1.

Table 1. Demographics of the ADNI dataset.

Biomarker	Category	CN	SMC	EMCI	LMCI	AD
Cortical Thickness	# of subjects	359	181	437	180	166
	Gender (M/F)	198/161	69/112	249/188	71/109	102/64
	Age (Mean ± Std)	72.8±1.4	72.1±4.9	71.7±0.9	70.9±6.1	74.8±8.7
FDG	# of subjects	345	186	461	231	162
	Gender (M/F)	173/172	66/120	262/199	152/79	102/60
	Age (Mean ± Std)	73.0±1.3	71.7±5.2	72.1±0.9	71.1±7.0	74.9±8.2

Implementation Details. As for the revised VAR architecture, we adopt the architecture of standard decoder-only transformers with adaptive normalization (AdaLN). Unlike large language models, we did not incorporate advanced techniques such as rotary position embeddings (RoPE), SwiGLU MLP, or RMS Norm [31,32]. Instead, we adopted a simple scaling rule similar to that in [15], where the width w, number of heads n_h, and dropout rate dr are linearly scaled with the depth factor n_d. Specifically, we set w to $64n_d$, n_h to n_d, and dr to $0.1\frac{n_d}{24}$. Our model was trained for 2000 epochs with a base learning rate of 10^{-4} using the AdamW optimizer and a batch size of 256.

Evaluation and Baselines. Following [21], we adopt three evaluation metrics to assess the quality of generated samples. Root Mean Squared Error (RMSE) focuses on the sample-wise differences by measuring the squared error between generated and real samples. In contrast, Wasserstein Distance (WD) and Jensen-Shannon Divergence (JSD) measure the distribution-wise distance, capturing global structural differences and distributional shape discrepancies. For all experiments, we use 80% samples for training and the remaining 20% for testing, and sampled the same number of data points as the test set while maintaining label distribution. For comparison, we adopt two representative graph generation models synthesizing node signals with the autoregressive approach, GraphRNN [35], and the diffusion-based approach, GDSS [14]. All baselines and our model were evaluated with three different parameter initializations, reporting the average results with standard deviation.

4.2 Performance Evaluation

Generation Performance. Table 2 demonstrates the generation performance of the SGAR model on the CT and FDG test sets. SGAR consistently outperformed baselines in WD and JSD. In CT, WD and JSD decreased by approximately 46% and 50%, respectively, compared to the second-best method. For FDG, SGAR lowered WD by about 54% and JSD by 34%. These results demonstrate SGAR's superior distributional alignment and enhanced sample diversity, confirming its effectiveness in generating realistic data.

Qualitative Analysis of Generation Results. Figure 2 illustrates qualitative results from SGAR on the CT experiments, highlighting generation quality across different labels. The visualization clearly demonstrates that the generated data effectively capture region-specific characteristics crucial for Alzheimer's disease. In particular, the Parahippocampal gyrus ($G\_oc\text{-}temp\_med\_Parahip$) located in the medial temporal-lobe shows progressive deterioration along the disease progression, which plays a critical role in spatial memory and scene recognition [26]. Notably, the Parahippocampal gyrus exhibits early neural degeneration and atrophy, as highlighted by the marked differences observed between CN and EMCI groups [12]. Similarly, the Middle frontal gyrus (G_front_middle), is involved in attention, decision-making, and various cognitive functions [13]. In the context of AD, a reduction in cortical thickness is observed in the Middle frontal gyrus—a change that becomes more pronounced as the disease progresses, reflecting widespread alterations across the brain.

Augmentation Performance. To evaluate the efficacy of our augmentation approach for downstream tasks, we assessed classification performance using Multi-Layer Perceptrons (MLPs) with 2 and 4 layers as classifiers. These MLPs were trained on concatenated node features, and performance was rigorously

Table 2. Generation performance on ADNI dataset's CT and FDG test sets, with averages and standard deviations from three replicates. The best results are in **bold**, and the second-best are underlined.

Methods	CT			FDG		
	RMSE ↓	WD ↓	JSD ↓	RMSE ↓	WD ↓	JSD ↓
GraphRNN [35]	**0.2017**$^{\pm 0.0078}$	0.1370$^{\pm 0.0120}$	0.2212$^{\pm 0.0188}$	**0.1976**$^{\pm 0.0028}$	0.1325$^{\pm 0.0052}$	0.2378$^{\pm 0.0140}$
GDSS [14]	0.2525$^{\pm 0.0001}$	0.4159$^{\pm 0.0341}$	0.0988$^{\pm 0.0009}$	0.2533$^{\pm 0.0012}$	0.3773$^{\pm 0.0560}$	0.0981$^{\pm 0.0006}$
SGAR (Ours)	0.2104$^{\pm 0.0039}$	**0.0742**$^{\pm 0.0079}$	**0.0500**$^{\pm 0.0153}$	0.2474$^{\pm 0.0364}$	**0.0604**$^{\pm 0.0098}$	**0.0646**$^{\pm 0.0368}$

Table 3. Classification performance on CT and FDG test sets. The best results are in **bold**, and the second-best are underlined.

Model	MLP (2 layers)				MLP (4 layers)			
	Acc ↑	Prec ↑	Rec ↑	F1 ↑	Acc ↑	Prec ↑	Rec ↑	F1 ↑
CT								
NoAug	0.781 ±0.016	0.771 ±0.026	0.781 ±0.015	0.775 ±0.021	0.439 ±0.103	0.396 ±0.292	0.407 ±0.183	0.341 ±0.213
GraphRNN [35]	**0.857** ±0.007	0.862 ±0.012	0.852 ±0.007	0.856 ±0.008	0.814 ±0.006	0.811 ±0.004	0.816 ±0.013	0.812 ±0.005
GDSS [14]	0.831 ±0.013	0.843 ±0.016	0.835 ±0.015	0.836 ±0.002	0.847 ±0.011	**0.855** ±0.019	0.849 ±0.016	0.850 ±0.013
SGAR (Ours)	0.850 ±0.010	**0.867** ±0.004	**0.860** ±0.008	**0.863** ±0.005	**0.863** ±0.005	0.852 ±0.005	**0.852** ±0.007	**0.851** ±0.004
FDG								
NoAug	0.785 ±0.006	0.794 ±0.016	0.798 ±0.005	0.795 ±0.010	0.485 ±0.164	0.422 ±0.328	0.447 ±0.223	0.399 ±0.281
GraphRNN [35]	0.856 ±0.003	0.866 ±0.009	**0.862** ±0.005	**0.869** ±0.002	0.844 ±0.009	0.847 ±0.006	0.847 ±0.008	0.847 ±0.007
GDSS [14]	0.836 ±0.043	0.832 ±0.040	0.835 ±0.049	0.831 ±0.044	0.842 ±0.024	0.849 ±0.017	0.840 ±0.024	0.843 ±0.020
SGAR (Ours)	**0.859** ±0.010	**0.869** ±0.009	0.861 ±0.016	0.864 ±0.013	**0.854** ±0.008	**0.857** ±0.014	**0.859** ±0.011	**0.857** ±0.011

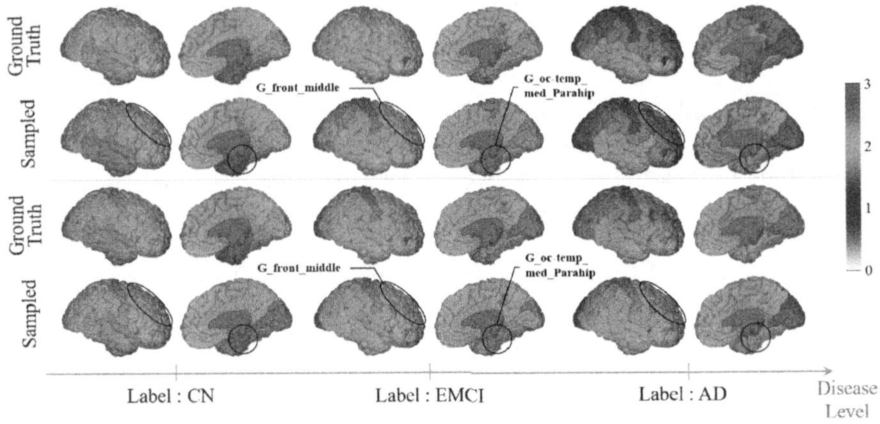

Fig. 2. Qualitative results from the CT test dataset. The first and third rows display the ground truth of three subjects with different labels (i.e., CN, EMCI, AD), while the second and fourth rows show the generated results by SGAR with the same labels. Throughout the label progression, vivid atrophy is consistently observed in the Parahippocampal gyrus ($G\_oc\_temp\_med\_Parahip$) and the Middle frontal gyrus [10] (G_front_middle), which are well-known AD-specific regions.

evaluated across accuracy, precision, recall, and F1-score, averaged over three replicates with varying parameter initializations for robust comparison.

In Table 3, we benchmarked SGAR against baselines and a No Augmentation control (NoAug). As shown in Table 3, the incorporation of augmented samples yielded a discernible improvement in classification performance compared to the NoAug, demonstrating the contribution of generated node features for AD classification. SGAR consistently outperformed all baselines across all evaluation metrics, ranking first in most cases and second otherwise. This consistency validates the effectiveness of our approach in generating high-quality augmented data that enhances downstream classification tasks.

5 Conclusion

We presented SGAR, a novel graph augmentation framework that leverages stable structural connectivity of brain networks and a conditional autoregressive spectral generation approach to synthesize continuous node signals. Our method utilize the common spectral space of brain networks to model pathological progression of AD biomarkers and further generate synthetic samples that highly resemble the real ones. Our method overcomes the limitations of traditional autoregressive and diffusion-based techniques, addressing data scarcity and imbalance challenges in AD research. Experiments on the ADNI dataset demonstrate that SGAR preserves global graph structure and improves downstream AD classification performance.

References

1. Achiam, J., et al.: GPT-4 technical report. arXiv preprint arXiv:2303.08774 (2023)
2. Anil, R., et al.: Palm 2 technical report. arXiv preprint arXiv:2305.10403 (2023)
3. Bergmeister, A., et al.: Efficient and scalable graph generation through iterative local expansion. arXiv preprint arXiv:2312.11529 (2023)
4. Brown, T., et al.: Language models are few-shot learners. In: Advances in Neural Information Processing Systems (NeurIPS), vol. 33, pp. 1877–1901 (2020)
5. Bullmore, E., Sporns, O.: Complex brain networks: graph theoretical analysis of structural and functional systems. Nat. Rev. Neurosci. **10**(3), 186–198 (2009)
6. Chawla, N.V., et al.: SMOTE: synthetic minority over-sampling technique. J. Artif. Intell. Res. **16**, 321–357 (2002)
7. Chung, F.R.: Spectral Graph Theory, vol. 92. American Mathematical Society (1997)
8. Dathathri, S., et al.: Plug and play language models: a simple approach to controlled text generation. arXiv preprint arXiv:1912.02164 (2019)
9. Defferrard, M., Bresson, X., Vandergheynst, P.: Convolutional neural networks on graphs with fast localized spectral filtering. In: Advances in Neural Information Processing Systems (NeurIPS), vol. 29, pp. 3844–3852 (2016)
10. Destrieux, C., et al.: Automatic parcellation of human cortical gyri and sulci using standard anatomical nomenclature. Neuroimage **53**(1), 1–15 (2010)
11. Dosovitskiy, A., et al.: An image is worth 16x16 words: transformers for image recognition at scale. arXiv preprint arXiv:2010.11929 (2020)
12. Echávarri, C., et al.: Atrophy in the parahippocampal gyrus as an early biomarker of Alzheimer's disease. Brain Struct. Funct. **215**, 265–271 (2011)
13. Germann, J., Petrides, M.: Area 8A within the posterior middle frontal gyrus underlies cognitive selection between competing visual targets. Eneuro **7**(5) (2020)
14. Jo, J., Lee, S., Hwang, S.J.: Score-based generative modeling of graphs via the system of stochastic differential equations. In: International Conference on Machine Learning (ICML). PMLR (2022)
15. Kaplan, J., et al.: Scaling laws for neural language models. arXiv preprint arXiv:2001.08361 (2020)
16. Karras, T., Laine, S., Aila, T.: A style-based generator architecture for generative adversarial networks. In: IEEE/CVF Conference on Computer Vision and Pattern Recognition (CVPR), pp. 4401–4410 (2019)

17. Karras, T., et al.: Analyzing and improving the image quality of styleGAN. In: IEEE/CVF Conference on Computer Vision and Pattern Recognition (CVPR), pp. 8110–8119 (2020)
18. Karras, T., et al.: Alias-free generative adversarial networks. In: Advances in Neural Information Processing Systems (NeurIPS), vol. 34, pp. 852–863 (2021)
19. Keskar, N.S., et al.: CTRL: a conditional transformer language model for controllable generation. arXiv preprint arXiv:1909.05858 (2019)
20. Kipf, T.N., Welling, M.: Semi-supervised classification with graph convolutional networks. arXiv preprint arXiv:1609.02907 (2016)
21. Kotelnikov, A., et al.: TabDDPM: modelling tabular data with diffusion models. In: International Conference on Machine Learning (ICML). PMLR (2023)
22. Liao, R., et al.: Efficient graph generation with graph recurrent attention networks. In: Advances in Neural Information Processing Systems (NeurIPS), vol. 32 (2019)
23. Lu, J., et al.: Unified-IO: a unified model for vision, language, and multi-modal tasks. arXiv preprint arXiv:2206.08916 (2022)
24. Lu, J., et al.: Unified-IO2: scaling autoregressive multimodal models with vision, language, audio, and action. arXiv preprint arXiv:2312.17172 (2023)
25. Mueller, S.G., et al.: The Alzheimer's disease neuroimaging initiative. Neuroimag. Clin. **15**(4), 869–877 (2005)
26. Owen, A.M., et al.: A specific role for the right parahippocampal gyrus in the retrieval of object-location: a positron emission tomography study. J. Cogn. Neurosci. **8**(6), 588–602 (1996)
27. Radford, A., et al.: Language models are unsupervised multitask learners. OpenAI Blog **1**(8), 9 (2019)
28. Shuman, D.I., et al.: The emerging field of signal processing on graphs: extending high-dimensional data analysis to networks and other irregular domains. IEEE Signal Process. Mag. **30**(3), 83–98 (2013)
29. Sporns, O., Tononi, G., Kötter, R.: The human connectome: a structural description of the human brain. PLoS Comput. Biol. **1**(4), e42 (2005)
30. Tian, K., et al.: Visual autoregressive modeling: scalable image generation via next-scale prediction. In: Advances in Neural Information Processing Systems (NeurIPS), vol. 37, pp. 84839–84865 (2025)
31. Touvron, H., et al.: Llama 2: open foundation and fine-tuned chat models. arXiv preprint arXiv:2307.09288 (2023)
32. Touvron, H., et al.: Llama: open and efficient foundation language models. arXiv preprint arXiv:2302.13971 (2023)
33. Vignac, C., et al.: DiGress: discrete denoising diffusion for graph generation. In: International Conference on Learning Representations (ICLR) (2023)
34. Xu, L., et al.: Modeling tabular data using conditional GAN. In: Advances in Neural Information Processing Systems (NeurIPS), vol. 32 (2019)
35. You, J., et al.: GraphRNN: generating realistic graphs with deep auto-regressive models. In: International Conference on Machine Learning (ICML), pp. 5294–5303. PMLR (2018)

HFR: Hemodynamic Feature Regression for Physically Constrained Pressure Drop Estimation

Jakub Chojnacki[✉], Szymon Kopeć, Konrad Duraj, and Maciej Zamorski

Hemolens Diagnostics Sp. z o.o., Legnicka 48G, 54-202 Wroclaw, Poland
jakub.chojnacki@hemolens.com

Abstract. Computational fluid dynamic (CFD) simulation is a leading approach to accurately model state of physical environment, e.g. blood pressure in cardiovascular system. However, due to the high level of detail of simulated system that CFD methods require they tend to be computationally expensive creating a need for ml-assisted solutions. While methods incorporating deep learning for pressure drop and fractional flow reserve (FFR) estimation are advancing rapidly, they suffer from need for large amount of real data, which is scarcely available, simulated under many possible scenarios presenting a challenge from computational point of view. In this work we present Hemodynamic Feature Regression (HFR)–method for pressure drop estimation that by incorporating hemodynamical modeling is able generalize from synthetic to real samples. We test the pressure drop predictions on real coronary geometries compared to CFD simulations.

Keywords: Coronary arteries · FFR · Pressure drop · Reduced order modelling · Machine learning

1 Introduction

Coronary artery disease (CAD) is one of the major cause of death worldwide [16], characterized by the presence of atherosclerotic plaques in coronary arteries (CA), resulting in the narrowing of the arterial lumen area and decrease in blood pressure incrising risk of adverse cardiac events [24]. The diagnosis of CAD is accomplished by imaging methods such as invasive coronary angiography (ICA), which may not result in correct determination of plaques [24]. Consequently, clinical practice exams fractional flow reserve (FFR) - measuring drop in artery pressure under induced hyperemia, providing lesion-specific severity of stenoses, complementing ICA [12]. However, due to invasiveness and high cost of the procedure, alternative methods are being explored [7,10].

Computational Fluid Dynamics (CFD) become essential for modelling the cardiovascular system [9] – e.g. as non-invasive approximation of FFR. Despite its promise, the routine use of CFD simulations is constrained by: (i) the need for accurate, patient-specific boundary conditions, (ii) difference of CA geometry between the real and simulated states, (iii) the low resolution of imaging method use to create CA model for simulation (iv) computational resources needed for satisfactory precision [13].

The computational requirements may be lowered by applying simplified approaches, e.g. reduced order models (ROMs) [19], in particular 0-D and 1-D lumped-parameter models. While offering superior performance, they may make the model unable to accurately model detailed 3D geometries [3,19,26]

To address ROMs limitations, deep learning methods have been proposed as an alternative [2], such as physics-informed neural networks (PINNs) [21] and graph neural networks (GNN) [18]. Nevertheless, the development of ML models require meticulously curated datasets and substantial computational resources, while their black-box nature pose challenge in terms of interpretability and regulatory compliance [25].

In this work, we propose a novel approach called Hemodynamic Feature Regression (HFR), which incorporates CatBoost [4] and hydraulic-based modelling as physical constraints to achieve strong generalization to real-world data using a limited number of synthetically generated samples.

The primary contributions of this paper are threefold:

1. We present that our method has very robust generalization ability. We show that it can effectively match the CFD results obtained from real data after being trained solely on a limited amount of synthetic data.
2. We show that by first learning correction factors from the hydraulic coefficient and then using them to estimate pressure drop, we can effectively stabilize model training.
3. We demonstrate fully interpretable approach to predict pressure drops in the coronary arteries.

2 Related Work

The complex task of predicting the pressure drop in coronary arteries is an active field of research. In [17] authors propose a modification of MeshGraphNet [18] which learns to simulate blood flow at the centerline nodes. Other work [5] take advantage of hybrid approach that combine physics with neural networks. The network incorporate ROM's predictions and learn their CFD residuals to attain CFD-quality FFR with limited training data. The work [27] proposes a method based on neural networks to address the complex parameter estimation challenge in cardiovascular hemodynamics. Subsequently, they demonstrate the application of this method by utilizing a reduced-order hemodynamic model to conduct a preliminary patient-specific assessment. Other approach [6] leverages PINN by simultaneously calibrates Windkessel boundary conditions and reconstructs 3D aortic pressure and velocity fields from sparse 2D phase-contrast MRI, achieving hemodynamic predictions in both steady and pulsatile flow scenarios. Alternative approaches for estimating pressure drop in CAs employ intricate geometric representations, such as point clouds [22] or meshes [15]. However, to the best of the author's knowledge, there is no fully interpretable approach that can generalize effectively to out-of-distribution datasets while adhering to physical constraints.

3 Methodology

In this section, we provide formulation of the pipeline depicted in Fig. 1 and 2. In Sect. 3.1, we delve into the construction process of the segments. We describe derivation of the intermediate state our model is based upon in Sect. 3.2, the concept of Hjorth parameters in Sect. 3.3 and finally the training process in Sect. 3.4.

3.1 Vessel Segments

The presented method utilizes centerline graphs as vessel representations. Subsequently, these centerline graphs are further subdivided into branches, which are defined as the distance from the inlet to the outlet. In accordance with best practices of performing CFD simulations we truncate (place the last measurement plane) a 5 mm before the end of the extracted geometry. These branches are further segmented into vessel segments, as illustrated in Fig. 1 in step 4, and defined as the range of points between splitting places, which are performed at the coronary artery bifurcations. In contrast to the standard usage of hydraulic coefficients, which treats a tube with stenosis as three distinct segments defined as proximal (before stenosis), stenotic, and distal (after stenosis), we treat it as a singular segment and describe it using Eq. 1.

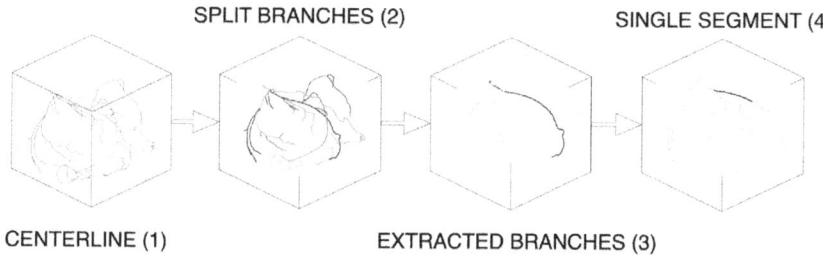

Fig. 1. Vascular geometry processing. **1)** Coronary artery centerline extraction from the segmented lumen. **2)** Root-to-leaf path tracing to obtain anatomically consistent branches. **3)** Partition the branches into inletbifurcation, bifurcationbifurcation, and bifurcationoutlet segments **4)** Extract segments for further calculations.

3.2 Hydraulic Coefficients

We compute the hydraulic coefficients of the polynomial expansion of the pressure base on its geometry and patient-specific parameters, including volumetric

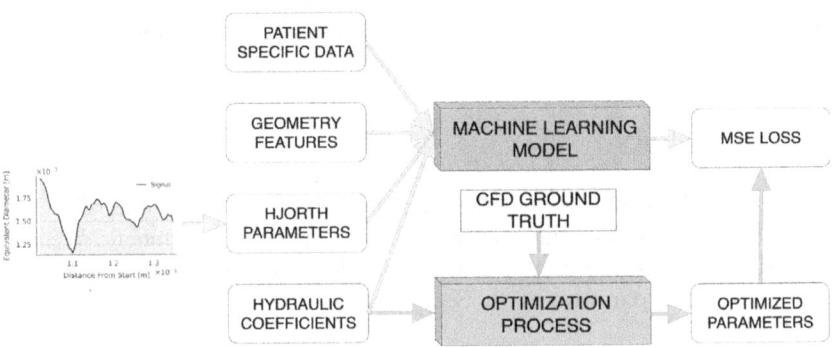

Fig. 2. Schematics of HFR. A 23-dimensional descriptor—hydraulic coefficients, Hjorth parameters, geometry features and patient specific data—is fed to a CatBoost regressor trained on CFD-labelled trees. The model predicts multiplicative corrections $(1 + \delta_k)$ for the coefficient relevant to the stenosis status. The corrected coefficient is substituted back into the analytical law to estimate the segmental pressure drop.

blood flow. The first coefficient a_1 is the HagenPoiseuille [20] equation for the straight pipe, while the second coefficient a_2 follows the Seeley & Young [23] formula:

$$a_1 = \frac{8\pi\mu L}{A_{\text{med}}}, \quad a_2 = \frac{1.52\rho}{2}\left(\frac{1}{A_{min}} - \frac{1}{A_{max}}\right)^2 \quad (1)$$

where $\mu = 3.5 \times 10^{-3}\,[\text{Pa} \cdot \text{s}]$ – viscosity, $\rho = 1060\,[\text{kg} \cdot \text{m}^{-3}]$ – density, L – segment length and $A_{med}, A_{min}, A_{max}$ – median, minimal and maximal cross-section area. We utilize the obtained coefficients to compute initial pressure drop for all analyzed segments:

$$\Delta p_{\text{base}} = a_1\tilde{Q} + a_2\tilde{Q}^2. \quad (2)$$

where $\tilde{Q}$ represents the segment-specific flow, which is calculated by multiplying the initial flowrate by the Murray [14] coefficient factor specific to a particular vessel segment.

3.3 Hjorth-Compressed Geometry

Conversion of 3D mesh into 1D centerline, entails the loss of spatial information, which we adress by we can effectively address the impact of noise, the difficulty in selecting the optimal number of points, and the model complexity. To address the aforementioned concerns, we have resolved to treat equivalent diameter, cross-sectional area, and Murray coefficients as signals and summarize them employing the Hjorth [8] parameters. We distinguish equivalent diameters and cross-sectional areas due to the disparities in their modeling relationships: linear and squared, which we believe exhibit a correlation with hydraulic coefficients. This approach aims to provide machine learning model with information regarding the subtle changes in vessel segments and to succinctly encapsulate the overall amplitude variability, average slope, and waveform roughness.

3.4 Training

Furthermore, inference frequently yields non-physical solutions, either in isolated or multiple segments of the blood vessel network. We have resolved to incorporate an intermediate state that facilitates the regularization of training. This involves the initial calculation of hydraulic flow coefficients, denoted as a_1 and a_2. Subsequently, the initial state of the system is determined by computing the pressure drop derived from these coefficients, as specified in Eq. 2.

The coronary arteries induced higher-order effects that were incorporated into Eq. 1. However, as previously mentioned, they can serve as effective early predictors of pressure drop. Therefore, instead of directly predicting the pressure drop using the model, we aimed to correct the hydraulic coefficients. To compensate for their limitations, we incorporated feature vectors that encompass various properties. For each segment, the model is with a 23-D feature vector comprising:

- Hydraulic flow coefficients: a_1 and a_2, initial pressure drop,
- Geometry features: segment length, minimal, maximal and median area, equivalent diameter, stenosis information,
- Hjorth parameters for eq. diameter, Murray coeffs., cross-section areas,
- Equations-estimated pressure drop and Murray coefficients.

We observe that a_1 parameter tends to have a greater impact in healthy regions and a_2 in stenotic segments.

$$a_1^\star = \frac{\Delta p - a_2 Q^2}{Q}, \quad a_2^\star = \frac{\Delta p - a_1 Q}{Q^2}, \tag{3}$$

where Δp is the ground-truth CFD pressure drop.

We observed that the model has a tendency to overcorrect the initial coefficients. Consequently, we decided to incorporate an additional optimization step utilizing Optuna [1] to compel the fit between two distributions, which represent the final model pressure drop prediction and the CFD ground truth:

$$\mathcal{L}(\widehat{\Delta p}, \Delta p^{\mathrm{CFD}}) = \left(\frac{|\widehat{\Delta p}|}{|\Delta p^{\mathrm{CFD}}| + \varepsilon} - 1\right)^2, \quad \varepsilon = 10^{-2}\,\mathrm{Pa}. \tag{4}$$

This method is particularly useful when our model exhibits a small mean squared error, yet its error on stenotic segments is relatively larger compared to healthy segments or vice versa.

4 Dataset

We consider two datasets for evaluation:

- **Synthetic data**: 100 generated cardiovascular patients' geometries (Fig. 3a), prepared as follows:

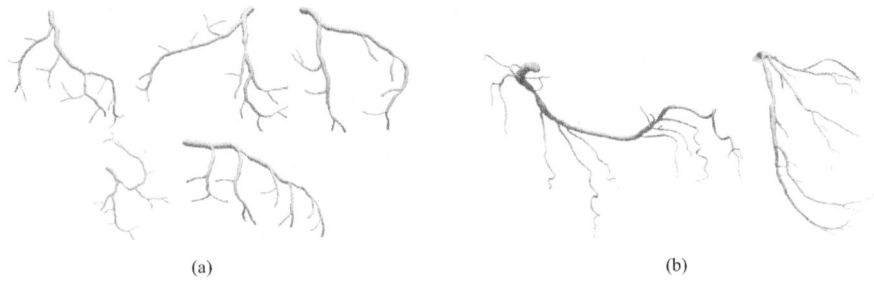

Fig. 3. Cardiovascular models derived from a synthetic data generation pipeline (a) and clinical CCTA images (b).

1. **Coronary Tree Synthesis**: A rooted centerline graph is expanded recursively based on anatomical priors (radius, taper, bifurcation angle, path length) until distal radii are less than 0.4 mm. Trees with intersecting branches are excluded.
2. **Lesion Modeling**: Each segment undergoes concentric stenoses, with radial severity sampled from a C^1-continuous polynomial basis, adhering to the CAD-RADS 1–4 statistics.
3. **Spatial embedding**: Two-dimensional centerlines are enveloped within random-radius spheres to achieve realistic curvature. Subsequently, these spheres are transformed into signed distance fields using the VMTK library, and the resulting data is subsequently meshed into watertight surfaces employing the marching cubes algorithm.

– **Real data**: Randomly selected 140 geometries from private dataset of coronary artery models. Segmentations were performed manually by skilled experts using CT Angiography imaging data collected from various institutions in Poland. The exemplary geometry is presented in Fig. 3b.

5 Results

In this section, we present the results obtained on the datasets with real patients described in Sect. 4. All of the experiments were performed on a local machine equipped with AMD Ryzen 9 5900X and NVIDIA GeForce RTX 3090.

5.1 Comparison Against Direct ML Pressure Drop Prediction

We evaluated the performance of the HFR in comparison to the catboost prediction, excluding the Hjorth parameters and hydraulic coefficients, and directly predicting pressure drop (as shown in Table 1). All input datasets underwent the following preprocessing: duplicated segments were removed, and cases with physically improbable values of hydraulic coefficients were excluded (indicating geometry errors). We computed the pressure drop per segment and aggregated

Table 1. Mean absolute errors in pressure drop for varying flowrates of HFR. Pressure drops were aggregated per patient and subsequently summed. All improvements were statistically significant (p-value < 0.05)

Flow (ml·s^{-1})	3	4	5	6	7
Base	12.02	16.78	26.52	36.37	47.21
HFR	**5.97**	**10.02**	**15.43**	**22.39**	**32.60**

these values for each specific patient, ensuring that the calculations were performed for all flowrates utilized in the simulation. As the flowrate increases, both the gap between the machine learning model and the actual data and the mean squared error between them increase. Our results also indicate that the generalization gap between ML models decreases when we add intermediate states to the models.

In Table 2, we present a comparative analysis of mean absolute errors between pressure drop derived from the HFR and BASE models in relation to coronary artery branches. It is important to note that different vascular regions exhibit varying ranges for key quantities, including pressure, flow rate, cross-sectional area, and equivalent diameter. Since the HFR was trained on synthetic data, we aim to investigate whether certain regions have better generalizations ability than others in case of real data.

Table 2. Mean absolute errors in millimeters of mercury (mmHg) of pressure drop predictions for each of the inspected coronary artery branches. Δ denotes the absolute and relative change (HFR - Base).

Branch	lad6	lad7	lad8	lad9	lad9a	lcx11	lcx12a	lcx12b	lcx13	lcx14	lcx15	rca3	rca4	rca16
Base	7.98	8.83	9.39	3.55	3.67	2.21	6.23	5.67	4.66	9.99	7.48	8.42	7.55	7.43
HFR	5.50	5.04	6.61	1.14	1.04	1.36	3.65	4.46	1.92	5.27	6.17	5.67	7.86	2.90
Δ (abs)	−2.48	−3.79	−2.78	−2.41	−2.63	−0.85	−2.58	−1.21	−2.74	−4.72	−1.31	−2.75	+0.31	−4.53
Δ (%)	−31%	−43%	−30%	−68%	−72%	−38%	−41%	−21%	−59%	−47%	−18%	−33%	+4%	−61%

We also conducted an analysis of the sensitivity of HFR to each feature. We utilized SHAP [11] to evaluate the impact of these features on specific predictions. Features associated with higher sensitivity values have a more significant impact on the result accuracy compared to those associated with lower values. Our findings are summarized in Fig. 4, which confirms that the model does not overfit. This is because, based on our expectations, the highest impact has been achieved by stenosis presence, Hjorth parameters (equivalent diameter complexity and area activity), volumetric flowrate, and precomputed parameters—dominated by a_2 (capturing pressure drops caused by stenosis) with only minor contribution from a_1.

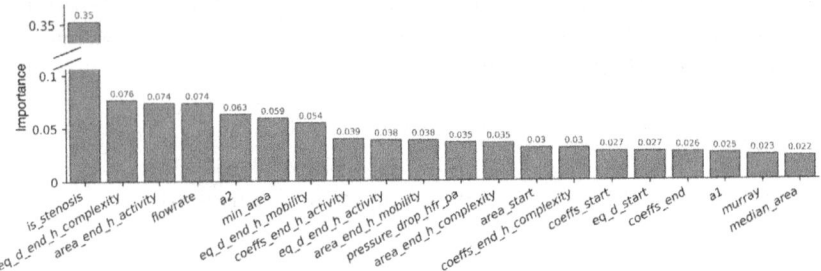

Fig. 4. SHAP-based sensitivity analysis for the HFR pressure drop prediction. Bars indicate the mean absolute SHAP value of each feature, representing its average marginal contribution to the prediction across all test instances. Higher bars indicate greater global importance.

5.2 Ablation Study

We investigate the impact of a) restricting the model to use only hydraulic coefficients and b) supplementing model with just Hjorth parameters on the final prediction of pressure drop. The ablation study results are presented in Table 3. We observe that enhancing the process with both ML model step and with additional features beneficially impacts the model predictive capabilities. The statistical analysis show significance with p-values < 0.05.

Table 3. Mean absolute error in mmHg of pressure drop predictions for each of the considered model. All improvements yield statistical significant ($p < 0.05$) drop in MAE.

Flowrate	3	4	5	6	7
Equations	8.28	13.25	19.55	27.37	38.00
+Model	7.45	11.50	16.63	23.63	33.87
+Model + Hjorth	5.97	10.02	15.43	22.39	32.60

6 Conclusions

We presented a HFR model to estimate pressure drop in coronary arteries in one-dimensional approximations of three-dimensional simulations. Our architecture is a modified version of the traditional approach to predicting pressure drop values based on hydraulic coefficients. We demonstrated the model's generalizability on a real dataset, evaluating its performance across diverse flowrates, geometries and particular anatomical coronary branches. We performed a sensitive analysis to determine which features are more important to correctly approximating the pressure drop. In our experiments, the most influential features were Hjorth parameters, geometric quantities like stenosis indicator and min area, hydraulic features like a_2 and initial pressure drop estimation. In Sect. 5.1, we conducted a direct comparison with the base model, which directly estimates pressure drop without any intermediate steps, showing superior performance of HFR. We also evaluated various approaches to estimate pressure drop, including methods that incorporate the Hjorth effect and solely utilize hydraulic coefficients (as detailed in Sect. 5.2). Our findings demonstrate that incorporating Hjorth parameters is crucial for obtaining reliable results. Furthermore, hydraulic features can serve as an efficient predictor of pressure drop, even without the need for any machine learning model.

References

1. Akiba, T., Sano, S., Yanase, T., Ohta, T., Koyama, M.: Optuna: a next-generation hyperparameter optimization framework. In: Proceedings of the 25th ACM SIGKDD International Conference on Knowledge Discovery & Data Mining, pp. 2623–2631 (2019)
2. Brivio, S., Fresca, S., Manzoni, A.: Ptpi-dl-roms: pre-trained physics-informed deep learning-based reduced order models for nonlinear parametrized pdes. Comput. Methods Appl. Mech. Eng. **432**, 117404 (2024)
3. Cai, L., Zhong, Q., Xu, J., Huang, Y., Gao, H.: A lumped-parameter model for evaluating coronary artery blood-supply capacity. Math. Biosci. Eng. **21**(4), 5838–5862 (2024). https://doi.org/10.3934/mbe.2024258
4. Dorogush, A.V., Ershov, V., Gulin, A.: Catboost: gradient boosting with categorical features support. arXiv preprint arXiv:1810.11363 (2018)
5. Fossan, F.E., et al.: Machine learning augmented reduced-order models for ffr-prediction. Comput. Methods Appl. Mech. Eng. **384**, 113892 (2021)
6. Garay, J., Dunstan, J., Uribe, S., Costabal, F.S.: Physics-informed neural networks for blood flow inverse problems. arXiv preprint arXiv:2308.00927 (2023)
7. Gulati, M., Levy, P.D., Mukherjee, D., et al.: 2021 aha/acc guideline for the evaluation and diagnosis of chest pain. Circulation **144**(22), e368–e454 (2021). https://doi.org/10.1161/CIR.0000000000001029
8. Hjorth, B.: The physical significance of time domain descriptors in eeg analysis. Electroencephalogr. Clin. Neurophysiol. **34**(3), 321–325 (1973)
9. Kim, H.J., Vignon-Clementel, I., Figueroa, C., Jansen, K., Taylor, C.: Developing computational methods for three-dimensional finite element simulations of coronary blood flow. Finite Elem. Anal. Des. **46**(6), 514–525 (2010)
10. Knuuti, J., Windecker, S., Bax, J.J., et al.: 2024 esc guidelines for the management of chronic coronary syndromes. Eur. Heart J. **45**(36), 3415–3547 (2024). https://doi.org/10.1093/eurheartj/ehae177
11. Lundberg, S., Lee, S.I.: A unified approach to interpreting model predictions (2017). https://arxiv.org/abs/1705.07874
12. Mangiacapra, F., Bressi, E., Sticchi, A., Morisco, C., Barbato, E.: Fractional flow reserve (ffr) as a guide to treat coronary artery disease. Expert Rev. Cardiovasc. Ther. **16**(7), 465–477 (2018)
13. Morris, P.D., et al.: Computational fluid dynamics modelling in cardiovascular medicine. Heart **102**(1), 18–28 (2016)

14. Murray, C.D.: The physiological principle of minimum work: I. the vascular system and the cost of blood volume. Proc. Natl. Acad. Sci. **12**(3), 207–214 (1926)
15. Nannini, G., et al.: Learning hemodynamic scalar fields on coronary artery meshes: a benchmark of geometric deep learning models. arXiv preprint arXiv:2501.09046 (2025)
16. Okrainec, K., Banerjee, D.K., Eisenberg, M.J.: Coronary artery disease in the developing world. Am. Heart J. **148**(1), 7–15 (2004). https://doi.org/10.1016/j.ahj.2003.11.027. https://www.sciencedirect.com/science/article/pii/S0002870304002042
17. Pegolotti, L., et al.: Learning reduced-order models for cardiovascular simulations with graph neural networks. Comput. Biol. Med. **168**, 107676 (2024)
18. Pfaff, T., Fortunato, M., Sanchez-Gonzalez, A., Battaglia, P.: Learning mesh-based simulation with graph networks. In: International Conference on Learning Representations (2020)
19. Pfaller, M.R., Pegolotti, L., Pham, J., Rubio, N.L., Marsden, A.L.: Reduced-order modeling of cardiovascular hemodynamics. In: Biomechanics of the Aorta, pp. 449–476. Elsevier (2024)
20. Pfitzner, J.: Poiseuille and his law. Anaesthesia **31**(2), 273–275 (1976)
21. Raissi, M., Perdikaris, P., Karniadakis, G.E.: Physics informed deep learning (part i): data-driven solutions of nonlinear partial differential equations. arXiv preprint arXiv:1711.10561 (2017)
22. Rygiel, P., Pluszka, P., Zięba, M., Konopczyński, T.: Centerlinepointnet++: a new point cloud based architecture for coronary artery pressure drop and vffr estimation. In: International Conference on Medical Image Computing and Computer-Assisted Intervention, pp. 781–790. Springer, Heidelberg (2023). https://doi.org/10.1007/978-3-031-43990-2_73
23. Seeley, B.D., Young, D.F.: Effect of geometry on pressure losses across models of arterial stenoses. J. Biomech. **9**(7), 439–448 (1976)
24. Shahjehan, R.D., Sharma, S., Bhutta, B.S.: Coronary artery disease. In: StatPearls [Internet]. StatPearls Publishing (2024)
25. Teng, Q., Liu, Z., Song, Y., Han, K., Lu, Y.: A survey on the interpretability of deep learning in medical diagnosis. Multimedia Syst. **28**(6), 2335–2355 (2022)
26. Yin, M., Yazdani, A., Karniadakis, G.E.: One-dimensional modelling of fractional flow reserve in coronary artery disease: Uncertainty quantification and bayesian optimisation. Comput. Methods Appl. Mech. Eng. **353**, 66–85 (2019). https://doi.org/10.1016/j.cma.2019.05.005
27. Zhou, Y., He, Y., Wu, J., Cui, C., Chen, M., Sun, B.: A method of parameter estimation for cardiovascular hemodynamics based on deep learning and its application to personalize a reduced-order model. Int. J. Numer. Methods Biomed. Eng. **38**(1), e3533 (2022)

WANCDR: Wasserstein Adversarial Network for Cancer Drug Response

Hanjun Choi and Mansu Kim(✉)

Department of AI convergence, Gwangju Institute of Science and Technology, Gwangju, Korea
mansu.kim@gist.ac.kr

Abstract. Predicting patient-specific drug responses from preclinical cell-line data remains challenging due to significant heterogeneity between preclinical (cell-line) and clinical (patient) gene expression profiles. In this study, we propose WANCDR, a novel adversarial neural network framework designed to improve the generalization of drug-response predictions by aligning latent representations across preclinical and clinical domains. Specifically, we introduce a domain alignment module trained adversarially, which enforces the encoder to generate domain-invariant latent embeddings. Extensive experiments conducted on preclinical (GDSC) and clinical (TCGA) datasets demonstrate that WANCDR achieves robust predictive performance on preclinical data, while substantially outperforming existing approaches in clinical generalization, particularly when classifying responses for previously unseen drugs. Qualitative analyses via UMAP visualization further validate the superior domain alignment capability of WANCDR. Collectively, these results highlight the potential of WANCDR to bridge the translational gap from preclinical insights to clinical applications.

Keywords: Graph neural network · Cancer drug response · Wasserstein Adversarial Network

1 Introduction

Cancer drug response (CDR) prediction aims to identify how effectively a tumor will respond to a given anticancer treatment, and is a critical component of precision oncology [5]. Accurate CDR predictions help clinicians to select proper therapies based on individual patient profiles, improving clinical outcomes and reducing adverse side effects. However, substantial genetic and transcriptomic variability among tumors makes reliable predictions challenging [8]. To address this issue, it is critical to develop robust computational methods that capture complex biological variations.

Recent studies on CDR prediction have adopted deep learning-based approaches due to their effectiveness in handling complex biological data. For example, DeepCDR integrates multi-omics data with drug structural information using graph convolutional networks (GCNs). PANCDR employs adversarial learning techniques to reduce the domain gap between preclinical and clinical datasets [9,11]. Despite these advancements, existing methods still face limitations when applied to clinical patient data due to differences between training (preclinical) and testing (clinical) domains.

A significant limitation of current methods is that they are trained on preclinical datasets (e.g., Genomics of Drug Sensitivity in Cancer, GDSC), resulting in decreased prediction performance when evaluated on clinical cohorts (e.g., The Cancer Genome Atlas, TCGA) [4,15]. Clinical data typically contain fewer samples, more diverse patient profiles, and distinct biological characteristics compared to preclinical datasets, leading to significant distributional shifts. For example, PANCDR employs a conventional adversarial framework to address these issues; however, it often suffers from instability during training and issues such as mode collapse, which limit its practical effectiveness [9]. Similarly, popular domain adaptation techniques borrowed from computer vision, including Gradient Reversal Layers (DANN) and correlation alignment (Deep CORAL), often struggle to effectively model the nuanced biological differences between preclinical and clinical data [4,15].

To overcome these limitations, we propose a Wasserstein Adversarial Network-based model for CDR prediction (WANCDR), specifically designed to align latent feature distributions between preclinical and clinical data robustly. Our main scientific contributions are summarized as follows: (1) We propose a Wasserstein Generative Adversarial Network based CDR prediction model that effectively aligns latent feature distributions between preclinical and clinical data. (2) By employing a critic network to minimize the Wasserstein distance, our model ensures stable adversarial training, thus addressing common issues such as mode collapse and vanishing gradients found in conventional adversarial learning frameworks [6]. (3) We demonstrate the effectiveness of our domain alignment approach through extensive seen-unseen CDR experiments.

2 Material and Methodology

2.1 Data Preparation

Source Dataset: GDSC. We use preclinical data obtained from the GDSC database [8]. Raw gene expression profiles for cancer cell lines are obtained from ArrayExpress (E-MTAB-3610), and drug response annotations (Sensitive or Resistant) are computed by binarizing IC_{50} values using LOBICO [10]. After removing samples without valid drug response annotations or drug identifiers (PubChem ID or SMILES), the final dataset contains 112,575 instances spanning 950 cell lines and 151 drugs, with an approximate Sensitive-to-Resistant ratio of 1:7.5. For drug representation, SMILES strings are converted into molecular graphs using RDKit (https://www.rdkit.org), and feature and adjacency matrices are computed using DeepChem [13].

Target Dataset: TCGA. Gene expression profiles are downloaded from the GDAC Broad Institute (http://gdac.broadinstitute.org/), and clinical drug response annotations are acquired from Ding et al. [3]. Specifically, drug responses are categorized as Sensitive (Complete Response/Partial Response) or Resistant (Progressive Disease/Stable Disease). Following MOLI's preprocessing strategy [14], cases involving multi-drug treatments are excluded, leaving only single-agent treatments. As a result, the final external test set comprises 666 instances from 569 patients treated with 69 distinct drugs, with a Sensitive-to-Resistant ratio of approximately 1:1. Additionally, we use 9,424 unlabeled primary solid tumor samples from TCGA for adversarial training, which helps to align feature distributions between preclinical and clinical datasets [9].

2.2 Wasserstein Adversarial Network for Cancer Drug Response (WANCDR)

The proposed WANCDR model comprises two primary components: a Wasserstein GAN (WGAN) for domain alignment and a Graph Convolutional Network (GCN)-based predictor for cancer drug response prediction. First, WANCDR utilizes a WGAN framework [1,7] to robustly align latent genomic feature distributions between the source (cell-line) and target (patient) domains. Second, structural drug information is embedded through a GCN and integrated with the aligned genomic representation to accurately estimate drug response, quantified by IC_{50} values (Fig. 1).

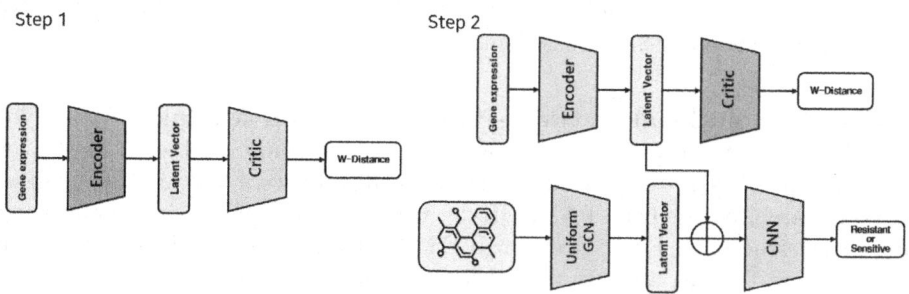

Fig. 1. WANCDR architecture: An MLP-based encoder generates latent representations of gene expression, while a GCN-based predictor captures drug features. A WGAN critic distinguishes between cell-line and patient-derived latent vectors. The training alternates between Step 1: freezing encoder/predictor and training the critic, and Step 2: freezing critic and jointly training encoder and predictor.

WGAN-Based Domain Alignment Module. We employ a Wasserstein Generative Adversarial Network (WGAN) framework [1,7] to effectively address the domain distribution mismatch between preclinical (cell-line) and clinical (patient) datasets. The WGAN framework is particularly suitable for our task, as it robustly aligns latent distributions by mitigating common adversarial training challenges such as mode collapse and gradient instability through enforcing Lipschitz continuity constraints via gradient penalties. This approach provides enhanced training stability compared to traditional GAN-based methods.

Formally, let $E(\cdot)$ denote the MLP encoder network mapping gene expression profiles into latent feature representations. Specifically, given input profiles x_{GDSC} and x_{TCGA}, we have latent embeddings $z_g = E(x_{\text{GDSC}})$ and $z_t = E(x_{\text{TCGA}})$, which follow latent distributions $z_g \sim \mathbb{P}_z^{\text{GDSC}}$ and $z_t \sim \mathbb{P}_z^{\text{TCGA}}$, respectively. To quantify the discrepancy between these distributions, we introduce a critic network $C : z \mapsto \mathbb{R}$, which approximates the Wasserstein-1 distance:

$$W_1(\mathbb{P}_z^{\text{TCGA}}, \mathbb{P}_z^{\text{GDSC}}) = \sup_{\|C\|_L \leq 1} \left[\mathbb{E}_{z_t \sim \mathbb{P}_z^{\text{TCGA}}}[C(z_t)] - \mathbb{E}_{z_g \sim \mathbb{P}_z^{\text{GDSC}}}[C(z_g)] \right].$$

To enforce the critic's 1-Lipschitz constraint effectively, we adopt the gradient penalty regularization term from WGAN-GP [7], defined as:

$$\Omega_{\text{GP}} = \mathbb{E}_{\hat{z} \sim P_{\hat{z}}} \left[(\|\nabla_{\hat{z}} C(\hat{z})\|_2 - 1)^2 \right], \quad \hat{z} = \varepsilon \cdot z_t + (1 - \varepsilon) z_g, \ \varepsilon \sim U(0,1).$$

Thus, the final objective for training the critic network becomes:

$$L_{\text{critic}} = \mathbb{E}_{z_t \sim P_z^{\text{TCGA}}}[C(z_t)] - \mathbb{E}_{z_g \sim P_z^{\text{GDSC}}}[C(z_g)] + \lambda \Omega_{\text{GP}} \qquad (1)$$

where λ denotes a hyperparameter controlling the gradient penalty strength. By optimizing this objective, WANCDR effectively aligns the latent feature distributions of cell-line and patient data, thereby enhancing model generalization and predictive robustness across domains.

Graph Convolutional Network-Based Prediction Module. We employ a GCN-based prediction module that integrates genomic and structural drug information to accurately estimate CDR. Specifically, we adapt the GCN+MLP predictor architecture from PANCDR [9], which jointly models genomic features and drug structures. First, drug SMILES strings are converted into molecular graphs using RDKit, and atom-level features generated by DeepChem [13] are processed through a GCN to update node embeddings. A global average pooling layer is then applied to aggregate these node embeddings into a fixed-length drug representation h_{drug}.

Formally, let $GCN(\cdot)$ represent the GCN-based embedding network for structural drug information. Given a molecular graph (G), constructed from SMILES strings using RDKit, GCN generates a fixed-length embedding vector:

$$h_{\text{drug}} = GCN(G).$$

We then combine two latent embeddings into a unified representation:

$$h_{\text{combined}} = h_{\text{drug}} \oplus z_i, i \in \{t, g\}$$

which is further processed by a fully-connected layer followed by a 1D convolutional layer, as described previously in PAN-CDR [9], to capture high-level interactions between drug and genomic features. Finally, a sigmoid activation function predicts the probability of drug sensitivity ($\hat{y}$):

$$\hat{y} = \sigma(\text{Conv1D}(\text{FC}(h_{\text{combined}}))).$$

The prediction module is trained by minimizing the binary cross-entropy loss:

$$\mathcal{L}_{\text{pred}} = -\left[y \log \hat{y} + (1-y)\log(1-\hat{y})\right]. \tag{2}$$

By integrating the domain-aligned genomic representations, z_{expr}, with structural drug embeddings, h_{drug}, we accurately predict drug responses and enhance the generalization performance across both clinical and preclinical populations.

Loss Function. We alternately minimize two losses to optimize our model. First, we update the *Domain Alignment Module* by minimizing the critic loss $\mathcal{L}_{\text{critic}}$, defined in Eq. 1. Subsequently, we optimize the *prediction module* by minimizing the joint predictive and adversarial loss:

$$\mathcal{L}_{\text{joint}} = \mathcal{L}_{\text{pred}} + \lambda \mathcal{L}_{\text{adv}}, \tag{3}$$

where the predictive loss $\mathcal{L}_{\text{pred}}$ is defined in Eq. 2, and λ is a hyperparameter balancing predictive accuracy and domain alignment. The adversarial loss is given by:

$$\mathcal{L}_{\text{adv}} = -\mathbb{E}_{x_{GDSC} \sim \mathbb{P}_{\text{GDSC}}}[C(E(x_{GDSC}))],$$

which encourages the encoder $E(\cdot)$ to generate domain-invariant latent representations by restricting the critic's capability to discriminate between preclinical (cell-line) and clinical (patient) data. This joint optimization ensures robust predictive performance and effective alignment of latent features across domains.

3 Experiments and Results

3.1 Experiments Setup

To assess the classification performance of WANCDR, we conduct experiments on the preclinical GDSC dataset using a stratified 5-fold cross-validation approach. Specifically, among the 5 folds, four are assigned as the training set, from which 5% is set as an internal validation set for early stopping; the remaining fold serves as the test set. The classification performance is evaluated in terms of Area Under the Curve (AUC), F1-score, Precision, and Recall. To ensure the robustness of the results, we repeat this entire cross-validation process 10 times and report the average outcomes.

Furthermore, to assess the model's generalization to clinical data, we perform additional evaluations using the clinical TCGA dataset. Briefly, we train WANCDR on 95% of the GDSC preclinical dataset, reserving 5% as an internal validation set to determine optimal hyperparameters via random search [2]. The model trained on GDSC is then directly applied to the external TCGA clinical dataset. Similar to the preclinical experiment setup, we measure the model's ability to classify Sensitive and Resistant patient-drug pairs using AUC, F1-score, Precision, and Recall. This procedure is also repeated 10 times to ensure robust and reliable estimates of clinical generalization performance.

3.2 Results on Preclinical (GDSC) Dataset

We evaluate the predictive performance of our proposed model on the preclinical GDSC dataset. The comparative results are summarized in Table 1. In detail, WANCDR achieve an AUC of 0.815 ± 0.010, accuracy (ACC) of 0.753 ± 0.023, precision of 0.284 ± 0.019, recall of 0.723 ± 0.032, and F1-score of 0.407 ± 0.016. Although DeepCDR slightly outperform WANCDR in terms of AUC (0.832 ± 0.002), ACC (0.762 ± 0.017), precision (0.295 ± 0.014), recall (0.737 ± 0.022), and F1-score (0.421 ± 0.011), WANCDR shows competitive performance on preclinical data, demonstrating the effectiveness of the proposed approach.

Table 1. Comparison of classification performance between WANCDR and DeepCDR on preclinical (GDSC) and clinical (TCGA) datasets. Results are reported in the format of mean (standard deviation).

Dataset	Model	AUC	ACC	Precision	Recall	F1
GDSC	WANCDR	0.815(0.010)	0.753(0.023)	0.284(0.019)	0.723(0.032)	0.407(0.016)
	DeepCDR	**0.832(0.002)**	**0.762(0.017)**	**0.295(0.014)**	**0.737(0.022)**	**0.421(0.011)**
1TCGA	WANCDR	**0.672(0.032)**	**0.644(0.024)**	**0.656(0.015)**	**0.561(0.096)**	**0.601(0.059)**
	DeepCDR	0.550(0.030)	0.574(0.028)	0.609(0.070)	0.416(0.143)	0.496(0.054)

3.3 Results on Clinical (TCGA) Dataset

To evaluate the clinical generalization capability of WANCDR, we train the model on the preclinical GDSC dataset and assess its predictive performance directly on the external clinical TCGA dataset. As summarized in Table 2, WANCDR significantly outperforms DeepCDR across all performance metrics, achieving an AUC of 0.672 ± 0.032, accuracy (ACC) of 0.644 ± 0.024, precision of 0.656 ± 0.015, recall of 0.561 ± 0.096, and F1-score of 0.601 ± 0.059.

Furthermore, we evaluate the model's classification performance specifically on unseen drug-gene expression pairs. As presented in Table 2, WANCDR consistently demonstrates superior results for both previously seen and unseen

Table 2. Comparison of classification performance for seen and unseen drug-gene expression pairs on the clinical TCGA dataset. Results are reported in the format of mean (standard deviation).

Dataset	Model	AUC	ACC	Precision	Recall	F1
Seen	WANCDR	**0.690(0.048)**	**0.660(0.029)**	**0.671(0.027)**	**0.569(0.144)**	**0.604(0.087)**
	DeepCDR	0.554(0.044)	0.595(0.045)	0.639(0.097)	0.431(0.121)	0.496(0.054)
Unseen	WANCDR	**0.632(0.017)**	**0.622(0.015)**	**0.641(0.031)**	0.623(0.143)	**0.622(0.059)**
	DeepCDR	0.534(0.040)	0.568(0.022)	0.597(0.058)	0.614(0.239)	0.573(0.107)

drug classifications. Specifically, we observe the robust predictive performance achieved on unseen drugs (AUC: 0.632±0.017, ACC: 0.622±0.015, and F1-score: 0.622 ± 0.059), highlighting WANCDR's promising potential for generalization to novel therapeutic agents.

These findings clearly demonstrate that WANCDR provides robust predictive performance and enhanced clinical generalization, effectively bridging the domain gap between preclinical (cell-line) and clinical (patient) data.

3.4 Domain Alignment Analysis

To qualitatively assess the effectiveness of domain alignment achieved by WANCDR, we visualize latent embeddings of the GDSC (blue) and TCGA (orange) datasets using UMAP [12] (Fig. 2). As shown in Fig. 2-(a), latent embeddings from the two domains form clearly separated clusters before training, indicating a considerable domain gap between preclinical (GDSC) and clinical (TCGA) data. After training with DeepCDR (Fig. 2-(b)), the embeddings within each domain become more structured; however, they still remain distinctly separated, highlighting DeepCDR's limited capacity for cross-domain generalization. In contrast, WANCDR (Fig. 2-(c)) generates a cohesive latent space wherein the two domains substantially overlap, forming a continuous manifold that effectively bridges GDSC and TCGA samples. This significant intermingling demonstrates the successful domain-invariant representation learning capability of WANCDR. The qualitative observation aligns closely with the quantitative evaluations, where WANCDR shows comparable performance with DeepCDR on the preclinical GDSC dataset and significantly superior performance on the clinical TCGA dataset.

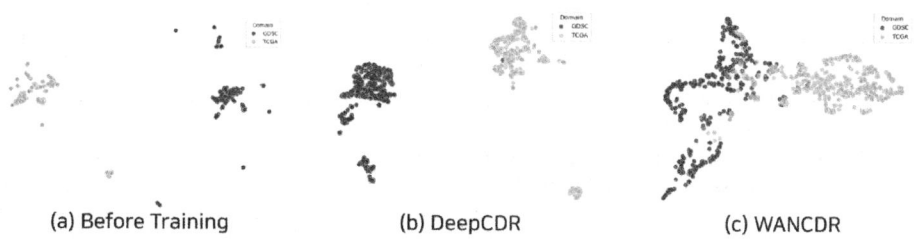

(a) Before Training   (b) DeepCDR   (c) WANCDR

Fig. 2. UMAP of latent gene-expression embeddings for GDSC (blue) and TCGA (orange): (a) Before training, (b) DeepCDR, (c) WANCDR. (Color figure online)

4 Conclusion

In this study, we propose WANCDR, improving cross-domain generalization in cancer drug response prediction through latent representation alignment between preclinical (GDSC) and clinical (TCGA) gene expression profiles. Experimental results demonstrate that WANCDR achieves competitive predictive performance on the preclinical dataset while substantially outperforming baseline methods in clinical generalization, particularly for unseen therapeutic agents. Qualitative analysis using UMAP visualizations further confirms the effective alignment of latent features across domains, highlighting WANCDR's capability to bridge the domain gap between preclinical cell-line data and clinical patient data. Our findings collectively indicate that WANCDR represents a promising approach towards translating preclinical insights into clinical practice.

Acknowledgment. This work was supported by the National Research Foundation of Korea(NRF) grant funded by the Korea government(MSIT) (RS-2022-NR073137, RS-2025-00521250). This work was partly supported by Institute of Information & communications Technology Planning & Evaluation (IITP) grant funded by the Korea government(MSIT) (No. RS-2021-II212068, Artificial Intelligence Innovation Hub; No.2019-0-01842, Artificial Intelligence Graduate School Program (GIST)), and Ministry of Trade, Industry, and Energy (MOTIE) Korea, under the "Infrastructure program for industrial innovation" supervised by the Korea Institute for Advancement of Technology (KIAT) (No. RS-2024-00434342).

References

1. Arjovsky, M., Chintala, S., Bottou, L.: Wasserstein gan. In: International Conference on Machine Learning (ICML) (2017)
2. Bergstra, J., Bengio, Y.: Random search for hyper-parameter optimization. J. Mach. Learn. Res. **13**(1), 281–305 (2012)
3. Ding, Z., Zu, S., Gu, J.: Evaluating the molecule-based prediction of clinical drug responses in cancer. Bioinformatics **32**(19), 2891–2895 (2016)
4. Ganin, Y., et al.: Domain-adversarial training of neural networks. J. Mach. Learn. Res. **17**(59), 1–35 (2016)
5. Garnett, M.J., et al.: Systematic identification of genomic markers of drug sensitivity in cancer cells. Nature **483**(7391), 570–575 (2012)
6. Goodfellow, I.J., et al.: Generative adversarial nets. In: Advances in Neural Information Processing Systems, vol. 27 (2014)
7. Gulrajani, I., Ahmed, F., Arjovsky, M., Dumoulin, V., Courville, A.C.: Improved training of wasserstein gans. In: Advances in Neural Information Processing Systems (NeurIPS) (2017)
8. Iorio, F., et al.: A landscape of pharmacogenomic interactions in cancer. Cell **166**(3), 740–754 (2016)
9. Kim, J., Park, S.H., Lee, H.: Pancdr: precise medicine prediction using an adversarial network for cancer drug response. Brief. Bioinf. **25**(2), bbae088 (2024)
10. Knijnenburg, T.A., et al.: Logic models to predict continuous outputs based on binary inputs with an application to personalized cancer therapy. Sci. Rep. **6**(1), 36812 (2016)
11. Liu, Q., Hu, Z., Jiang, R., Zhou, M.: Deepcdr: a hybrid graph convolutional network for predicting cancer drug response. Bioinformatics **36**(Supplement_2), i911–i918 (2020)

12. McInnes, L., Healy, J., Melville, J.: Umap: uniform manifold approximation and projection for dimension reduction. arXiv preprint arXiv:1802.03426 (2018)
13. Ramsundar, B., Eastman, P., Walters, P., Pande, V.: Deep learning for the life sciences: applying deep learning to genomics, microscopy, drug discovery, and more. O'Reilly Media (2019)
14. Sharifi-Noghabi, H., Zolotareva, O., Collins, C.C., Ester, M.: Moli: multi-omics late integration with deep neural networks for drug response prediction. Bioinformatics **35**(14), i501–i509 (2019)
15. Sun, B., Saenko, K.: Deep CORAL: correlation alignment for deep domain adaptation. In: Hua, G., Jégou, H. (eds.) ECCV 2016. LNCS, vol. 9915, pp. 443–450. Springer, Cham (2016). https://doi.org/10.1007/978-3-319-49409-8_35

Author Index

A
Abraham, Daniel Raz 118
Ahn, Hayoung 209
Aillet, Albert Sund 162
Aktar, Mumu 55
Ashraf, Saad 55

B
Bae, Jaehyeok 118
Baugh, Matthew 172
Bayer, Siming 76
Beauferris, Youssef 3
Beliveau, Vincent 108
Bento, Mariana 3, 55
Broersen, Alexander 131

C
Cacace, Paolo 162
Carlier, Thomas 195
Castonguay, Samuel 65
Cattin, Philippe C. 87
Cechnicka, Sarah 172
Cheng, Xinxing 44
Cherkaoui, Oumeymah 87
Cho, Hyuna 209
Choi, Hanjun 229
Chojnacki, Jakub 219

D
De Bosch, Marc Molina Van 162
Deshpande, Gouri 3
Dijkstra, Jouke 131
Dirix, Pietro 65
Dombrowski, Mischa 172

E
Duan, Jinming 44
Duraj, Konrad 219

E
Eichhorn, Hannah 23

F
Flouris, Kyriakos 65

G
Ganz, Melanie 108
Gao, Yuzhen 141, 151
Gatidis, Sergios 97
Georgiev, Hristo 108
Ghoul, Aya 97
Giusti, Lorenzo 162
Gu, Mingxuan 76

H
Hadramy, Sidaty El 87
Halter, Moritz 65
Hammernik, Kerstin 97
Hargreaves, Brian Andrew 118
Huang, Wenqi 23, 118
Hummel, Friedhelm 162
Hwang, Yechan 209

I
Idoko, Jacob 3
Ikuma, Takeshi 34

J
Jacobs, Luuk 65

K
Kainz, Bernhard 172
Kim, Mansu 229
Kim, Won Hwa 209
Kist, Andreas M. 34

Konukoglu, Ender 65
Kopeć, Szymon 219
Kopp, Markus 34
Kozerke, Sebastian 65
Krumm, Patrick 97
Kuebler, Jens 97
Kunduk, Melda 34
Küstner, Thomas 97

L

Laribi, Hakima 195
Larsen, Deirdre 34
Lee, Seungjoo 209
Lee, Yolanne Y. R. 65
Li, Zhe 172
Lin, Yimeng 118
Lingg, Andreas 97
Liu, Mingxia 141, 151

M

Maas, Kirsten W. H. 13
Maier, Andreas 76
Mamun, Md Afif Al 55
Mateus, Diana 195
Mei, Siyuan 76
Meng, Qingjie 44
Menzel, Marion I. 23

N

Nestmann, Jonathan 65
Neubig, Luisa 34
Niessen, Natascha 23
Nurdinova, Aizada 118

O

O'Regan, Declan 44
Omidi, Abbas 3

P

Paetzold, Johannes C. 172
Pang, Yan 44
Petersen, Jens 108
Pezzotti, Nicola 13
Pintea, Silvia L. 131
Pirkl, Carolin M. 23

Potter, Guy G. 151
Protani, Andrea 162

R

Rekik, Islem 184
Reynaud, Hadrien 44
Rizkallah, Mira 195
Roufosse, Candice 172
Rueckert, Daniel 97
Ruijters, Danny 13

S

Schnabel, Julia A. 23
Schneider, Linda-Sophie 76
Schwarz, Annette 76
Sengupta, Prajit 184
Serio, Luigi 162
Setsompop, Kawin 118
Shamaei, Amirmohammad 3
Sheye, Jakob 108
Silva, Heloisa Barbosa Da 162
Sim, Jaeyoon 209
Solana, Ana Beatriz 23
Souza, Roberto 3, 55
Spieker, Veronika 23
Sprenger, Tim 23
Stephens, David C. 151
Sun, Yipeng 76
Sun, Yongheng 141

T

Thiery, Oriane 195

V

Vallières, Martin 195
van Erp, Gonnie C. M. 131
Vilanova, Anna 13

W

Wang, Cui 141
Wang, Li 151
Wang, Lihong 151
Wu, Guorong 209
Wu, Mengqi 151
Wyburd, Madeleine 108

Author Index

X
Xu, Siying 97

Y
Ye, Chengze 76
Yin, Jiahui 44

Yu, Minhui 141

Z
Zamorski, Maciej 219
Zhang, Weitong 172
Zhang, Xiaotong 131

Made in the USA
Monee, IL
03 May 2026

49438642R00142